Emerging Antibody Engineering Strategies and Applications for Immunotherapy of Cancer

Emerging Antibody Engineering Strategies and Applications for Immunotherapy of Cancer

Guest Editors

Christian Klein
Sophia Karagiannis
Ann White

MDPI

Basel • Beijing • Wuhan • Barcelona • Belgrade • Novi Sad • Cluj • Manchester

Guest Editors

Christian Klein
Faculty of Chemistry and Pharmacy
Ludwig Maximilians University of Munich
Munich
Germany

Sophia Karagiannis
School of Basic & Medical Biosciences
King's College London
London
UK

Ann White
Translational Science
Mestag Therapeutics
Cambridge
UK

Editorial Office
MDPI AG
Grosspeteranlage 5
4052 Basel, Switzerland

This is a reprint of the Special Issue, published open access by the journal *Antibodies* (ISSN 2073-4468), freely accessible at: https://www.mdpi.com/journal/antibodies/special_issues/F8278093W1.

For citation purposes, cite each article independently as indicated on the article page online and as indicated below:

Lastname, A.A.; Lastname, B.B. Article Title. *Journal Name* **Year**, *Volume Number*, Page Range.

ISBN 978-3-7258-6770-7 (Hbk)
ISBN 978-3-7258-6771-4 (PDF)
https://doi.org/10.3390/books978-3-7258-6771-4

Contents

About the Editors

Christian Klein

Christian Klein specializes in the discovery, engineering and preclinical development of (bispecific) mAbs for immunotherapy. He is CSO and co-founder of a biotech startup. Previously, he was CXO and Drug Maker at Curie.Bio where he worked with founders of seed stage biotechs. As a distinguished scientist, research project/program leader and mentor he made essential contributions during 22 years at Roche pRED oncology to more than 30 drug candidates entering the clinic of which GAZYVA, VABYSMO, and COLUMVI are approved. He led teams developing CrossMab®, 2+1 TCB, 4-1BB/CD28 costims, (PD1-cis) immunocytokines, next gen biologics, and contributed to research on the IGF-1R mAb TEPEZZA and Nutlin MDM2i. In 1998, he received a diploma in biochemistry from the University of Tübingen as a scholar of the Studienstiftung working at the MPI for Biology, and in 2002 a doctorate from TU Munich working on p53/MDM2 at the MPI for Biochemistry. Also in 2002, he joined Roche Pharma Research Penzberg's Therapeutic Protein Initiative as a Postdoctoral researcher, where he became Lab Head in 2003 and Department Head of Discovery Oncology in 2007. Later on, in 2010, he joined Roche Innovation Center Zurich as Head of Oncology Programs, taking on additional roles as Department Head of CIT-3 in 2015 and Site Head of RICZ in 2019. Since his habilitation in 2017, he is a lecturer (PD) for Biochemistry at LMU Munich.

Sophia Karagiannis

Sophia Karagiannis, Professor of Translational Cancer Immunology and Immunotherapy at King's College London, is a translational B cell and cancer immunologist with academic and biotechnology experience in the UK and USA. She heads a cancer antibody discovery team focusing on dissecting the crosstalk between human immunity and solid tumors and understanding the impact of antibody isotype on therapeutic efficacy. The Karagiannis group evaluate cancer patient-derived B cells and their expressed antibodies, antibody-engaging immune effector cells such as monocytes/macrophages, NK cells and basophils, and engineer antibodies of different isotypes, Fc-optimized antibodies and antibody-drug conjugates (ADCs) for cancer therapy. Her group was the first to design and translate IgE class antibodies from concept to clinical testing and to commercialization. Professor Karagiannis is the author of several patents on antibody technologies and academic founder of Epsilogen Ltd, the first immuno-oncology company dedicated to developing IgE-based cancer immunotherapies. She is also a founding member and elected Chair of the AllergoOncology Working Group in the European Academy of Allergy and Clinical Immunology (EAACI), an international interdisciplinary group focused on exploring the interface between Th2 immunity, allergy, IgE biology and cancer. Her groundbreaking work has been recognized with an EAACI Research Fellow Award in 2025.

Ann White

Ann White, Head of Translational Science at Mestag Therapeutics in Cambridge, has worked for over 20 years on antibody cancer therapeutics, encompassing multiple roles across academia and industry. Her research at the University of Southampton helped define the crucial role of antibody isotype and the antibody Fc in therapeutic antibody function. During her time in industry, at UCB Pharma and now Mestag Therapeutics, she has leveraged this knowledge to drive therapeutic antibody projects from conception to the clinic. Her current focus is the Mestag M300 program, an FAP-dependent LTBR agonist designed to stimulate anti-tumor immunity through the induction of tertiary lymphoid structures in solid tumors.

antibodies

MDPI

Brief Report

Characterization of a Trispecific PD-L1 Blocking Antibody That Exhibits EGFR-Conditional 4-1BB Agonist Activity

Laura Rubio-Pérez [1,2,3,4], Susana Frago [5], Marta Compte [5], Rocío Navarro [5], Seandean L. Harwood [6], Rodrigo Lázaro-Gorines [1,2,3], Marina Gómez-Rosel [1,2,3], Oana Hangiu [1,2,5], Noelia Silva-Pilipich [7,8], Lucía Vanrell [9,10], Cristian Smerdou [7,8] and Luis Álvarez-Vallina [1,2,3,4,*]

1 Cancer Immunotherapy Unit (UNICA), Department of Immunology, Hospital Universitario12 de Octubre (H12O), 28041 Madrid, Spain; laura.rubio.imas12@h12o.es (L.R.-P.); rlazaro@cnio.es (R.L.-G.); marinagomezrosel.imas12@h12o.es (M.G.-R.); oana.hangiu@leadartis.com (O.H.)
2 Immuno-Oncology and Immunotherapy Group, Instituto de Investigacion Sanitaria 12 de Octubre (imas12), 28041 Madrid, Spain
3 H12O-CNIO Cancer Immunotherapy Clinical Research Unit, Spanish National Cancer Research Centre (CNIO), 28029 Madrid, Spain
4 Chair for Immunology UFV/Merck, Universidad Francisco de Vitoria (UFV), Pozuelo de Alarcón, 28223 Madrid, Spain
5 Department of Antibody Engineering, Leadartis SL, QUBE Technology Park, Tres Cantos, 28760 Madrid, Spain; susana.frago@leadartis.com (S.F.); marta.compte@leadartis.com (M.C.); rocio.navarro@leadartis.com (R.N.)
6 Department of Molecular Biology and Genetics, Aarhus University, 8000 Aarhus C, Denmark; sdlh@mbg.au.dk
7 Division of DNA and RNA Medicine, Cima Universidad de Navarra, 31008 Pamplona, Spain; nsilva.1@alumni.unav.es (N.S.-P.); csmerdou@unav.es (C.S.)
8 Instituto de Investigación Sanitaria de Navarra (IdISNA) and CCUN, 31008 Pamplona, Spain
9 Facultad de Ingeniería, Universidad ORT Uruguay, 11100 Montevideo, Uruguay; lvanrell@unav.es
10 Nanogrow Biotech, Montevideo 11500, Uruguay
* Correspondence: lav.imas12@h12o.es

Abstract: Immune checkpoint blockade has changed the treatment paradigm for advanced solid tumors, but the overall response rates are still limited. The combination of checkpoint blockade with anti-4-1BB antibodies to stimulate tumor-infiltrating T cells has shown anti-tumor activity in human trials. However, the further clinical development of these antibodies has been hampered by significant off-tumor toxicities. Here, we generated an anti-4-1BB/EGFR/PD-L1 trispecific antibody consisting of a triple-targeting tandem trimerbody (TT) fused to an engineered silent Fc region. This antibody (IgTT-4E1-S) was designed to combine the blockade of the PD-L1/PD-1 axis with conditional 4-1BB costimulation specifically confined to the tumor microenvironment (TME). The antibody demonstrated simultaneous binding to purified EGFR, PD-L1, and 4-1BB in solution, effective blockade of the PD-L1/PD1 interaction, and potent 4-1BB-mediated costimulation, but only in the presence of EGFR-expressing cells. These results demonstrate the feasibility of IgTT-4E1-S specifically blocking the PD-L1/PD-1 axis and inducing EGFR-conditional 4-1BB agonist activity.

Keywords: cancer immunotherapy; trispecific antibody; epithelial growth factor receptor; immune checkpoint blockade; 4-1BB costimulation

1. Introduction

One of the most promising strategies for enhancing anti-tumor immune responses is the blockade of inhibitory immune checkpoints, such as cytotoxic T lymphocyte antigen 4 (CTLA-4), programmed cell death protein 1 (PD-1), or PD-1 ligand (PD-L1) [1]. Immune checkpoint blockers have transformed cancer treatment for a wide range of tumor types, but the overall response rates are still limited, as many patients have no response or only a transient response [2,3]. As of January 2024, seven PD-L1 blockers and eight PD-1 blockers have been approved for clinical use in the United States and Europe, with atezolizumab being the first anti-PD-L1 monoclonal antibody (mAb) on the market (2017) [4]. Another strategy involves targeting costimulatory pathways, such as 4-1BB, also known as CD137, a member of the TNF receptor (TNFR) superfamily (TNFRSF9), which is an activation-induced surface receptor that provides antigen-primed T cells with augmented survival, proliferation and effector functions, as well as metabolic advantages [5]. Anti-4-1BB agonistic mAbs have shown considerable potential in promoting tumor rejection in preclinical cancer models [6]. However, the clinical development of full-length anti-4-1BB antibodies has been hampered by off-tumor toxicity, which is mainly due to Fc-FcγR interactions [6–8]. Therefore, to fully exploit their therapeutic potential, novel approaches are being developed that generally aim to confine 4-1BB costimulation to the TME and draining lymph nodes by adding tumor-specific moieties to generate bispecific 4-1BB agonistic antibodies [8–10]. Tumor-associated antigens (TAAs), such as epidermal growth factor receptor (EGFR), fibroblast activation protein (FAP), CD19, B7-H3 (CD276), carcinoembryonic antigen (CEA), and EGFR 2 (HER2) have been targeted to develop 4-1BB bispecifics [8–15]. Recently, a range of bispecific constructs targeting 4-1BB-mediated T cell costimulation to PD-L1-overexpressing tumor cells and simultaneously blocking the PD-1/PD-L1 axis have been generated [16–22] and are being clinically evaluated.

Here, we generated and characterized a multispecific antibody by fusing a trispecific 4-1BB/EGFR/PD-L1 tandem trimerbody (TT) [23] with the human IgG_1 hinge and Fc regions based on a previously described IgTT platform [24]. We used an engineered silenced Fc region to inhibit the binding of FcγR but retain the binding of FcRn for IgG-like pharmacokinetics [25]. The trispecific and hexavalent antibody was designed to simultaneously modulate two key pathways to enhance anti-tumor immune responses: PD-L1 blockade and tumor-specific 4-1BB costimulation. Our choice of TAA was the EGFR, a well-characterized tyrosine kinase receptor whose dysregulation promotes cancer cell proliferation, inhibits apoptosis, and promotes invasion [26].

2. Materials and Methods

2.1. Cell Lines and Culture Conditions

Dulbecco's modified Eagle's medium (DMEM) (Life Technologies, Carlsbad, CA, US; cat# 10313021) supplemented with antibiotics (100 units/mL of penicillin, 100 µg/mL of streptomycin; both from Life Technologies), 10% (v/v) heat-inactivated fetal bovine serum (FBS) (Merck Life Science, Darmstadt, Germany; cat# F7524-500 ML), and 2 mmol/L L-glutamine was used to culture HEK-293 (CRL-1573), NIH/3T3 (CRL-1658), CHO-K1 (CCL-61), and MDA-MB-231 (HTB-26) cells at 37 °C in 5% CO_2. All these cell lines were obtained from the American Type Culture Collection. NIH/3T3 cells expressing human EGFR ($3T3^{EGFR}$) were kindly provided by Dr. A. Villalobo (Instituto de Investigaciones Biomédicas "Alberto Sols", IIBm CSIC-UAM, Madrid, Spain). Expi293F cells (from Gibco, Thermo Fisher Scientific, Waltham, MA, USA) were cultured in Expi293 expression medium in a humidified, 8% CO_2 incubator rotating at 95 rpm at 37 °C. Jurkat T cells (TIB-152) were cultured in RPMI-1640 (Lonza Bioscience, Basel, Switzerland; cat# 12-702Q) supplemented with 10% (v/v) heat-inactivated FBS, 2 mmol/L of L-glutamine, and antibiotics. Jurkat T cells stably expressing human PD-1 and NFAT-induced luciferase ($Jurkat^{NFAT\text{-}PD\text{-}1}$) and CHO-K1 cells stably expressing human PD-L1 (PD-L1 aAPC/CHO-K1) were obtained from Promega (Madison, WI, USA; cat# J1250). Jurkat T cells stably expressing human 4-1BB and NFAT-induced luciferase ($Jurkat^{NFAT\text{-}4\text{-}1BB}$) were obtained from Promega (cat# JA2351). CHO-K1 cells stably expressing human PD-L1 ($CHO^{PD\text{-}L1}$) were obtained from Genlantis (xCELLerateTM PD-L1 Stable Cell Line, XCL-PDL1), and CHO-K1 cells stably expressing human EGFR (CHO^{EGFR}) were generated using human EGFR-encoding lentiviral particles (G&P Biosciences, Santa Clara, CA, US; cat# LTV0169). Jurkat T cells expressing GFP-tagged 4-1BB ($Jurkat^{4\text{-}1BB}$) were generated by lentiviral transduction using commercial lentiviral particles (Origene, Rockville, MD, USA; cat# RC200664L4V). The ORF present in the lentivirus contains 4-1BB sequence followed by mGFP sequence in the C-terminus, a

P2A cleavage site, and a puromycin resistance gene, all in frame. Briefly, 1×10^4 cells were incubated with the lentiviral particles (titer of 5.5×10^7 TU/mL) at a MOI (multiplicity of infection) of 20 for 8 h in retronectin-coated 24-well plates (20 µg/cm^2). The cells were allowed to recover for 48 h and subsequently selected in 2 µg/mL of puromycin. The cells were maintained in culture in the presence of 1 µg/mL of puromycin. All cell lines were routinely screened for mycoplasma contamination by PCR using the Mycoplasma Plus TM Primer Set (Biotools B&M Labs, Madrid, Spain; cat# 90022).

2.2. Construction of the Expression Vectors

To generate the 4-1BB/EGFR/PD-L1 trispecific IgTT Fc^{wt} (IgTT-4E1) expression vector, the plasmid pCR3.1-FLAG/Strep-α4-1BB-αEGFR-αPD-L1-hFc^{wt}-Myc-His was generated by cloning the insert OncoM-FLAG/Strep-α4-1BB scFv SAP3.28 [13] flanked by *Hind*III-*Not*I sites (GeneArt AG, Thermo Fischer, Waltham, MA, USA) and the insert αPD-L1 V$_{HH}$ (Nb6p generated in llama immunized with human PD-L1 ectodomain [27]) flanked by *Afe*I-*Sac*II sites (GeneArt AG, Thermo Fischer) into the plasmid from pCR3.1-FLAG/Strep-αPD-L1-αEGFR-hFc^{wt}-Myc-His encoding the bispefic IgTT-1E [24]. Then, to generate the 4-1BB/EGFR/PD-L1 trispecific IgTT Fc^{silent} (IgTT-4E1-S), the insert hFc^{silent}-Myc-His flanked by *Sac*II-*Xba*I sites (GeneArt AG, Thermo Fischer) was subcloned into the plasmid pCR3.1-FLAG/Strep-α41BB-αEGFR-αPD-L1-hFc^{wt}-Myc-His. Finally, to generate the 4-1BB/PD-L1/EGFR trispecific IgTT Fc^{silent} (IgTT-41E-S), the insert αPD-L1 V$_{HH}$/αEGFR V$_{HH}$ flanked by *Xho*I-*Sac*II sites (GeneArt AG, Thermo Fischer) was cloned into the plasmid pCR3.1-FLAG/Strep-α41BB-αEGFR-αPD-L1-hFc^{silent}-Myc-His. All the sequences were verified using the FwCMV and RvBGH oligonucleotide primers (Table S1).

2.3. Expression and Purification of the Recombinant Antibodies

HEK-293 cells (4×10^5 cells/well) were transfected with different vectors (2.5 µg of DNA/well) using a Lipofectamine 3000 kit (Fisher Scientific, Waltham, MA, USA; cat# 15292465) and cultured at 37 °C. The cells were then selected in complete DMEM supplemented with 500 µg/mL of G418 to generate stable cell lines. To produce large amounts of protein, 5.8×10^8 Expi293F cells were transfected with 180 µg of DNA using the Expifectamine 293 transfection reagent (Life Technologies) in 180 mL of medium and cultured at 37 °C for four days. The conditioned media were collected and processed using the Strep-Tactin purification system (IBA Lifesciences, Göttingen, Germany) in an ÄKTA Prime plus system (Life Technologies). The purified antibodies were then dialyzed at 4 °C with PBS pH 7.4 supplemented with 150 mM NaCl, analyzed by sodium dodecyl sulphate (SDS)–polyacrylamide gel electrophoresis (PAGE) under reducing conditions, and assayed for endotoxin levels using a limulus amebocyte lysate (LAL) endotoxin quantitation kit according to the manufacturer's specifications (Thermo Fisher Scientific, Waltham, MA, USA; cat# A39552S). The endotoxin levels were <0.25 EU/mL.

2.4. Enzyme-Linked Immunosorbent Assay

To determine the binding activity of the purified or secreted protein, human EGFR-Fc (hEGFR-Fc, R&D Systems, Minneapolis, MN, USA; cat# 344-ER), human PD-L1-Fc (hPD-L1-Fc, R&D Systems, cat# 156-B7-100), or human 4-1BB-Fc (h4-1BB-Fc, R&D Systems, cat# 838-4B-100) chimeras were immobilized (2.5 µg/mL in PBS) on Maxisorp 96-well plates (NUNC Brand Products, Thermo Fisher Scientific; cat# 44240) overnight at 4 °C. After washing and blocking with PBS–5% BSA, the conditioned medium or the purified protein solution (1 µg/mL) was added, and the plates were incubated for 1 h at room temperature. After washing, proteins were detected with HRP-conjugated anti-FLAG mAb (M2 clone, Sigma-Aldrich, Burlington, MA, USA; cat# A8592) (1 µg/mL) for 45 min at room temperature. The plates were then developed with 100 µL of 3,3′,5,5′-tetramethylbenzidine (TMB) (Sigma-Aldrich, cat# T0440), and the reaction was stopped with 100 µL of 1 N H_2SO_4. Absorbance was read at 450–620 nm using a Multiskan FC photometer (Thermo Scientific, Waltham, MA, USA).

2.5. Western Blotting

Conditioned medium samples containing 0.1% (v/v) FBS were analyzed by 10–20% tris-glycine SDS-PAGE under reducing conditions and transferred onto nitrocellulose membranes (Thermo Fisher Scientific, cat# IB23002). They were incubated with mouse anti-FLAG mAb (clone M2, Sigma-Aldrich, cat# F3165) (1 μg/mL) and probed with HRP-conjugated goat anti-mouse polyclonal antibody (GAM-HRP) (Sigma-Aldrich, Burlington, MA, USA; cat# A2554) (1:10,000 dilution). The visualization of protein bands was performed with Pierce ECL Plus Western Blotting Substrate (Thermo Scientific, cat# 32132) using the ChemiDoc MP Imaging System and Image Lab software (version 6.0.1, both from BioRad, Hercules, CA, USA).

2.6. Size-Exclusion Chromatography

Size-exclusion chromatography (SEC) was performed on a Superdex 200 Increase 10/300 GL column (Cytiva) on an ÄKTA go chromatography system (Cytiva). The column was equilibrated in 0.22 μm filtered PBS. Analytical runs of the antibody were performed at room temperature at a flow rate of 0.75 mL/min, injecting 20 μg of protein. The column was calibrated using the Gel Filtration HMW Calibration Kit from Cytiva (cat# 28403842). Data acquisition and analysis were performed using ASTRA software (version 7.6, Wyatt Technology, Santa Barbara, CA, USA). The reported molar mass corresponds to the center of the chromatography peaks.

2.7. Molecular Modeling

The three-dimensional representation of IgTT-4E1 was built by comparative homology modeling using MODELLER [28]. The model was created by using different templates for the different domains. The V_{HH} monospecific TT model was generated utilizing as a template the monospecific anti-CEA TT from a prior study [23]. The Fc region was modeled based on the human IgG_1 B12 structure (pdb:1HZH.H) [29], obtained from the Protein Data Bank [30] (99% sequence identity, e-value of 1×10^{-169} as calculated by BLAST (version 2.2.24) [31]). The 1HZH structure also guided the dimer modeling. The scFv domain was modeled using the interleukin 18 receptor antagonist scFv (pdb:6NK9.D) [32] as a template (83% sequence identity, e-value of 2×10^{-138} as calculated by BLAST). Ultimately, the final representation was generated by merging the models.

2.8. Biolayer Interferometry

The simultaneous binding of all three cognate antigens by IgTT-4E1 was investigated using biolayer interferometry (BLI) on an Octet RED96 system (Fortebio, Fremont, CA, USA). We immobilized 5 g/L of hEGFR-Fc onto AR2G biosensors (Fortebio) at pH 5.0 using amine-reactive coupling. Then, IgTT-4E1 in HEPES-buffered saline (HBS; 20 mM HEPES, 150 mM NaCl, pH 7.4) at 20 nM was allowed to associate with the hEGFR-Fc-coated biosensors for 20 min, after which the dissociation of the antibody from the biosensors was measured for 5 min in HBS buffer only. The biosensors were then moved into HBS containing 50 nM hPD-L1-Fc for 20 min and subsequently into HBS containing 50 nM h4-1BB-Fc for 20 min, followed by 5 min of dissociation in HBS.

2.9. Flow Cytometry

The purified antibody IgTT-4E1 or control monospecific mAbs (6.67 nM) were incubated with CHO, CHO^{EGFR}, $CHO^{PD\text{-}L1}$, 3T3, $3T3^{EGFR}$, Jurkat, or $Jurkat^{4\text{-}1BB}$ T cells (1×10^5 cells/well) for 30 min on ice. After washing, PE-conjugated F(ab')$_2$ goat anti-human (GAH) polyclonal antibody (Jackson ImmunoResearch, West Grove, PA, USA; cat# 109-116-170) was added and incubated for 30 min. Rituximab (Fritz Hoffmann-La Roche, Basel, Switzerland), urelumab (MedChemExpress, Monmouth Junction, NJ, USA; cat# HY-p99055), atezolizumab (Fritz Hoffmann-La Roche), and cetuximab (Merck KGaA, Darmstadt, Germany) (6.67 nM) were used as monospecific controls. To immunophenotype MDA-MB-231 cells and peripheral blood mononuclear cells (PBMC), the following mAbs were used: PE-conjugated anti-EGFR (Becton Dickinson, Franklin Lakes, NJ, USA; cat# 555997), APC-conjugated anti-PD-L1 (BD, cat# 563741), FITC-conjugated anti-PD-1 (BD, cat# 557860), PE-conjugated anti-CD137 (BD, cat# 555956), PE-conjugated isotype IgG_1 (BD, cat# 554680), APC-conjugated isotype IgG_1 (BD, cat# 554681), and FITC-conjugated isotype IgG_1 (ImmunoTools; Friesoythe, Germany; cat# 21275513) and incubated for 30 min. Finally, after washing, DAPI (Sigma Aldrich, cat# D9542) was added, and the samples were analyzed with a FACSCanto II flow cytometer (Becton Dickinson).

2.10. ADCC Reporter Bioassay

The ADCC reporter bioassay (Promega, cat# G7010) was performed according to the manufacturer's instructions. Briefly, 3T3 or $3T3^{EGFR}$ cells (1.2×10^4/well) were seeded in 96-well white plates in DMEM-10% FBS and incubated overnight at 37 °C. Then, the medium was removed, and different amounts of rituximab, IgTT-4E1, or IgTT-4E1-S (final concentrations 66.7; 6.67; 0.667 nM) were added in 40 µL of RPMI-1% FBS/well. Then, $Jurkat^{NFAT\text{-}CD16}$ effector cells (7.5×10^4/well) were added to RPMI-1% FBS (40 µL/well) and incubated for 6 h at 37 °C. Finally, 80 µL/well of BioGlo Reagent (Promega) was added, and bioluminescence was measured in a Tecan Infinite F200 Fluorescence Microplate Reader (Life Sciences, St. Petersburg, FL, USA; Tecan).

2.11. PD-1/PD-L1 Blockade Bioassay

The PD-1/PD-L1 Bioassay (Promega, cat# J1250) was used following the manufacturer's instructions. Briefly, PD-L1 aAPC/CHO-K1 cells (2.5×10^4/well) were seeded in 96-well white plates in DMEM-10% FBS and incubated overnight at 37 °C. Then, the medium was removed, and different amounts of atezolizumab, cetuximab, or IgTT-4E1-S (final concentrations 66.7; 6.67; 0.667 nM) were added in RPMI-1% FBS (40 µL well). Then, $Jurkat^{NFAT\text{-}PD\text{-}1}$ cells (1.25×10^5/well) were added to RPMI-1% FBS (40 µL/well) and incubated for 6 h at 37 °C. Finally, 80 µL/well of BioGlo Reagent was added, and bioluminescence was measured in a Tecan Infinite F200 Fluorescence Microplate Reader.

2.12. Antigen-Dependent Jurkat 4-1BB Activation Assay

The 4-1BB Bioassay (Promega, cat#. J2332) was used following the manufacturer´s instructions. Briefly, CHO, CHO^{EGFR}, or $CHO^{PD\text{-}L1}$ cells (3×10^4/well) were seeded in 96-well white plates in DMEM-10% FBS and incubated overnight at 37 °C. Then, the medium was removed, and different amounts of urelumab, rituximab, purified IgTT-4E1-S (final concentrations 66.7; 6.67; 0.667 and 0.0667 nM), or conditioned media containing IgTT-1E, IgTT-4E1-S, and IgTT-41E-S were added. Then, $Jurkat^{NFkB\text{-}4\text{-}1BB}$ cells (1.25×10^5/well) were added to RPMI-1% FBS (40 µL/well) and incubated for 6 h at 37 °C. Finally, 80 µL/well of BioGlo Reagent was added, and bioluminescence was measured in a Tecan Infinite F200 Fluorescence Microplate Reader.

2.13. IFNγ Secretion Analysis

Human peripheral blood mononuclear cells (PBMCs) were isolated from healthy donors by density gradient centrifugation (Lymphoprep, Axis–Shield, Dundee, UK; cat# AXS-1114544). All donors provided a written informed consent in accordance with the Declaration of Helsinki. PBMCs were activated overnight with 0.3 µg/mL of anti-CD3 mAb (OKT3; BD, cat#. 566685) and co-cultured with MDA-MB-231 cells (5×10^4/well) at an effector/target (E/T) ratio of 1:1 (5×10^4 cells/well) and 2:1 (1×10^5 cells/well) in the presence of 6.67 nM control polyclonal human IgG (Grifols, Barcelona, Spain), urelumab, atezolizumab, or IgTT-4E1-S for 48 h. The supernatants were collected, and IFNγ levels were analyzed by ELISA (Bionova, Fremont, CA, USA; cat# 851.560.010) following the manufacturer's instructions. The experiment was performed in triplicate.

2.14. Statistical Analysis

All graphs and statistical analyses were performed using GraphPad Prism 9.0. Most in vitro experiments were performed in triplicate, and the values are presented as mean ± SD. Significant differences (*p* value) were identified using one-way analysis of variance (ANOVA), adjusted by Dunnett's test for multiple comparisons, as indicated. *p* values are shown in the corresponding figures. Two-way ANOVA was used to analyze experiments that evaluated the interaction of two variables, such as cell type and therapy, following multiple comparison testing using either Dunnett's or Tukey's test, as appropriate.

3. Results

3.1. Generation of IgTT-4E1 and IgTT-4E1-S

In this study, we generated a trispecific 4-1BBxEGFRxPD-L1 IgTT by replacing the N-terminal scFv from IgTT-1E [24] with an anti-4-1BB agonistic scFv [13], and the C-terminal anti-EGFR V_{HH} with an anti-PD-L1 blocking V_{HH} [27] (IgTT-4E1, Figure 1a). This antibody was further modified by replacing the parental wild-type IgG_1 Fc region with an engineered Fc region containing P329G L234A/L235A (PGLALA) mutations attenuating FcγR binding [33] (IgTT-4E1-S, Figure S1a). As shown in Figure 1b,c, the middle V_{HH} (green) and the C-terminal V_{HH} (blue) were each connected to the trimerization domains by two linkers, whereas the N-terminal scFv (magenta) had a single linker, providing the binding domains with variable distances and broad areas of action around the central Fc region. Both IgTT-4E1 and IgTT-4E1-S antibodies were purified from the conditioned medium of transfected Expi293 cells by Strep-Tactin affinity chromatography and eluted as a single and an symmetrical peak with protein yields of approximately 13 mg/L. The analysis of the purified IgTT-4E1-S by SEC showed a major symmetric peak with an elution volume corresponding to an experimental molecular mass of 305 kDa, in accordance with the calculated value for the dimeric species in solution (240 kDa) (Figure S1b). A major band corresponding to the IgTT-4E1-S monomer was identified by visual inspection of the reducing SDS-PAGE analysis after SEC (Figure S1c). Both IgTT-4E1 and IgTT-4E1-S specifically recognized human 4-1BB-Fc (4-1BB), human EGFR-Fc (EGFR), and human PD-L1 Fc (PD-L1) (Figure S1d).

BLI experiments were used to investigate whether IgTT-4E1 could simultaneously bind to all three antigens. First, IgTT-4E1 was found to interact with EGFR immobilized onto biosensors. Then, the IgTT-4E1-loaded biosensors were incubated with PD-L1 and subsequently with 4-1BB, which resulted in cumulative increases in the signal (Figure 1d). The signal increases were not observed when the antigens were incubated with EGFR-coated biosensors not loaded with IgTT-4E1, demonstrating that they in fact represented complex formation with IgTT-4E1 and not non-specific interactions with EGFR or the biosensor surface. Complex formation between IgTT-4E1 and all three antigens showed that IgTT-4E1 is functionally trispecific, demonstrating a lack of steric hindrance between its three cognate interactions in this experimental context.

The ability to specifically detect antigens in a cellular context was studied by flow cytometry, using urelumab, cetuximab, and atezolizumab as binding controls, and rituximab as a negative control (Figure 1e). With $Jurkat^{4\text{-}1BB}$ cells, a specific displacement in fluorescence intensity was observed with IgTT-4E1 and urelumab compared to untransfected Jurkat cells. A specific increase in fluorescence intensity was observed upon the addition of both IgTT-4E1 and cetuximab to EGFR-positive cells ($3T3^{EGFR}$ and CHO^{EGFR}) when compared to EGFR-negative cells (3T3 and CHO). For $CHO^{PD\text{-}L1}$ cells, a shift is shown with IgTT-4E1 and atezolizumab, with respect to CHO cells. Moreover, with IgTT-4E1 and atezolizumab, a shift was observed in both Jurkat and $Jurkat^{4\text{-}1BB}$ cells with respect to the control, due to PD-L1 expression in Jurkat cells.

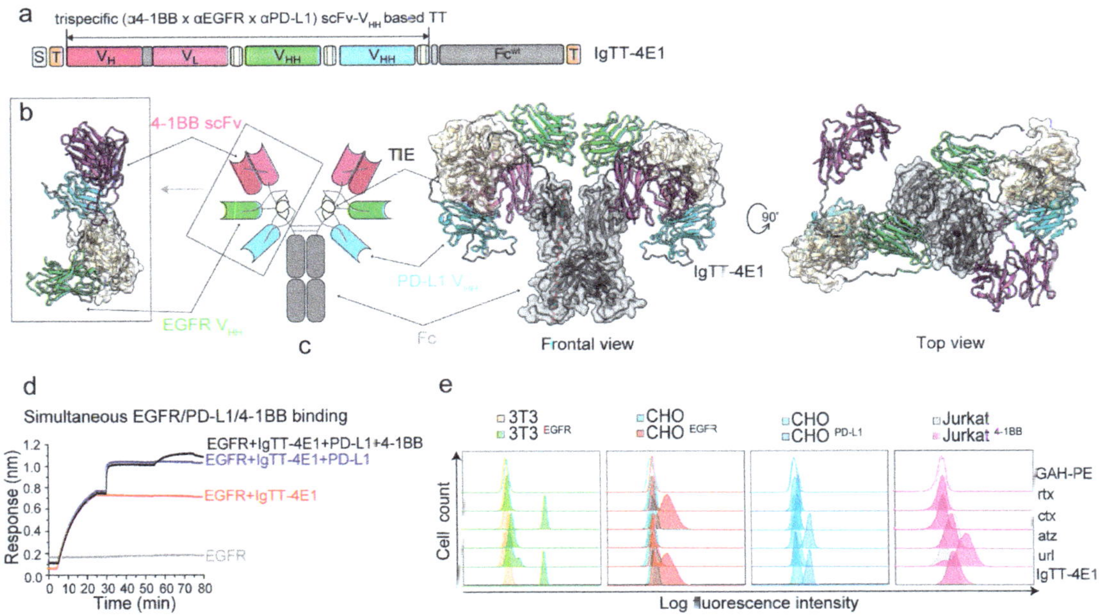

Figure 1. Gene construct, molecular model, and binding characteristics of the IgTT-4E1 antibody. (**a**) Gene layout of the IgTT-4E1 antibody bearing a signal peptide from oncostatin M (white box), one anti-4-1BB scFv (magenta boxes), one anti-EGFR V_{HH} (green box), one anti-PD-L1 V_{HH} (blue box), three collagen-derived trimerization (TIE) domains (yellow boxes) flanked by peptide linkers, and the Fc-encoding element (gray box). N-terminal FLAG-Strep and C-terminal Myc-His tags (orange boxes) were appended for purification and immunodetection purposes. (**b**) Schematic diagram showing the three-dimensional model of the anti-4-1BB/EGFR/PD-L1 tandem trimerbody (TT). (**c**) Molecular diagram of the IgTT-4E1 antibody and its three-dimensional modelizations, in front and top views. (**d**) Human EGFR-Fc (EGFR) was immobilized onto four different biosensors; three biosensors were incubated in 20 nM IgTT-4E1 for 20 min (red trace), while the fourth was kept in HBS as a negative control (gray trace). Then, 50 nM human PD-L1-Fc (PD-L1) was added to two of the IgTT-4E1-loaded biosensors (blue trace) and the control biosensor for 20 min. Also, 50 nM human 4-1BB-hFc (4-1BB) was then added to one PD-L1- and IgTT-4E1-treated biosensor (black trace) and the control biosensor. The cumulative signal increases on the biosensor receiving IgTT-4E1, PD-L1, and 4-1BB demonstrate that all three antigens can be bound by the IgTT-4E1 antibody simultaneously. (**e**) The binding to human EGFR, PD-L1, and 4-1BB on the cell surface of 3T3, $3T3^{EGFR}$, CHO, CHO^{EGFR}, $CHO^{PD\text{-}L1}$, Jurkat, and $Jurkat^{4\text{-}1BB}$ cells by rituximab (rtx), cetuximab (ctx), atezolizumab (atz), urelumab (url), and IgTT-4E1 at 6.67 nM was measured by FACS. Cells incubated with PE-conjugated GAH antibody (GAH-PE) are shown as non-filled histogram. The *y*-axis shows the relative cell number, and the *x*-axis represents the intensity of fluorescence expressed on a logarithmic scale.

3.2. Determination of Fc-Mediated Effector Functions

To measure differences in ADCC activity, Jurkat cells constitutively expressing human FcγRIIIa (CD16) on the cell surface ($Jurkat^{NFAT\text{-}CD16}$) and a luciferase reporter driven by an NFAT response element were co-cultured with $3T3^{EGFR}$ cells or with 3T3 cells as a negative control. IgTT-4E1 induced the activation of $Jurkat^{NFAT\text{-}CD16}$ cells in co-cultures with $3T3^{EGFR}$ cells, as evidenced by a significant increase in luciferase activity ($p < 0.0001$; Figure 2a), while the IgTT-4E1-S protein and rituximab (negative control) did not show an increase in luciferase activity. In the absence of EGFR-mediated interactions (co-cultures with 3T3 cells), rituximab and both IgTT-4E1 and IgTT-4E1-S showed no induction on Jurkat^{CD16} cells.

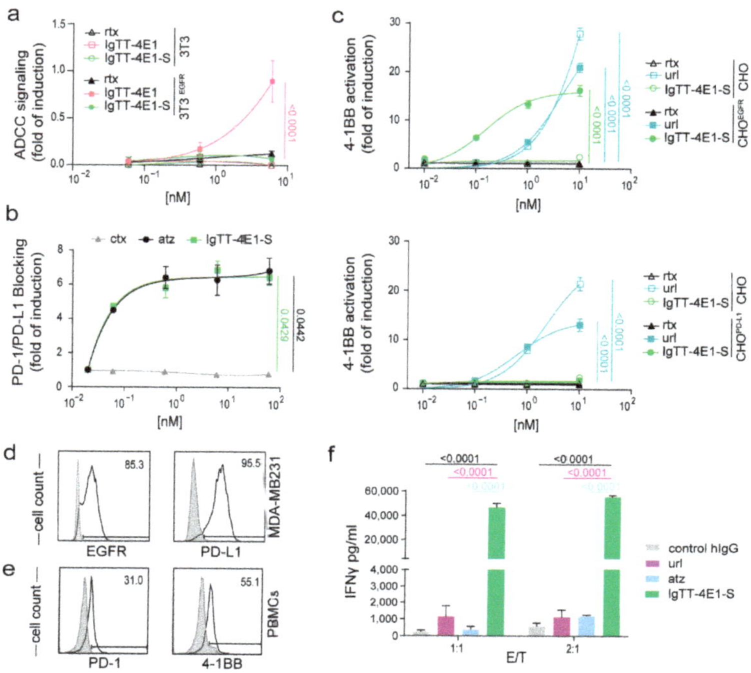

Figure 2. Effect of IgTT-4E1-S on PD-L1/PD-1 blockade, 4-1BB costimulation, and IFNγ secretion. (**a**) ADCC reporter bioassay response to rituximab (rtx), IgTT-4E1, and IgTT-4E1-S, using ADCC bioassay effector Jurkat$^{NFAT\text{-}CD16}$ cells co-cultured with 3T3 or 3T3^{EGFR} target cells at a 6:1 E/T ratio for 6 h at 37 °C. After incubation, Bio-Glo™ Luciferase Assay Reagent was added for luminescence determination. Data shown represent the mean ± standard deviation of triplicates. Data are presented as the mean ± SD (n = 3). Significance was determined by two-way ANOVA with Tukey's multiple comparison test. (**b**) PD-L1/PD-1 blockade bioassay assesses the inhibitory activities of IgTT-4E1-S. Y-axis represents reporter gene fold induction. Cetuximab (ctx) was used as a negative control, and atezolizumab (atz) as a positive control. Fold induction relative to negative control (ctx)-incubated cells. Results are expressed as mean ± SD (n = 3). Significance was measured by one-way ANOVA with Dunnett's multiple comparison test. (**c**) Costimulation of Jurkat$^{NFkB\text{-}4\text{-}1BB}$ cells co-cultured with CHO cells, CHO^{EGFR}, or CHO$^{PD\text{-}L1}$ cells in the presence of 10-fold increasing concentrations of IgTT-4E1-S, urelumab (url), and rituximab (rtx) antibodies for 6 h at 37 °C. After incubation, luminescence was determined. Data are presented as fold induction relative to the values obtained from unstimulated Jurkat$^{NFkB\text{-}4\text{-}1BB}$ cells. Representative dose–concentration curves are presented and expressed as mean ± SD (n = 3). Significance was determined by two-way ANOVA with Tukey's multiple comparison test. (**d**) Flow cytometry analysis of EGFR and PD-L1 expression in MDA-MB-231 cancer cells. Cells incubated with PE-conjugated and APC-conjugated isotypes are shown as gray-filled histogram. (**e**) Flow cytometry analysis of PD-1 and 4-1BB expression in PBMCs preactivated with anti-CD3 mAb for 24 h. PBMCs incubated with FITC-conjugated and PE-conjugated isotypes are shown as gray-filled histogram. (**f**) Anti-CD3 preactivated PBMCs were co-cultured with MDA-MB231 target cells at effector/target (E/T) ratios of 1:1 and 2:1. Control human polyclonal IgG (control hIgG), urelumab (url), atezolizumab (atz), and IgTT-4E1-S were added at 6.67 nM. The IFNγ concentrations in supernatant after 48 h were analyzed and expressed as mean ± SD (n = 3). Significance was determined by two-way ANOVA with Tukey's multiple comparison test.

3.3. Effect of IgTT-4E1-S on PD-1/PD-L1 Blockade

To demonstrate the ability of IgTT-4E1-S to block PD-1/PD-L1 interactions, Jurkat cells expressing human PD-1 and a luciferase reporter driven by an NFAT response element ($Jurkat^{NFAT\text{-}PD\text{-}1}$) were co-cultured with APC/CHO-K1 cells, which express human PD-L1, and an engineered cell surface protein designed to activate cognate TCRs in an antigen-independent manner. As shown in Figure 2b, IgTT-4E1-S efficiently blocked PD-1/PD-L1 interactions, leading to a significant induction of luciferase activity in $Jurkat^{NFAT\text{-}PD\text{-}1}$ cells ($p = 0.0429$), similar to that observed with the PD-L1 blocking antibody atezolizumab ($p = 0.0442$). In contrast, no PD-1/PD-L1 blocking activity was observed in the presence of cetuximab.

3.4. Costimulatory Activity of IgTT-4E1-S

The agonist activities of IgTT-4E1-S and urelumab were assessed using Jurkat cells stably transfected with NFκB-inducible luciferase and 4-1BB ($Jurkat^{NFkB\text{-}4\text{-}1BB}$), co-cultured with target cells expressing either EGFR (CHO^{EGFR}) or PD-L1 ($CHO^{PD\text{-}L1}$), and untransfected CHO cells as a negative control. In the absence of EGFR- or PD-L1-mediated cross-linking at the target cell surface (CHO cells), IgTT-4E1-S showed no induction in untreated $Jurkat^{NFkB\text{-}4\text{-}1BB}$ cells at all tested concentrations, whereas urelumab showed an approximately 25-fold induction (EC_{50} = 0.3115 nM, Figure 2c). In the presence of EGFR-mediated cross-linking (i.e., using CHO^{EGFR} as target cells), IgTT-4E1-S induced an NFκB dose-dependent activation with a 15-fold induction ($p < 0.0001$). Effective costimulation was achieved at concentrations that were an order of magnitude lower than those of urelumab (EC_{50} = 0.02635 nM) (Figure 2c). In contrast, in co-cultures with PD-L1-positive cells ($CHO^{PD\text{-}L1}$), IgTT-4E1-S showed no induction of luciferase at any concentration tested (Figure 2d). The negative control rituximab showed no activation (Figure 2c,d).

Thus, IgTT-4E1-S demonstrated potent and conditional costimulation that was dependent on EGFR but not on PD-L1 expression. To further investigate this issue, we generated an additional IgTT silent molecule with the order of the V_{HH} domains reversed (Figure S1a). This construct (IgTT-41E-S) was also efficiently secreted, and Western blot analysis under reducing conditions showed a migration pattern consistent with the molecular weight calculated from their amino acid sequence (Figure S2a). ELISA analysis demonstrated that IgTT-41E-S specifically recognized 4-1BB, EGFR, and PD-L1 (Figure S2b). In the presence of PD-L1-expressing cells, IgTT-41E-S also showed no induction of luciferase in $Jurkat^{NFkB\text{-}4\text{-}1BB}$ cells. However, in the presence of EGFR cells, IgTT-41E-S showed a fold induction similar to that observed with IgTT-4E1-S (Figure S2c). These results show that EGFR but not PD-L1 enabled cross-linking-mediated 4-1BB signaling with two different configurations of the PD-L1- and EGFR-binding V_{HHs}, indicating that the issue did not arise from steric hindrance of the PD-L1-binding domain.

3.5. IgTT-4E1-S Enhances the Activation of Primary Human T Cells

The triple-negative breast cancer MDA-MB-231 cells expressing both EGFR and PD-L1 (Figure 2d) were co-cultured for 48 h at two different effector/target ratios (1:1 and 2:1) with anti-CD3-prestimulated PBMCs expressing PD-1 and 4-1BB (Figure 2e). The agonistic anti-4-1BB mAb (urelumab) and anti-PD-L1 blocking mAb (atezolizumab) had a modest effect on IFNγ production (Figure 2f). In contrast, the IgTT-4E1-S antibody strongly enhanced IFNγ secretion levels, which were significantly higher ($p < 0.0001$) than those observed in the single-treated co-cultures (Figure 2f).

4. Discussion

The design of multi-specific antibodies against a combination of immunomodulatory targets is a promising approach to enhance the clinical benefit of conventional checkpoint blockers. Here, we generated a trispecific IgG-like antibody by fusing an anti-4-1BB x anti-EGFR x anti-PD-L1 TT with an engineered Fc silent region to abrogate FcγR binding. This molecule was based on the previously described IgTT platform, which is a fusion of a mono- or multispecific TT with an Fc region to generate a hexavalent IgG-like antibody capable of bivalently recognizing up to three different antigens [24]. The binding domains of IgTT-4E1 are positioned around the human collagen homodimerization domain, and BLI studies showed that IgTT-4E1 could bind all three antigens simultaneously in solution. Furthermore, IgTT-4E1 was able to recognize the antigens in a cellular context. These results suggest that the binding domains are predominantly sterically unhindered and available for antigen binding.

Anti-4-1BB agonistic mAbs can be classified as either strong or weak agonists. Here, we confirmed in a 4-1BB-reporting cell line that urelumab, a strong agonist, could induce signaling activation without FcγR-mediated cross-linking. The anti-4-1BB antibody used to generate IgTT-4E1-S is a weak agonist [13], and therefore 4-1BB signaling was only induced with additional cross-linking. IgTT-4E1-S exhibited conditional 4-1BB agonist activity dependent on the cross-linking with EGFR but not with PD-L1. This finding was independent of the position of the anti-EGFR and anti-PD-L1 V_{HH} domains in the TT, suggesting that the agonist activity was not influenced by steric hindrance issues. These results are not consistent with those published for other anti-4-1BB × anti-PD-L1 antibodies [19–22,34,35], where conditional 4-1BB agonist activity dependent on PD-L1 cross-linking was reported. These differences in the ability of IgTT to induce 4-1BB-mediated costimulation may be related to structural determinants or to the epitopes recognized by the binding domains. The anti-4-1BB antibody binds to the N-terminal CRD1 domain [36]; anti-EGFR V_{HH} binds in the cleft formed between domains II and III [37], while the epitope recognized by anti-PD-L1 V_{HH} has not been characterized. In the IgTT molecule, each TT is trispecific, and the three binding domains are very close and spatially restricted, which may limit the potential for the simultaneous recognition of distally located epitopes and epitopes more proximal to the membrane [38], as might occur with 4-1BB (86 kDa) and PD-L1 (30 kDa) interactions. This would probably not be a limitation with conventional bispecific antibodies, where each arm recognizes a different antigen, and the area of influence is significantly greater. In the context of simultaneous interactions between 4-1BB and EGFR (134 kDa) mediated by IgTT-4E1-S, available structural data suggest that the epitopes may be in closer proximity to each other. The ability to block the PD-L1/PD-1 interaction in a CHO cell system was similar to that obtained with atezolizumab. This was also previously observed with a dimeric version of the same anti-PD-L1 V_{HH} by using an ELISA-type blocking assay [27]. Simultaneous targeting of PD-L1 and EGFR by bispecific antibodies promotes PD-L1 blockade selectively in the TME due to EGFR overexpression in malignant cells, while reducing potential off-tumor binding to normal PD-L1-expressing cells [35]. In addition, EGFR overexpression and activation enhances PD-L1 expression by tumor cells [39]. IgTT-4E1-S enhanced the activation and effector functions of human primary T cells co-cultured with $EGFR^+$/$PD\text{-}L1^+$ cancer cells. However, to determine the ability of IgTT-4E1-S to restrict 4-1BB costimulation to EGFR-expressing tumors while minimizing off-target costimulation, further studies are required in cancer cell lines expressing different levels of EGFR and PD-L1, as well as in humanized mouse models carrying patient-derived xenografts. In addition, IgTT-4E1-S carries a silenced Fc domain to avoid or minimize anti-4–1BB-induced hepatic damage, which is partly explained by Fc-FcγRs interactions [6–8], but studies in non-human primates are warranted to further explore its toxicity profile.

In summary, here we described the generation of IgTT-4E1-S, the first IgG-like trispecific antibody designed to provide EGFR-conditional 4-1BB stimulation to antigen-experienced T cells while constitutively blocking the PD-1/PD-L1 inhibitory axis. This apparent asymmetry between conditional 4-1BB agonism and PD-L1 blockade may be relevant when comparing our antibody to other bispecific antibodies already on the market [17,20,22,35], but further studies are needed to determine the efficacy and safety of the IgTT-4E1-S antibody.

Supplementary Materials: The following supporting information can be downloaded at: https://www.mdpi.com/article/10.3390/antib13020034/s1, Figure S1: Gene constructs and structural characterization of the IgTT-4E1-S antibody; Figure S2: Characterization of the IgTT-41E-S antibody; Table S1: Oligonucleotides used in this study.

Author Contributions: Conception and design of the study: L.Á.-V.; sample acquisition of the data: L.R.-P., S.F., R.N., S.L.H., M.G.-R., O.H., N.S.-P., L.V. and C.S.; analysis and interpretation of the data: L.R.-P., S.F., R.N., S.L.H. and M.C.; investigation: L.R.-P., S.F., R.N. and S.L.H.; writing—original draft preparation, L.R.-P. and R.L.-G.; writing—review and editing: all authors; visualization: L.Á.-V.; supervision: L.Á.-V.; project administration: L.Á.-V.; funding acquisition: C.S. and L.Á.-V. All authors have read and agreed to the published version of the manuscript.

Funding: L.Á.-V. was supported by grants from the MCIN/AEI/10.13039/501100011033 (PID2020-117323RB-100 and PDC2021-121711-100), the Instituto de Salud Carlos III (DTS20/00089), the Comunidad de Madrid (S2022/BMD-7225 Next Generation CART MAD), the CRIS Cancer Foundation (FCRIS-2021-0090), the Spanish Association Against Cancer (PROYE19084ALVA and PRYGN234844ALVA), the Fundación "La Caixa" (HR21-00761 project IL7R_LungCan), and the Fundación de Investigación Biomédica 12 de Octubre Programa Investiga (2022-0082). L.R.-P. was supported by a predoctoral fellowship from the Immunology Chair, Universidad Francisco de Vitoria/Merck. O.H. was supported by an industrial PhD fellowship from the Comunidad de Madrid (IND2020/BMD-17668). C.S. was supported by the Instituto de Salud Carlos III (PI20/00415) and the Gobierno de Navarra, Departamento de Salud, GN21-2022.

Institutional Review Board Statement: Not applicable.

Informed Consent Statement: Informed consent was obtained from all subjects involved in the study.

Data Availability Statement The datasets used and/or analyzed during the current study are available from the corresponding author [L.Á.-V.] on reasonable request.

Conflicts of Interest: M.C., R.N., S.F., and O.H. are all employees of Leadartis. L.Á.-V. is a co-founder of Leadartis. The authors declare no conflicts of interest.

References

1. Sharma, P.; Goswami, S.; Raychaudhuri, D.; Siddiqui, B.A.; Singh, P.; Nagarajan, A.; Liu, J.; Subudhi, S.K.; Poon, C.; Gant, K.L.; et al. Immune checkpoint therapy-current perspectives and future directions. *Cell* **2023**, *186*, 1652–1669. [CrossRef]
2. Wei, S.C.; Duffy, C.R.; Allison, J.P. Fundamental Mechanisms of Immune Checkpoint Blockade Therapy. *Cancer Discov.* **2018**, *8*, 1069–1086. [CrossRef]
3. Sharpe, A.H.; Pauken, K.E. The diverse functions of the PD1 inhibitory pathway. *Nat. Rev. Immunol.* **2018**, *18*, 153–167. [CrossRef] [PubMed]
4. Antibody Therapeutics Approved or in Regulatory Review in the EU or US. The Antibody Society. Available online: https://www.antibodysociety.org/resources/approved-antibodies/ (accessed on 10 January 2024).
5. Chester, C.; Sanmamed, M.F.; Wang, J.; Melero, I. Immunotherapy targeting 4-1BB: Mechanistic rationale, clinical results, and future strategies. *Blood* **2018**, *131*, 49–57. [CrossRef]
6. Melero, I.; Shuford, W.W.; Newby, S.A.; Aruffo, A.; Ledbetter, J.A.; Hellström, K.E.; Mittler, R.S.; Chen, L. Monoclonal antibodies against the 4-1BB T-cell activation molecule eradicate established tumors. *Nat. Med.* **1997**, *3*, 682–685. [CrossRef] [PubMed]
7. Li, F.; Ravetch, J.V. Antitumor activities of agonistic anti-TNFR antibodies require differential FcγRIIB coengagement in vivo. *Proc. Natl. Acad. Sci. USA* **2013**, *110*, 19501–19506. [CrossRef]
8. Compte, M.; Harwood, S.L.; Muñoz, I.G.; Navarro, R.; Zonca, M.; Perez-Chacon, G.; Erce-Llamazares, A.; Merino, N.; Tapia-Galisteo, A.; Cuesta, A.M.; et al. A tumor-targeted trimeric 4-1BB-agonistic antibody induces potent anti-tumor immunity without systemic toxicity. *Nat. Commun.* **2018**, *9*, 4809. [CrossRef] [PubMed]
9. Claus, C.; Ferrara, C.; Xu, W.; Sam, J.; Lang, S.; Uhlenbrock, F.; Albrecht, R.; Herter, S.; Schlenker, R.; Hüsser, T.; et al. Tumor-targeted 4-1BB agonists for combination with T cell bispecific antibodies as off-the-shelf therapy. *Sci. Transl. Med.* **2019**, *11*, eaav5989. [CrossRef]
10. Hinner, M.J.; Aiba, R.S.B.; Jaquin, T.J.; Berger, S.; Dürr, M.C.; Schlosser, C.; Allersdorfer, A.; Wiedenmann, A.; Matschiner, G.; Schüler, J.; et al. Tumor-Localized Costimulatory T-Cell Engagement by the 4-1BB/HER2 Bispecific Antibody-Anticalin Fusion PRS-343. *Clin. Cancer Res.* **2019**, *25*, 5878–5889. [CrossRef]
11. Compte, M.; Harwood, S.L.; Martínez-Torrecuadrada, J.; Perez-Chacon, G.; González-García, P.; Tapia-Galisteo, A.; Van Bergen En Henegouwen, P.M.P.; Sánchez, A.; Fabregat, I.; Sanz, L.; et al. Case Report: An EGFR-Targeted 4-1BB-agonistic Trimerbody Does Not Induce Hepatotoxicity in Transgenic Mice With Liver Expression of Human EGFR. *Front. Immunol.* **2020**, *11*, 614363. [CrossRef]
12. Hangiu, O.; Compte, M.; Dinesen, A.; Navarro, R.; Tapia-Galisteo, A.; Mandrup, O.A.; Erce-Llamazares, A.; Lázaro-Gorines, R.; Nehme-Álvarez, D.; Domínguez-Alonso, C.; et al. Tumor targeted 4-1BB agonist antibody-albumin fusions with high affinity to FcRn induce anti-tumor immunity without toxicity. *iScience* **2022**, *25*, 104958. [CrossRef]

13. Compte, M.; Harwood, S.L.; Erce-Llamazares, A.; Tapia-Galisteo, A.; Romero, E.; Ferrer, I.; Garrido-Martin, E.M.; Enguita, A.B.; Ochoa, M.C.; Blanco, B.; et al. An Fc-free EGFR-specific 4-1BB-agonistic Trimerbody Displays Broad Antitumor Activity in Humanized Murine Cancer Models without Toxicity. *Clin. Cancer Res.* **2021**, *27*, 3167–3177. [CrossRef]
14. Mikkelsen, K.; Harwood, S.L.; Compte, M.; Merino, N.; Mølgaard, K.; Lykkemark, S.; Alvarez-Mendez, A.; Blanco, F.J.; Álvarez-Vallina, L. Carcinoembryonic Antigen (CEA)-Specific 4-1BB-Costimulation Induced by CEA-Targeted 4-1BB-Agonistic Trimerbodies. *Front. Immunol.* **2019**, *10*, 1791. [CrossRef]
15. Shen, A.; Liu, W.; Wang, H.; Zeng, X.; Wang, M.; Zhang, D.; Zhao, Q.; Fang, Q.; Wang, F.; Cheng, L.; et al. A novel 4-1BB/HER2 bispecific antibody shows potent antitumor activities by increasing and activating tumor-infiltrating T cells. *Am. J. Cancer Res.* **2023**, *13*, 3246–3256.
16. Tapia-Galisteo, A.; Compte, M.; Álvarez-Vallina, L.; Sanz, L. When three is not a crowd: Trispecific antibodies for enhanced cancer immunotherapy. *Theranostics* **2023**, *13*, 1028–1041. [CrossRef]
17. Geuijen, C.; Tacken, P.; Wang, L.-C.; Klooster, R.; van Loo, P.F.; Zhou, J.; Mondal, A.; Liu, Y.; Kramer, A.; Condamine, T.; et al. A human CD137×PD-L1 bispecific antibody promotes anti-tumor immunity via context-dependent T cell costimulation and checkpoint blockade. *Nat. Commun.* **2021**, *12*, 4445. [CrossRef]
18. Cheng, L.-S.; Zhu, M.; Gao, Y.; Liu, W.-T.; Yin, W.; Zhou, P.; Zhu, Z.; Niu, L.; Zeng, X.; Zhang, D.; et al. An Fc-muted bispecific antibody targeting PD-L1 and 4-1BB induces antitumor immune activity in colorectal cancer without systemic toxicity. *Cell Mol. Biol. Lett.* **2023**, *28*, 47. [CrossRef]
19. Jeong, S.; Park, E.; Kim, H.-D.; Sung, E.; Kim, H.; Jeon, J.; Kim, Y.; Jung, U.; Son, Y.-G.; Hong, Y.; et al. Novel anti-4-1BB×PD-L1 bispecific antibody augments anti-tumor immunity through tumor-directed T-cell activation and checkpoint blockade. *J. Immunother. Cancer* **2021**, *9*, e002428. [CrossRef]
20. Peper-Gabriel, J.K.; Pavlidou, M.; Pattarini, L.; Morales-Kastresana, A.; Jaquin, T.J.; Gallou, C.; Hansbauer, E.-M.; Richter, M.; Lelievre, H.; Scholer-Dahirel, A.; et al. The PD-L1/4-1BB Bispecific Antibody-Anticalin Fusion Protein PRS-344/S095012 Elicits Strong T-Cell Stimulation in a Tumor-Localized Manner. *Clin. Cancer Res.* **2022**, *28*, 3387–3399. [CrossRef] [PubMed]
21. Muik, A.; Altintas, I.; Gieseke, F.; Schoedel, K.B.; Burm, S.M.; Toker, A.; Salcedo, T.W.; Verzijl, D.; Eisel, D.; Grunwitz, C.; et al. An Fc-inert PD-L1×4-1BB bispecific antibody mediates potent anti-tumor immunity in mice by combining checkpoint inhibition and conditional 4-1BB co-stimulation. *Oncoimmunology* **2022**, *11*, 2030135. [CrossRef] [PubMed]
22. Muik, A.; Garralda, E.; Altintas, I.; Gieseke, F.; Geva, R.; Ben-Ami, E.; Maurice-Dror, C.; Calvo, E.; LoRusso, P.M.; Alonso, G.; et al. Preclinical Characterization and Phase I Trial Results of a Bispecific Antibody Targeting PD-L1 and 4-1BB (GEN1046) in Patients with Advanced Refractory Solid Tumors. *Cancer Discov.* **2022**, *12*, 1248–1265. [CrossRef]
23. Alvarez-Cienfuegos, A.; Nuñez-Prado, N.; Compte, M.; Cuesta, A.M.; Blanco-Toribio, A.; Harwood, S.L.; Villate, M.; Merino, N.; Bonet, J.; Navarro, R.; et al. Intramolecular trimerization, a novel strategy for making multispecific antibodies with controlled orientation of the antigen binding domains. *Sci. Rep.* **2016**, *6*, 28643. [CrossRef]
24. Rubio-Pérez, L.; Lázaro-Gorines, R.; Harwood, S.L.; Compte, M.; Navarro, R.; Tapia-Galisteo, A.; Bonet, J.; Blanco, B.; Lykkemark, S.; Ramírez-Fernández, Á.; et al. A PD-L1/EGFR bispecific antibody combines immune checkpoint blockade and direct anti-cancer action for an enhanced anti-tumor response. *Oncoimmunology* **2023**, *12*, 2205336. [CrossRef]
25. Delidakis, G.; Kim, J.E.; George, K.; Georgiou, G. Improving Antibody Therapeutics by Manipulating the Fc Domain: Immunological and Structural Considerations. *Annu. Rev. Biomed. Eng.* **2022**, *24*, 249–274. [CrossRef]
26. Roskoski, R. The ErbB/HER family of protein-tyrosine kinases and cancer. *Pharmacol. Res.* **2014**, *79*, 34–74. [CrossRef]
27. Silva-Pilipich, N.; Blanco, E.; Lozano, T.; Martisova, E.; Igea, A.; Herrador-Cañete, G.; Ballesteros-Briones, M.C.; Gorraiz, M.; Sarrión, P.; González-Sapienza, G.; et al. Local delivery of optimized nanobodies targeting the PD-1/PD-L1 axis with a self-amplifying RNA viral vector induces potent antitumor responses. *Cancer Lett.* **2023**, *561*, 216139. [CrossRef]
28. Šali, A.; Blundell, T.L. Comparative protein modelling by satisfaction of spatial restraints. *J. Mol. Biol.* **1993**, *234*, 779–815. [CrossRef]
29. Saphire, E.O.; Parren, P.W.; Pantophlet, R.; Zwick, M.B.; Morris, G.M.; Rudd, P.M.; Dwek, R.A.; Stanfield, R.L.; Burton, D.R.; Wilson, I.A. Crystal structure of a neutralizing human IGG against HIV-1: A template for vaccine design. *Science* **2001**, *293*, 1155–1159. [CrossRef]
30. Berman, H.M.; Westbrook, J.; Feng, Z.; Gilliland, G.; Bhat, T.N.; Weissig, H.; Shindyalov, I.N.; Bourne, P.E. The Protein Data Bank. *Nucleic Acids Res.* **2000**, *28*, 235–242. [CrossRef]
31. Altschul, S.F.; Gish, W.; Miller, W.; Myers, E.W.; Lipman, D.J. Basic local alignment search tool. *J. Mol. Biol.* **1990**, *215*, 403–410. [CrossRef]
32. Liu, S.; Miersch, S.; Li, P.; Bai, B.; Liu, C.; Qin, W.; Su, J.; Huang, H.; Pan, J.; Sidhu, S.S.; et al. A Synthetic Human Antibody Antagonizes IL-18Rβ Signaling Through an Allosteric Mechanism. *J. Mol. Biol.* **2020**, *432*, 1169–1182. [CrossRef] [PubMed]
33. Lo, M.; Kim, H.S.; Tong, R.K.; Bainbridge, T.W.; Vernes, J.-M.; Zhang, Y.; Lin, Y.L.; Chung, S.; Dennis, M.S.; Zuchero, Y.J.Y.; et al. Effector-attenuating Substitutions That Maintain Antibody Stability and Reduce Toxicity in Mice. *J. Biol. Chem.* **2017**, *292*, 3900–3908. [CrossRef] [PubMed]
34. Zhai, T.; Wang, C.; Xu, Y.; Huang, W.; Yuan, Z.; Wang, T.; Dai, S.; Peng, S.; Pang, T.; Jiang, W.; et al. Generation of a safe and efficacious llama single-domain antibody fragment (vHH) targeting the membrane-proximal region of 4-1BB for engineering therapeutic bispecific antibodies for cancer. *J. Immunother. Cancer* **2021**, *9*, e002131. [CrossRef] [PubMed]

35. Warmuth, S.; Gunde, T.; Snell, D.; Brock, M.; Weinert, C.; Simonin, A.; Hess, C.; Tietz, J.; Johansson, M.; Spiga, F.M.; et al. Engineering of a trispecific tumor-targeted immunotherapy incorporating 4-1BB co-stimulation and PD-L1 blockade. *Oncoimmunology* **2021**, *10*, 2004661. [CrossRef]
36. Bitra, A.; Doukov, T.; Croft, M.; Zajonc, D.M. Crystal structures of the human 4-1BB receptor bound to its ligand 4-1BBL reveal covalent receptor dimerization as a potential signaling amplifier. *J. Biol. Chem.* **2018**, *293*, 9958–9969. [CrossRef] [PubMed]
37. Schmitz, K.R.; Bagchi, A.; Roovers, R.C.; van Bergen en Henegouwen, P.M.P.; Ferguson, K.M. Structural evaluation of EGFR inhibition mechanisms for nanobodies/VHH domains. *Structure* **2013**, *21*, 1214–1224. [CrossRef] [PubMed]
38. Bluemel, C.; Hausmann, S.; Fluhr, P.; Sriskandarajah, M.; Stallcup, W.B.; Baeuerle, P.A.; Kufer, P. Epitope distance to the target cell membrane and antigen size determine the potency of T cell-mediated lysis by BiTE antibodies specific for a large melanoma surface antigen. *Cancer Immunol. Immunother.* **2010**, *59*, 1197–1209. [CrossRef]
39. Zhang, W.; Pang, Q.; Yan, C.; Wang, Q.; Yang, J.; Yu, S.; Liu, X.; Yuan, Z.; Wang, P.; Xiao, Z. Induction of PD-L1 expression by epidermal growth factor receptor-mediated signaling in esophageal squamous cell carcinoma. *Onco. Targets Ther.* **2017**, *10*, 763–771. [CrossRef]

antibodies

MDPI

Article

^{177}Lu Anti-Angiogenic Radioimmunotherapy Targeting ATP Synthase in Gastric Cancer Model

Bok-Nam Park, Young-Sil An, Su-Min Kim, Su-Jin Lee, Yong-Jin Park and Joon-Kee Yoon *

Department of Nuclear Medicine & Molecular Imaging, Ajou University School of Medicine, Worldcup-ro 164, Suwon 16499, Republic of Korea; curies@ajou.ac.kr (B.-N.P.); aysays@aumc.ac.kr (Y.-S.A.); smversion@naver.com (S.-M.K.); suesj202@aumc.ac.kr (S.-J.L.); yjpark@aumc.ac.kr (Y.-J.P.)

* Correspondence: jkyoon3@ajou.ac.kr; Tel.: +82-31-219-7839

Abstract: This study investigated a novel radioimmunotherapy strategy for targeting tumor angiogenesis. We developed a radiopharmaceutical complex by labeling an anti-adenosine triphosphate synthase (ATPS) monoclonal antibody (mAb) with the radioisotope ^{177}Lu using DOTA as a chelating agent. ^{177}Lu-DOTA-ATPS mAb demonstrated high labeling efficiency (99.0%) and stability in serum. MKN-45 cancer cells exhibited the highest cellular uptake, which could be specifically blocked by unlabeled ATPS mAb. In mice, ^{177}Lu-DOTA-ATPS mAb accumulated significantly in tumors, with a tumor uptake of 16.0 ± 1.5%ID/g on day 7. ^{177}Lu-DOTA-ATPS mAb treatment significantly reduced the viability of MKN-45 cells in a dose-dependent manner. In a xenograft tumor model, this radioimmunotherapy strategy led to substantial tumor growth inhibition (82.8%). Furthermore, combining ^{177}Lu-DOTA-ATPS mAb with sunitinib, an anti-angiogenic drug, enhanced the therapeutic efficacy of sunitinib in the mouse model. Our study successfully developed ^{177}Lu-DOTA-ATPS mAb, a radioimmunotherapy agent targeting tumor blood vessels. This approach demonstrates significant promise for inhibiting tumor growth, both as a single therapy and in combination with other anti-cancer drugs.

Keywords: radioimmunotherapy; ATP synthase; ^{177}Lu; angiogenesis; gastric cancer

1. Introduction

Angiogenesis, the formation of new blood vessels, is a crucial process for normal tissue growth and tumor expansion. Disrupting this process has become a major focus for researchers aiming to develop effective cancer therapies. Tumors hijack angiogenesis to fuel their growth and metastasis [1]. This uncontrolled vessel formation is countered by endogenous inhibitors like angiostatin. Angiostatin maintains a balanced angiogenic environment by suppressing the effects of proangiogenic factors like vascular endothelial growth factor and fibroblast growth factor [2]. While typically residing within the mitochondrial inner membrane, adenosine triphosphate synthase (ATPS), an enzyme responsible for ATP generation, has been found on the surface of endothelial and tumor cells, called "ectopic" ATPS [3]. Specific subunits (α/β) of this ectopic ATPS serve as binding sites for angiostatin [4]. Antibodies developed against specific ATPS subunits can directly inhibit the enzyme's activity on the endothelial cell surface, which, in turn, disrupts the formation of new blood vessels and directly hinders the proliferation and migration of cancer cells [5–7]. We also have shown that radiolabeled anti-ATPS antibodies are promising tools for radioimmunotherapy and immuno-positron emission tomography (PET) imaging. Radioiodine (^{131}I)-labeled anti-ATPS antibody effectively suppressed the tumor growth by 2.5-fold in a gastric cancer model [8]. Anti-ATPS antibodies labeled with Zirconium-89 (^{89}Zr) demonstrated significantly higher targeting specificity for MDA-MB-231 tumors with abundant ectopic ATPS expression compared to PC-3 tumors with low expression [9]. Given its role in tumor angiogenesis, ectopic ATPS emerges as a novel target for developing effective anti-angiogenic therapies.

Since the landmark report in 1981 on a successful radiolabeled antibody targeting carcinoembryonic antigen [10], radioimmunotherapy has remained a beacon of hope in the fight against cancer, offering a powerful tool for targeted therapy. Zevalin™ and Bexxar™, being used for the treatment of non-Hodgkin's lymphoma, are prime examples of radioimmunotherapy successfully employed in modern medicine. Lutetium-177 (^{177}Lu) has gained widespread popularity as a preferred radioisotope in recent years. This radioisotope demonstrates its versatility in cancer treatment. Studies have shown its effectiveness in treating two distinct cancers: unresectable metastatic neuroendocrine tumors [11] and metastatic castration-resistant prostate cancer [12]. In both cases, it is used as a targeted therapy approach (peptide receptor radionuclide therapy and radioligand therapy). ^{177}Lu has favorable characteristics for radioimmunotherapy. It emits both high-energy β-ray (Emax = 761 KeV) and γ-rays (113 in 6.4% and 208 KeV in 13%) and decays with a half-life of 6.7 days [13], which is favorable for radioimmunotherapy.

This study aims to develop a new radioimmunotherapy approach that specifically targets tumor angiogenesis. We synthesized ^{177}Lu-labeled anti-ATPS antibody to achieve this and evaluated its antitumor efficacy in a gastric cancer animal model.

2. Materials and Methods

2.1. Radiosynthesis of ^{177}Lu-DOTA-ATPS mAb

The ATPS mAb was purchased from Abcam (ab14730, MW 52 kDa; Cambridge, MA, USA) and stored as aliquots at −78 °C. $^{177}LuCl_3$ was obtained from PLATOM (National Centre for Nuclear Research, Poland). For conjugation, a 50-fold molar excess of p-SCN-Bn-tetraazacyclododecane-1,4,7,10-tetraacetic acid (DOTA, Futurechem, Seoul, Republic of Korea) in 30 μL dimethyl sulfoxide was added to the ATP mAb (100–200 μg in 20 μL 0.1 N $NaHCO_3$ buffer), and the reaction mixture was incubated at 37 °C for 30 min. After incubation, the conjugation mixture was purified by using Slide-A-Lyzer™ Dialysis Cassettes (2K MWCO, Thermo Fisher Scientific, Rockford, IL, USA) to remove unconjugated *p*-SCN-Bn-DOTA. $^{177}LuCl_3$ (37–111 MBq/10 μL) was buffered with 0.5 M NH_4OAc (pH 5.5), followed by 100–200 μg DOTA-mAb. The reaction mixture was incubated at 37 °C for 1 h [14]. After completion of synthesis, the reaction mixture was purified on a size-exclusive PD-10 column (GE Healthcare) and ^{177}Lu-DOTA-ATPS mAb was separated by eluting with PBS (Figure 1). For analysis of labeled ^{177}Lu-DOTA-ATPS mAb, it was spotted on silica gel impregnated aluminum sheets (Merck, Darmstadt, Germany), developed with 0.02 M citrate buffer, and confirmed by using radio-thin-layer chromatography (radio-TLC) (Bioscan, Eckert & Ziegler Radiopharma Inc., Wilmington, MA, USA). In vitro stability was measured in triplicates at 2 h, 24 h, and on the 7th day, in PBS or serum, at 4 °C, room temperature, or 37 °C.

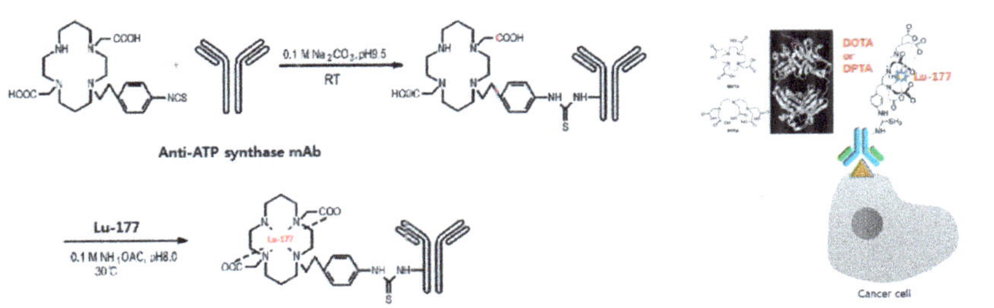

Figure 1. Schematic diagram for radiosynthesis of ^{177}Lu-DOTA-ATPS mAb.

2.2. Cancer Cell Culture and Xenograft Tumor Model

All cancer cell lines, human breast adenocarcinoma (MDA-MB-231), human fibrosarcoma (HT-1080), human lung adenocarcinoma (A549), human follicular thyroid carcinoma (FTC-133), human prostate adenocarcinoma (PC-3), and human gastric adenocarcinoma (MKN-45) were purchased from the Korean Cell Line Bank (Seoul, Republic of Korea).

All cells except FTC-133 (FTC-133 cells were cultured in DMEM/F-12, 1:1 mixture, 10% fetal bovine serum, WelGENE Inc., Daegu, Republic of Korea) were cultured with RPMI-1640 medium (WelGENE) supplemented with 10% FBS and 1% penicillin/streptomycin (WelGENE) at 37 °C and 5% fully humidified CO_2. Animal experiments were performed according to protocols approved by the Care of Experimental Animals Committee (IACUC No. 2021-0067). Six-week-old female Balb/c nude mice (Orient Bio, Seongnam, Republic of Korea) were maintained under specific pathogen-free conditions [15]. To create a tumor xenograft model, 5×10^6 tumor cells were mixed with phenol red-free Matrigel (Corning®, Bedford, MA, USA) and subcutaneous inoculation was injected into the right shoulder of each mouse. Experiments were performed about 10 to 14 days after injection of cells, when tumors reached a diameter of approximately 5 to 10 mm.

2.3. Cellular Uptake of ^{177}Lu-DOTA-ATPS mAb in Various Cancer Cells

Cellular uptake of ^{177}Lu-DOTA-ATPS mAb was measured in the six human cancer cell lines as described previously [8]. In brief, 5×10^5 cells were seeded per well in 12-well plates and cultured for 24 h. Upon attachment, 37 kBq ^{177}Lu-DOTA-ATPS mAb was added to freshly replaced culture media, followed by incubation of the cells for 1, 4, or 24 h at 37 °C and 5% CO_2. After incubation, the cells were washed twice with cold PBS and harvested with 0.1 N NaOH. Radioactivity of the cells was counted using a Gamma-HEs gamma counter (Shinjin Medics Inc., Goyang, Republic of Korea) and normalized to the cell protein content obtained using the Bradford method [16]. Data are presented as the percentage of 1 h uptake.

2.4. Specific Binding of ^{177}Lu-DOTA-ATPS mAb in MKN-45 Cells

MKN-45 cells were cultured in 12-well plates and incubated with 111 kBq free ^{177}Lu, ^{177}Lu-DOTA-ATPS mAb, or ^{177}Lu-DOTA-IgG for 24 h at 37 °C and 5% CO_2. After incubation, the cells were washed twice with cold PBS, harvested with 0.1 N NaOH, and the radioactivity was counted using a gamma counter. The cell protein content was determined using the Pierce 660™ Protein Assay Kit for normalization (Thermo Fisher Scientific, Rockford, IL, USA). Cellular uptake was expressed as a percentage of free ^{177}Lu uptake.

2.5. Competitive Binding of ^{177}Lu-DOTA-ATPS mAb in MKN-45 Cells

Competitive inhibition of ^{177}Lu-DOTA-ATPS mAb binding was examined using unlabeled ATPS mAb in MKN-45 cells. Cells were cultured in 12-well plates and treated with 10% FBS and 1% penicillin/streptomycin at 37 °C and 5% CO_2 (as described previously). The cells were pretreated with 6.4 μM unlabeled ATPS mAb for 1 h, while control cells were incubated with the vehicle. Then, 111 kBq ^{177}Lu-DOTA-ATPS mAb was added to the cells and incubated for 4 or 24 h under the same conditions. After incubation, cellular uptake was calculated and expressed as a percentage relative to that of untreated control. Cellular uptake, specific binding, and inhibition experiments were all performed in triplicate.

2.6. ^{177}Lu-DOTA-ATPS mAb Radioimmunotherapy in MKN-45 Cells

MKN-45 cells were cultured in 96-well plates and treated with 3.7 or 7.4 MBq ^{177}Lu-DOTA-ATPS mAb, unlabeled ATPS mAb, or left untreated for 24 h at 37 °C and 5% CO_2. To investigate the effect of combination therapy, MKN-45 cells were treated with either 7.4 MBq ^{177}Lu-DOTA-ATPS mAb or 5 mg/mL sunitinib (LC Laboratories®®, Woburn, MA, USA), both, or left untreated for 24 h at 37 °C and 5% CO_2 [17]. After incubation, cell viability was measured using an XTT assay kit (Cayman Chemical, Ann Arbor, MI, USA) with a further 2 h incubation at 37 °C. Absorbance was read at 450 nm using a microplate reader (Bio-Rad Laboratories Inc., Hercules, CA, USA).

2.7. Biodistribution Study of Wild-Type Mice and MKN-45 Tumor Xenograft Models

Wild-type mice and MKN-45 tumor-bearing mice (n = 4–5 per time point) were intravenously injected with 3.7 MBq ^{177}Lu-DOTA-ATPS mAb, ^{177}Lu-IgG, or free ^{177}Lu. The mice were then anesthetized, sacrificed, and dissected for organ radioactivity analysis at 1, 2, 4, and 7 days after injection. Major organs (heart, lung, liver, spleen, stomach, kidneys, intestine, muscle, and bone), blood, and tumors were dissected, weighed, and counted for radioactivity using a gamma counter. Uptake in the organs and tumors was expressed as the percentage of the injected dose per gram of tissue (%ID/g).

2.8. Blocking Study of ^{177}Lu-DOTA-ATPS mAb in MKN-45 Tumor-Bearing Mice

To investigate blocking efficacy, 50 μg of unlabeled ATPS mAb was co-injected with 3.7 MBq ^{177}Lu-DOTA-ATPS mAb (1 μg as mAb) through the tail vein (n = 2). Tumors and organs were then removed at 1 and 7 days after injection for subsequent radioactivity analysis. The organs were weighed and counted for radioactivity using a gamma counter. Results were expressed as %ID/g.

2.9. Radioimmunotherapy, Immunohistochemical Staining, and ^{18}F-FDG-PET Imaging in MKN-45 Tumor-Bearing Mice

To investigate radioimmunotherapy, tumor models were intravenously injected with 18.5 MBq ^{177}Lu-DOPA-ATPS mAb, 30 μg unlabeled ATPS mAb (10 times larger than the therapeutic dosage of ^{177}Lu-DOPA-ATPS mAb), 30 μg unlabeled IgG, and vehicle (normal saline), respectively, once a week for 4 weeks (n = 4 for each group) [8]. Tumor size was measured twice a week in two dimensions (length and width). Tumor volume was calculated using the formula, V = (length ×. width^2)/2, and compared among the groups [18]. Tumor growth inhibition (TGI, %) was calculated using the formula, TGI = (1 − mean volume of treated tumors/mean volume of control tumors) × 100.

The effect of combination therapy of ^{177}Lu-DOTA-ATPS mAb and sunitinib was evaluated in mice bearing MKN-45 tumors. Mice were divided into four groups: vehicle (0.9% NaCl), 18.5 MBq ^{177}Lu-DOTA-ATPS mAb alone, 40 mg/kg sunitinib alone, or a combination of both (n = 6 for each group). Mice received 18.5 MBq ^{177}Lu-DOTA-ATPS mAb and 40 mg/kg sunitinib once a week for 4 weeks [19]. Tumor size was measured twice a week as described above. Body weights were measured once a week.

For imaging, PET images were acquired 1 h after intravenous injection of 18.5 MBq 2-deoxy-2-[^{18}F]fluoro-D-glucose (^{18}F-FDG) using a PET scanner (SimPET, BRIGHTONIX IMAGING®, Seoul, Republic of Korea) before and after therapies. Mice were scanned for 20 min under anesthesia with isoflurane inhalation.

For immunohistochemistry, tumors were dissected immediately after the PET imaging at the 4th week. Slides were stained using an anti-CD31 antibody (ab28364, abcam) according to the manufacturer's standard procedure.

2.10. Statistical Analysis

All data are presented as means ± standard errors. The statistical comparison of cellular uptake and tumor size was evaluated by Student's *t*-test and Kruskal–Wallis test using statistical software (R, version 3.1.2), and the difference was considered significant at $p < 0.05$.

3. Results

3.1. Labeling Efficiency and In Vitro Stability of ^{177}Lu-DOTA-ATPS mAb

The ^{177}Lu-DOTA-ATPS mAb was successfully synthesized according to the schematic representation. The radiochemical yield of ^{177}Lu-DOTA-ATPS mAb was 99.0% (Figure 2A). The in vitro stabilities of ^{177}Lu-DOTA-ATPS mAb in serum were at least 95% on the 2nd day and 85% on the 7th day at all temperature conditions (Figure 2B). Similarly, the in-vitro stabilities of ^{177}Lu-DOTA-ATPS mAb in PBS were at least 94% on the 2nd day, regardless of temperature. However, on the 7th day, the stability dropped significantly to 65%, 63%, and 91% at 4 °C, room temperature, and 37 °C, respectively (all $p < 0.005$).

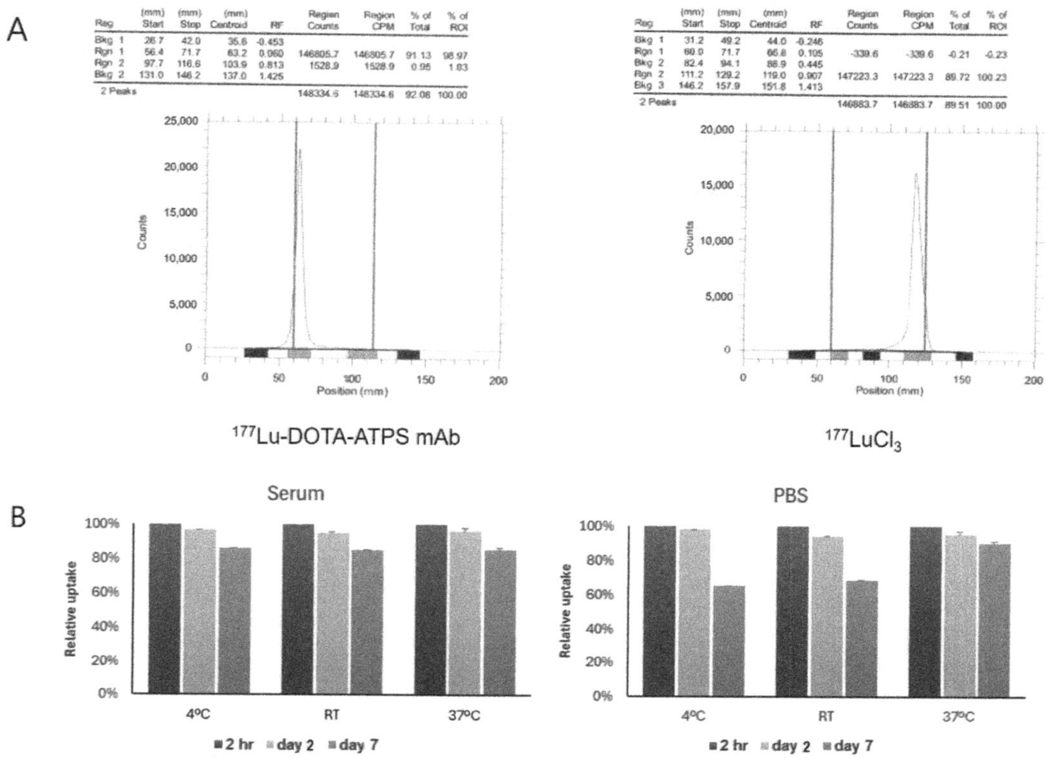

Figure 2. Labeling efficiency (**A**) and in vitro stability (**B**) of ^{177}Lu-DOTA-ATPS mAb. The Rf value of ^{177}Lu-DOTA-ATPS mAb was between 0.01 and 0.05, while that of $^{177}LuCl_3$ was between 0.6 and 1.0. The in vitro stabilities of ^{177}Lu-DOTA-ATPS mAb in serum remained unchanged up to 7 days. DOTA, tetraazacyclododecane-1,4,7,10-tetraacetic acid; ATPS, adenosine triphosphate synthase; mAb, monoclonal antibody; RT, room temperature; PBS, phosphate-buffered saline.

3.2. Cellular Uptake, Specific Binding, and Inhibition of ^{177}Lu-DOTA-ATPS mAb

Six cancer cell lines were evaluated for their cellular uptake of ^{177}Lu-DOTA-ATPS (Figure 3A). MKN-45 cells exhibited a time-dependent increase in uptake reaching 189.3% ± 9.8% and 450.8% ± 13.3% of the 1 h uptake at 4 and 24 h, respectively ($p < 0.0005$ and < 0.0001, respectively). PC-3 also showed increased uptake at both 4 h (121.9% ± 4.9%, $p < 0.05$) and at 24 h (190.7% ± 8.8%, $p < 0.0005$). Similarly, MDA-MB-231 (186.4% ± 7.7%, $p < 0.001$), HT-1080 (158.7% ± 9.5%, $p < 0.005$), A549 (128.0% ± 9.9%, $p < 0.05$), and FTC-133 (186.2% ± 4.9%, $p < 0.001$) showed a significant increase in cellular uptake of ^{177}Lu-DOTA-ATPS mAb at 24 h, while their 4 h uptake remained unchanged ($p >$ 0.05). Notably, MKN-45 cells demonstrated significantly higher uptake compared to other cell lines at 24 h (all $p < 0.0005$). This finding prompted further cellular and animal studies using MKN-45 cells.

In MKN-45 cells, the uptake of ^{177}Lu-DOTA-ATPS mAb was compared with that of free ^{177}Lu ($^{177}LuCl_3$) and ^{177}Lu-DOTA-IgG at 24 h (Figure 3B). The cellular uptake of ^{177}Lu-DOTA-ATPS mAb (128.7% ± 3.6%) was significantly higher than that of free ^{177}Lu and ^{177}Lu-DOTA-IgG (88.5% ± 7.8%) (all $p < 0.05$). There was no significant difference in cellular uptake between free ^{177}Lu and ^{177}Lu-DOTA-IgG ($p > 0.05$).

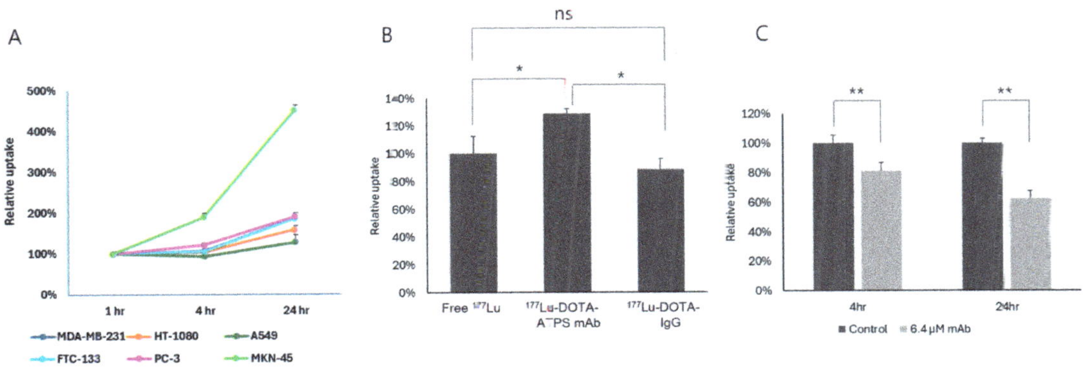

Figure 3. The cellular uptake (**A**), specific binding (**B**), and inhibition study (**C**) of ^{177}Lu-DOTA-ATPS mAb. MKN-45 cells showed the highest cellular uptake of ^{177}Lu-DOTA-ATPS mAb among the tested cancer cell lines. ^{177}Lu-DOTA-ATPS mAb uptake was specific and inhibited by unlabeled ATPS mAb in MKN-45 cells. DOTA, tetraazacyclododecane-1,4,7,10-tetraacetic acid; ATPS, adenosine triphosphate synthase; mAb, monoclonal antibody. $p < 0.05$ *, $p < 0.005$ **, ns: not significant.

The uptake of ^{177}Lu-DOTA-ATPS mAb was inhibited by a high dose of unlabeled ATPS mAb at both 4 h (81.0% ± 5.9%, $p < 0.005$) and at 24 h (62.2% ± 5.4%, $p < 0.005$) (Figure 3C). The inhibitory effect of unlabeled ATPS mAb was more pronounced at 24 h compared to 4 h ($p < 0.05$).

3.3. ^{177}Lu-DOTA-ATPS mAb Radioimmunotherapy in MKN-45 Cells

Radioimmunotherapy with ^{177}Lu-DOTA-ATPS mAb significantly reduced cell viability (Figure 4A) compared with vehicle-treated controls. Treatment with 3.7 MBq (78.9% ± 1.2%, $p < 0.005$) and 7.4 MBq (70.4% ± 1.7%, $p < 0.01$) of ^{177}Lu-DOTA-ATPS mAb resulted in a dose-dependent decrease in viable cells ($p < 0.005$). Unlabeled ATPS mAb also significantly reduced cell viability (89.8% ± 0.5%, $p < 0.001$ vs. control); however, the therapeutic effect of either dose of ^{177}Lu-DOTA-ATPS mAb was significantly greater (all $p < 0.0001$).

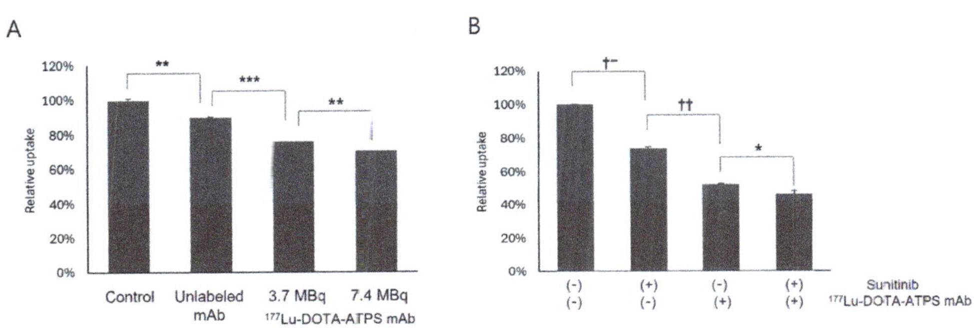

Figure 4. Radioimmunotherapy with ^{177}Lu-DOTA-ATPS mAb alone (**A**) and ^{177}Lu-DOTA-ATPS mAb in combination with sunitinib (**B**) in MKN-45 cells. DOTA, tetraazacyclododecane-1,4,7,10-tetraacetic acid; ATPS, adenosine triphosphate synthase; mAb, monoclonal antibody. $p < 0.05$ *, $p < 0.005$ **, $p < 0.001$ ***, $p < 0.0005$ $^{+}$, $p < 0.00005$ $^{++}$.

An anti-angiogenic therapy with sunitinib (5 mg/mL) significantly decreased the number of viable MKN-45 cells (73.8% ± 1.0% of controls, $p < 0.00005$) (Figure 4B). Combination therapy with 7.4 MBq of ^{177}Lu-DOTA-ATPS mAb and sunitinib showed a greater reduction in cell viability (46.5% ± 1.5%) compared to single therapy with ^{177}Lu-DOTA-ATPS mAb

(52.5% ± 0.2%, $p < 0.05$) or sunitinib ($p < 0.00005$). Notably, among single therapies, 7.4 MBq of ^{177}Lu-DOTA-ATPS mAb exhibited a greater therapeutic effect than 5 mg/mL sunitinib ($p < 0.00005$).

3.4. Biodistribution of ^{177}Lu-DOTA-ATPS mAb in Wild-Type Mice and MKN-45 Tumor Xenograft Models

The biodistribution of ^{177}Lu-DOTA-ATPS mAb was evaluated in wild-type mice (Figure 5) and mice bearing MKN-45 tumors (Figure 6) on days 1, 2, 4, and 7. In wild-type mice, bone marrow uptake of ^{177}Lu-DOTA-ATPS mAb reached 20.4 ± 1.3%ID/g on day 1 and remained stable from day 2 to day 7. Renal uptake was highest on day 1 (25.4 ± 0.6%ID/g) and decreased slightly from day 2 to day 7 (13.0 to 17.4%ID/g). Hepatic uptake ranged from 17.9 to 20.6%ID/g.

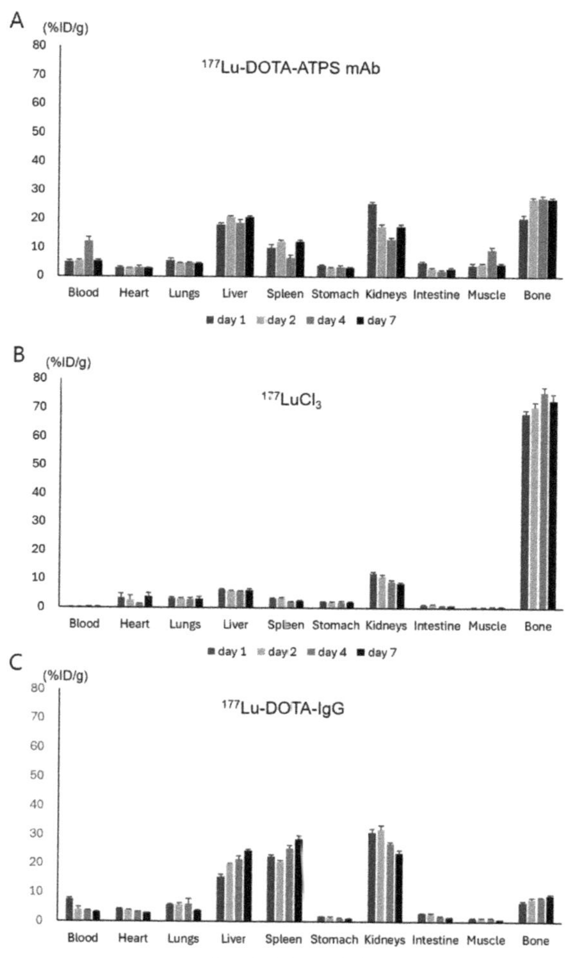

Figure 5. Biodistribution of ^{177}Lu-DOTA-ATPS mAb (**A**), ^{177}LuCl$_3$ (**B**), and ^{177}Lu-DOTA-IgG (**C**) in wild-type mice on days 1, 2, 4, and 7. DOTA, tetraazacyclododecane-1,4,7,10-tetraacetic acid; ATPS, adenosine triphosphate synthase; mAb, monoclonal antibody.

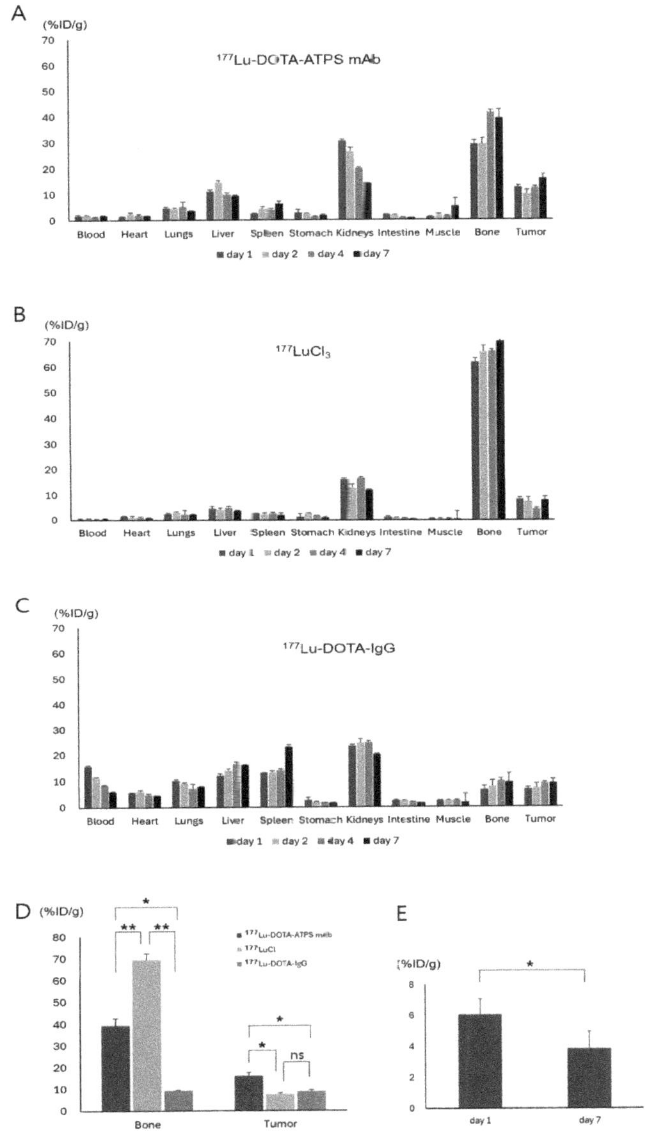

Figure 6. Biodistribution of ^{177}Lu-DOTA-ATPS mAb (**A**), $^{177}LuCl_3$ (**B**), and ^{177}Lu-DOTA-IgG (**C**) in mice bearing MKN-45 tumors on day 1, 2, 4, and 7. Comparison of bone marrow and tumor uptake among radiopharmaceuticals (**D**). Inhibition of ^{177}Lu-DOTA-ATPS mAb uptake in tumors by unlabeled ATPS mAb (**E**). DOTA, tetraazacyclododecane-1,4,7,10-tetraacetic acid; ATPS, adenosine triphosphate synthase; mAb, monoclonal antibody. $p < 0.05$ *, $p < 0.005$ **, ns: not significant.

In contrast, $^{177}LuCl_3$ primarily accumulated in the bone marrow (68.1 ± 1.2%ID/g, 70.4 ± 1.6%ID/g, 75.6 ± 1.7%ID/g, and 72.7 ± 2.2%ID/g on days 1, 2, 4, and 7, respectively). Renal uptake peaked at 11.8 ± 0.7%ID/g on day 1 and decreased slowly thereafter. The liver uptake ranged from 5.6 to 6.2%ID/g.

For ^{177}Lu-DOTA-IgG, bone marrow uptake remained lower than that of ^{177}Lu-DOTA-ATPS mAb and $^{177}LuCl_3$ throughout the study (all < 10%ID/g). Uptake of ^{177}Lu-DOTA-IgG was similar among the liver, spleen, and kidneys. Renal uptake reached a peak on day 1, while hepatic and splenic uptake increased over time.

Tumor uptake of ^{177}Lu-DOTA-ATPS mAb reached 16.0 ± 1.5%ID/g on day 7, which was significantly higher than that of $^{177}LuCl_3$ (7.6 ± 0.5%ID/g, $p < 0.05$) and ^{177}Lu-DOTA-IgG (8.9 ± 0.5%ID/g, $p < 0.05$). Similarly, on day 4, ^{177}Lu-DOTA-ATPS mAb showed greater tumor uptake (12.4 ± 0.4%ID/g) compared to $^{177}LuCl_3$ (4.2 ± 0.5%ID/g, $p < 0.005$) and ^{177}Lu-DOTA-IgG (8.8 ± 0.5%ID/g, $p < 0.00005$). There was no significant difference in tumor uptake between $^{177}LuCl_3$ and ^{177}Lu-DOTA-IgG on day 7. The biodistribution patterns of ^{177}Lu-DOTA-ATPS mAb, $^{177}LuCl_3$, and ^{177}Lu-DOTA-IgG in the liver, spleen, kidney, and bone marrow of tumor-bearing mice mirrored those observed in wild-type mice.

In an inhibition study, a high dose of unlabeled ATPS mAb significantly reduced the tumoral uptake of ^{177}Lu-DOTA-ATPS mAb from 6.08 ± 1.0%ID/g on day 1 to 3.8 ± 1.1%ID/g ($p < 0.05$).

3.5. *Radioimmunotherapy, Immunohistochemical Staining, and* ^{18}*F-FDG-PET Imaging in MKN-45 Tumor-Bearing Mice*

All animals survived until the end of the experiment regardless of treatment (single agent or combination). The therapeutic efficacy of single agents is shown in Figure 7A. No significant difference in initial tumor volume was observed among the groups (^{177}Lu-DOTA-ATPS mAb, 92.0 ± 15.5 mm^3; unlabeled ATPS mAb, 93.6 ± 3.5 mm^3; IgG,
104.6 ± 6.2 mm^3; vehicle, 97.1 ± 8.6 mm^3, $p > 0.05$). After four weeks, tumors treated with ^{177}Lu-DOTA-ATPS mAb (269.1 ± 130.4 mm^3) were significantly smaller than those treated with unlabeled ATPS mAb (836.4 ± 53.1 mm^3, $p < 0.01$), IgG (1117.1 ± 364.5 mm^3, $p < 0.05$), or vehicle (1561.4 ± 420.4 mm^3, $p < 0.05$). While tumors treated with unlabeled ATPS mAb or IgG displayed smaller volumes compared to controls, these differences were not statistically significant. %TGI after four weeks of treatment was 82.8% for ^{177}Lu-DOTA-ATPS mAb, 46.4% for unlabeled ATPS mAb, and 28.5% for IgG treatment. Tumor volume in the ^{177}Lu-DOTA-ATPS mAb treated group did not significantly change from baseline to the 4th week ($p > 0.05$).

Immunohistochemistry using an anti-CD31 antibody revealed moderate to strong staining in the small vessels of the tumors treated with unlabeled ATPS mAb, IgG, or vehicle (Figure 7B). Conversely, tumors treated with ^{177}Lu-DOTA-ATPS mAb exhibited minimal staining.

Representative ^{18}F-FDG PET images of mice after four weeks of treatment are shown in Figure 7C. Tumor volume increased in mice treated with unlabeled ATPS mAb, IgG, or vehicle at the 4th week. These tumors also displayed central metabolic defects, indicative of necrotic change. In contrast, tumor volume remained stable in mice treated with ^{177}Lu-DOTA-ATPS mAb.

The efficacy of combination therapy is shown in Figure 8A. Initial tumor volume did not differ significantly among groups (^{177}Lu-DOTA-ATPS mAb + sunitinib, 209.3 ± 31.3 mm^3; ^{177}Lu-DOTA-ATPS mAb, 210.3 ± 4.2 mm^3; sunitinib, 207.6 ± 22.9 mm^3; vehicle, 203.7 ± 18.9 mm^3, $p > 0.05$). After four weeks, tumors treated with ^{177}Lu-DOTA-ATPS mAb (2644.4 ± 703.7 mm^3, $p < 0.05$), sunitinib (3619.3 ± 1114.0 mm^3, $p < 0.05$), or the combination (1727.6 ± 793.5 mm^3, $p < 0.01$) were significantly smaller than those in the vehicle group. Combination therapy with ^{177}Lu-DOTA-ATPS mAb and sunitinib demonstrated a greater therapeutic effect compared to either single agent (^{177}Lu-DOTA-ATPS mAb, $p < 0.05$; sunitinib, $p < 0.005$). %TGI after four weeks of treatment was 70.3% for the combination, 54.6% for ^{177}Lu-DOTA-ATPS mAb alone, and 37.8% for sunitinib alone.

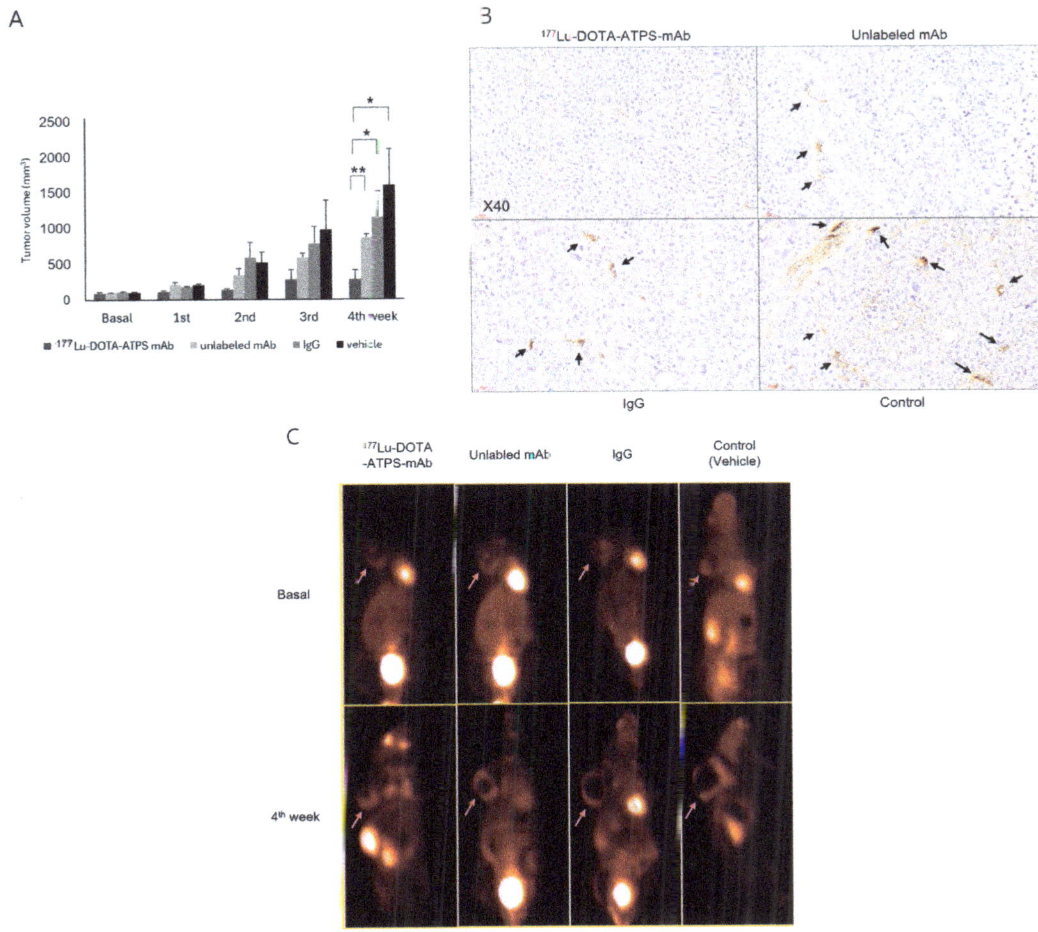

Figure 7. Radioimmunotherapy with ^{177}Lu-DOTA-ATPS mAb. (**A**) Tumor growth curve during the 4-week treatment with ^{177}Lu-DOTA-ATPS mAb, unlabeled ATPS mAb, IgG, and vehicle. (**B**) Immunohistochemical staining with anti-CD31 antibody for MKN-45 tumors after 4 weeks of treatment. (**C**) ^{18}F-FDG PET imaging in mice bearing MKN-45 tumors at baseline and at 4th week of treatment. DOTA, tetraazacyclododecane-1,4,7,10-tetraacetic acid; ATPS, adenosine triphosphate synthase; mAb, monoclonal antibody. $p < 0.05$ *, $p < 0.01$ **, arrows indicate positive staining.

Immunohistochemistry using an anti-CD31 antibody revealed strong staining in the small vessels of tumors from the vehicle group only (Figure 8B). In contrast, tumors treated with ^{177}Lu-DOTA-ATPS mAb, sunitinib or the combination exhibited minimal staining. These findings indicate an anti-angiogenic effect of the therapeutic approaches.

Representative ^{18}F-FDG PET images of mice after four weeks of treatment are shown in Figure 8C. Tumor volume increased in all groups at the 4th week. Tumors treated with single agents or vehicle were larger than those treated with the combination and displayed central metabolic defects.

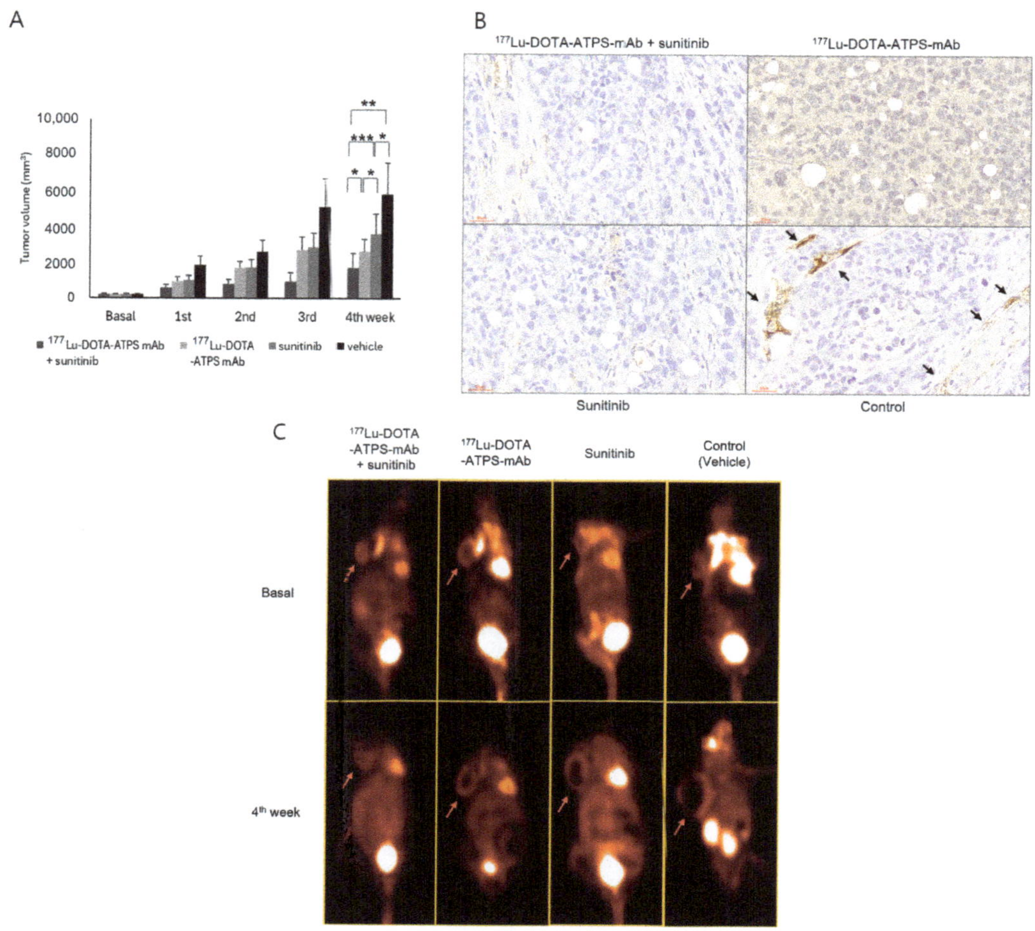

Figure 8. Combination chemo-radioimmunotherapy with sunitinib and ^{177}Lu-DOTA-ATPS mAb. (**A**) Tumor growth curve during the 4-week treatment with ^{177}Lu-DOTA-ATPS mAb, sunitinib, combination, and vehicle. (**B**) Immunohistochemical staining with anti-CD31 antibody for MKN-45 tumors after 4 weeks of treatment. (**C**) ^{18}F-FDG PET imaging in mice bearing MKN-45 tumors at baseline and 4th week of treatment. DOTA, tetraazacyclododecane-1,4,7,10-tetraacetic acid; ATPS, adenosine triphosphate synthase; mAb, monoclonal antibody. $p < 0.05$ *, $p < 0.01$ **, $p < 0.005$ ***, arrows indicate positive staining.

4. Discussion

In the present study, we aimed to develop a radioimmunotherapy agent targeting tumor angiogenesis. We achieved this by coupling anti-ATPS mAb to ^{177}Lu using DOTA as a chelator. Based on cellular uptake results, MKN-45 gastric cancer cells were selected for further development and the evaluation of the therapeutic efficacy of ^{177}Lu-DOTA-ATPS mAb. The uptake of ^{177}Lu-DOTA-ATPS mAb was specific and inhibited by unlabeled ATPS mAb in both in vitro and in vivo experiments. ^{177}Lu-DOTA-ATPS mAb demonstrated a superior therapeutic effect compared to unlabeled ATPS mAb against MKN-45 cells, both in vitro and in vivo. Furthermore, the combination of ^{177}Lu-DOTA-ATPS mAb with sunitinib significantly enhanced the therapeutic effect in mice bearing MKN-45 tumors. This enhanced efficacy was also evident in ^{18}F-FDG PET imaging and immunohistochemistry analysis. These results suggest that radioimmunotherapy using ^{177}Lu-DOTA-ATPS mAb has potential for application in cancer therapy targeting tumor angiogenesis.

^{177}Lu is a long-lived (half-life of 6.7 days), β-ray emitting (Emax = 0.49 MeV, range = 670 μm in soft tissue) radioisotope suitable for therapy. The success of ^{177}Lu-based radioligand therapy and peptide receptor radionuclide therapy has contributed to the growing popularity of ^{177}Lu as a radioisotope for radioimmunotherapy [20]. ^{177}Lu-PSMA has shown promising results as a treatment option for metastatic castration-resistant prostate cancer following chemotherapy and hormonal therapy [21]. ^{177}Lu-DOTA-TATE effectively reduces tumor growth and stabilizes disease in patients with gastroenteropancreatic well-differentiated neuroendocrine tumors, leading to its establishment as a second- or third-line treatment option [22]. ^{177}Lu-labeled trastuzumab, an antibody that targets the HER2 receptor protein, exemplifies radioimmunotherapy using ^{177}Lu. This approach demonstrates specific uptake in HER2-positive primary breast cancers and their metastatic sites [23]. Radioimmunotherapy with ^{177}Lu has also been explored in anti-angiogenesis cancer treatment. ^{177}Lu-labeled TRC105, an antibody targeting CD105, demonstrated tumor uptake in mice with breast cancer. The uptake was 14.3 ± 2.3%ID/g on day 1 and 11.6 ± 6.1%ID/g on day 7, similar to our findings. This approach also inhibited tumor growth and improved survival [24]. These results provide strong support for the use of ^{177}Lu-labeled radiopharmaceuticals in radioimmunotherapy.

Among the ^{177}Lu-labeled radiopharmaceuticals previously mentioned, ^{177}Lu-DOTA-TATE (Lutathera®) was approved by the FDA in 2018 for the treatment of somatostatin receptor-positive gastroenteropancreatic neuroendocrine tumors [25]. Additionally, ^{177}Lu -PSMA-617 (Pluvicto®) was approved in 2022 for the treatment of adult patients with prostate-specific membrane antigen-positive metastatic castration-resistant prostate cancer who have previously been treated with androgen receptor pathway inhibition and taxane-based chemotherapy [26]. Due to the success of ^{177}Lu-based therapy, it has recently garnered significant attention. In the near future, we can expect the development of more radiopharmaceutical therapies, not limited to those labeled with ^{177}Lu. This progress will provide clinicians with a broader range of treatment options for their cancer patients.

^{177}Lu, a radiometal isotope, requires a chelating agent to form a stable complex with antibodies. DOTA, diethylenetriamine pentaacetate (DTPA), and ethylene-diamine-tetraacetic acid (EDTA) are the most common chelators for radiometal isotopes [27,28]. In our preliminary study, ^{177}Lu-DOTA-ATPS mAb exhibited excellent labeling efficiency (around 99.0%) and stability in repeated experiments. Conversely, the labeling efficiency of ^{177}Lu-DTPA-ATPS mAb was significantly lower. This aligns with previous reports [29,30]. We opted for DOTA-based mAb on these findings.

DOTA, a macrocyclic chelator offers greater in vivo stability compared to acyclic chelators like DTPA and EDTA ("macrocyclic effect") [27]. This translates to its recommendation for labeling ^{177}Lu. As shown in this study, free ^{177}Lu accumulates significantly in bone marrow (72.7% for wild-type mice and 69.5% for tumor-bearing mice on day 7). In vivo dissociation of ^{177}Lu-DOTA-ATPS mAb can decrease therapeutic efficacy and increases bone marrow toxicity. Despite maintaining high in vitro stability (85.5% at 37 °C on day 7) in serum, ^{177}Lu-DOTA-ATPS mAb exhibited significant bone marrow uptake of free ^{177}Lu (27.0% for wild-type mice and 39.3% for tumor-bearing mice on day 7). This can be explained by two factors: first, the presence of various blood proteins like transferrin and albumin that strongly bind to ^{177}Lu and, furthermore, the dilution of ^{177}Lu-DOTA-ATPS mAb in vivo [28]. Overcoming this limitation is crucial, as researchers are actively developing new chelators [31]. Further studies to improve the in vivo stability of ^{177}Lu-DOTA-ATPS mAb are required.

ATPS is normally located in the inner mitochondrial membranes as part of the mitochondrial respiratory complex. It participates in ATP production using a proton gradient generated by mitochondrial respiratory complex I-IV [32]. Interestingly, ATPS can also be found on the surface of some cancer and endothelial cells, known as ectopic ATPS. This ectopic ATPS can be a binding site for angiostatin [3]. Therefore, ATPS could serve as a novel target for anti-angiogenic cancer therapies. As previously demonstrated, the anti-ATPS mAb used in this study can target tumor vasculature and cancer cells [9]. In this study, ^{177}Lu-DOTA-ATPS mAb showed a significant inhibitory effect on MKN-45 gastric cancers. TGI of ^{177}Lu-DOTA-ATPS mAb (82.8%) was greater than that of unlabeled ATPS mAb (46.6%). Additionally, immunohistochemistry with an anti-CD31 antibody revealed minimal staining in tumors treated with ^{177}Lu-DOTA-ATPS mAb (Figure 7B). Similarly, minimal anti-CD31 staining was observed in tumors treated with either single-agent sunitinib or combination therapy (Figure 8B). These findings indicate that the anticancer efficacy of these therapeutic strategies is likely mediated through targeting tumor angiogenesis.

Combination therapy offers a significant advantage by enhancing anticancer effects while potentially reducing side effects compared to single-agent treatment. In this study, combination therapy demonstrated a synergistic increase in therapeutic efficacy (TGI = 70.3%) compared to sunitinib alone (37.8%). Furthermore, ^{177}Lu-DOTA-ATPS mAb alone (TGI = 54.6%) also demonstrated a greater therapeutic effect than sunitinib alone. This finding suggests promising potential for the clinical application of ^{177}Lu-DOTA-ATPS mAb in the future. Chemotherapeutic agents, such as tyrosine kinase inhibitors, have been known to show a broad spectrum of adverse effects in both the hematologic system and nonhematologic systems [33]. ^{177}Lu-based radiopharmaceutical therapies have also been associated with various side effects, some of which are serious hematologic diseases [34,35]. Such side effects could obstruct the appropriate treatment of cancer patients, leading to a reduction in dosage or discontinuation of therapeutic agents. Based on the results of this study, combining ^{177}Lu-radioimmunotherapy with conventional chemotherapy could decrease the therapeutic doses of each treatment, leading to fewer adverse effects than single-agent therapy. However, further clinical studies should be conducted.

^{177}Lu decays by emitting two γ-rays (208 and 113 KeV), which are ideal for gamma camera imaging. While we employed ^{18}F-FDG PET for tumor visualization in this study, whole-body gamma camera imaging could directly visualize or predict the biodistribution of ^{177}Lu-DOTA-ATPS mAb. A limitation of this study is the lack of a small animal gamma camera, which prevented us from performing this complementary imaging modality.

We were unable to achieve tumor-free survival as tumors persistently grew in all groups, whether they were subjected to single or combination treatments. In the combination treatment group, tumors from two mice did not exhibit growth until the fourth week of therapy, although they did not completely disappear. In contrast, all tumors in the other groups showed significant growth by the fourth week of treatment. In this study, we initiated treatments when the tumors reached approximately 200 mm^3 in size. For a more accurate evaluation of tumor-free survival, treatments should ideally be started earlier. This is another limitation of the study.

In our previous study [9], we categorized six types of cancer cells into two groups based on their membranous ATPS expression, as determined by Western blot analysis and immunofluorescence microscope findings. From these, we chose MDA-MB-231 (which has high ATPS expression) and PC-3 (which has low ATPS expression) for further comparison. We evaluated these cells using in vitro cellular uptake, binding, and in vivo tumor uptake with ^{89}Zr-labeled ATPS mAb and positron emission tomography imaging (utilizing the same mAb as in this study). The MDA-MB-231 cells exhibited significantly higher cellular uptake, binding capability (Kd), and in vivo tumor uptake than the PC-3 cells. Based on these results, we hypothesized that cells demonstrating higher in vitro cellular uptake would inevitably show higher in vivo tumor uptake, leading to increased therapeutic efficacy. While it is a well-established concept in clinical radioimmunotherapy to predict therapeutic outcomes using diagnostic radiopharmaceuticals with the same antibodies, it would be beneficial to evaluate the ATPS expression of cancer cells concurrently with cellular uptake. As an alternative, comparing the therapeutic effects between tumors with high ATPS expression and those with lower ATPS expression could aid in drawing more credible conclusions. This is also acknowledged as a limitation of this study.

Despite encouraging preclinical results, this study represents early-stage research. Further technical refinements are necessary before clinical trials with ^{177}Lu-DOTA-ATPS mAb can be initiated.

Author Contributions: Conceptualization, J.-K.Y. and B.-N.P.; methodology, J.-K.Y. and B.-N.P.; formal analysis, Y.-J.P. and B.-N.P.; investigation, B.-N.P. and S.-M.K.; writing—original draft preparation, J.-K.Y. and B.-N.P.; writing—review and editing, Y.-S.A. and S.-J.L.; funding acquisition, B.-N.P. and Y.-S.A. All authors have read and agreed to the published version of the manuscript.

Funding: This work was supported by the National Research Foundation of Korea (NRF) grant funded by the Korea government (MSIT) (2019R1A2C109081314).

Institutional Review Board Statement: The animal study protocol was approved by the Institutional Animal Care and Use Committee of Ajou University School of Medicine (IACUC No. 2021-0067, 18 January 2022).

Informed Consent Statement: Not applicable.

Data Availability Statement: The data on this study are available in the manuscript. Further inquiries can be directed to the corresponding author.

Conflicts of Interest: The authors declare no conflicts of interest.

References

1. Folkman, J. Tumor angiogenesis: Therapeutic implications. *N. Engl. J. Med.* **1971**, *285*, 1182–1186. [CrossRef]
2. Folkman, J. Role of angiogenesis in tumor growth and metastasis. *Semin. Oncol.* **2002**, *29*, 15–18. [CrossRef]
3. Moser, T.L.; Stack, M.S.; Asplin, I.; Enghild, J.J.; Højrup, P.; Everitt, L.; Hubchak, S.; Schnaper, H.W.; Pizzo, S.V. Angiostatin binds ATP synthase on the surface of human endothelial cells. *Proc. Natl. Acad. Sci. USA* **1999**, *96*, 2811–2816. [CrossRef]
4. Kenan, D.J.; Wahl, M.L. Ectopic localization of mitochondrial ATP synthase: A target for anti-angiogenesis intervention? *J. Bioenerg. Biomembr.* **2005**, *37*, 461–465. [CrossRef]
5. Moser, T.L.; Kenan, D.J.; Ashley, T.A.; Roy, J.A.; Goodman, M.D.; Misra, U.K.; Cheek, D.J.; Pizzo, S.V. Endothelial cell surface F1-F0 ATP synthase is active in ATP synthesis and is inhibited by angiostatin. *Proc. Natl. Acad. Sci. USA* **2001**, *98*, 6656–6661. [CrossRef]
6. Chi, S.-J.L.; Wahl, M.L.; Mowery, Y.M.; Shan, S.; Mukhopadhyay, S.; Hilderbrand, S.C.; Kenan, D.J.; Lipes, B.D.; Johnson, C.E.; Marusich, M.F.; et al. Angiostatin-like activity of a monoclonal antibody to the catalytic subunit of F1F0 ATP synthase. *Cancer Res.* **2007**, *67*, 4716–4724. [CrossRef]
7. Zhang, X.; Gao, F.; Yu, L.L.; Peng, Y.; Liu, H.H.; Liu, J.-K.Y.; Yin, M.; Ni, J. Dual functions of a monoclonal antibody against cell surface F1F0 ATP synthase on both HUVEC and tumor cells. *Acta Pharmacol. Sin.* **2008**, *29*, 942–950. [CrossRef]
8. Park, B.N.; Lee, S.J.; Roh, J.H.; Lee, K.H.; An, Y.S.; Yoon, J.K. Radiolabeled Anti-Adenosine Triphosphate Synthase Monoclonal Antibody as a Theragnostic Agent Targeting Angiogenesis. *Mol. Imaging* **2017**, *16*, 1536012117737399. [CrossRef]
9. Park, B.N.; Kim, G.H.; Ko, S.A.; Shin, G.H.; Lee, S.J.; An, Y.S.; Yoon, J.K. Zr-89 Immuno-PET Targeting Ectopic ATP Synthase Enables In-Vivo Imaging of Tumor Angiogenesis. *Int. J. Mol. Sci.* **2019**, *20*, 5928. [CrossRef]
10. Sharkey, R.M.; Goldenberg, D.M. Cancer radioimmunotherapy. *Immunotherapy* **2011**, *3*, 349–370. [CrossRef]
11. Saravana-Bawan, B.; Bajwa, A.; Paterson, J.; McEwan, A.J.B.; McMullen, T.P.W. Efficacy of 177Lu Peptide Receptor Radionuclide Therapy for the Treatment of Neuroendocrine Tumors: A Meta-analysis. *Clin. Nucl. Med.* **2019**, *44*, 719–727. [CrossRef]
12. Yechiel, Y.; Chicheportiche, A.; Keidar, Z.; Ben-Haim, S. Prostate Cancer Radioligand Therapy: Beta-labeled Radiopharmaceuticals. *PET Clin.* **2024**, *19*(3), 389–399. [CrossRef]
13. Menon, S.R.; Mitra, A.; Chakraborty, A.; Tawate, M.; Sahu, S.; Rakshit, S.; Gaikwad, S.; Dhotre, G.; Damle, A.; Banerjee, S. Clinical Dose Preparation of [(177)Lu]Lu-DOTA-Pertuzumab Using Medium Specific Activity [(177)Lu]LuCl(3) for Radioimmunotherapy of Breast and Epithelial Ovarian Cancers, with HER2 Receptor Overexpression. *Cancer Biother. Radiopharm.* **2022**, *37*, 384–402. [CrossRef]
14. Thakral, P.; Singla, S.; Yadav, M.P.; Vasisht, A.; Sharma, A.; Gupta, S.-M.K.; Bal, C.S.; Snehlata; Malhotra, A. An approach for conjugation of (177) Lu- DOTA-SCN- Rituximab (BioSim) & its evaluation for radioimmunotherapy of relapsed & refractory B-cell non Hodgkins lymphoma patients. *Indian J. Med. Res.* **2014**, *139*, 544–554.
15. Jensen, R.L.; Leppla, D.; Rokosz, N.; Wurster, R.D. Matrigel augments xenograft transplantation of meningioma cells into athymic mice. *Neurosurgery* **1998**, *42*, 130–135; discussion 135–136. [CrossRef]
16. Bradford, M.M. A rapid and sensitive method for the quantitation of microgram quantities of protein utilizing the principle of protein-dye binding. *Anal. Biochem.* **1976**, *72*, 248–254. [CrossRef]
17. Lahti, S.; Ludwig, J.M.; Xing, M.; Sun, L.; Zeng, D.; Kim, H.S. In vitro biologic efficacy of sunitinib drug-eluting beads on human colorectal and hepatocellular carcinoma-A pilot study. *PLoS ONE* **2017**, *12*, e0174539. [CrossRef]

18. Bhagwat, S.V.; Gokhale, P.C.; Crew, A.P.; Cooke, A.; Yao, Y.; Mantis, C.; Kahler, J.; Workman, J.; Bittner, M.; Dudkin, L.; et al. Preclinical characterization of OSI-027, a potent and selective inhibitor of mTORC1 and mTORC2: Distinct from rapamycin. *Mol. Cancer Ther.* **2011**, *10*, 1394–1406. [CrossRef]
19. Jiang, X.L.; Zhang, Y.; Luo, C.L.; Wu, X.H. Targeting renal cell carcinoma with gambogic acid in combination with sunitinib in vitro and in vivo. *Asian Pac. J. Cancer Prev.* **2012**, *13*, 6463–6468. [CrossRef]
20. Niu, T.; Fan, M.; Lin, B.; Gao, F.; Tan, B.; Du, X. Current clinical application of lutetium-177 in solid tumors (Review). *Exp. Ther. Med.* **2024**, *27*, 225. [CrossRef]
21. Patell, K.; Kurian, M.; Garcia, J.A.; Mendiratta, P.; Barata, P.C.; Jia, A.Y.; Spratt, D.E.; Brown, J.R. Lutetium-177 PSMA for the treatment of metastatic castrate resistant prostate cancer: A systematic review. *Expert. Rev. Anticancer. Ther.* **2023**, *23*, 731–744. [CrossRef]
22. Becx, M.N.; Minczeles, N.S.; Brabander, T.; de Herder, W.W.; Nonnekens, J.; Hofland, J. A Clinical Guide to Peptide Receptor Radionuclide Therapy with (177)Lu-DOTATATE in Neuroendocrine Tumor Patients. *Cancers* **2022**, *14*, 5792. [CrossRef]
23. Bhusari, P.; Vatsa, R.; Singh, G.; Parmar, M.; Bal, A.; Dhawan, D.K.; Mittal, B.R.; Shukla, J. Development of Lu-177-trastuzumab for radioimmunotherapy of HER2 expressing breast cancer and its feasibility assessment in breast cancer patients. *Int. J. Cancer* **2017**, *140*, 938–947. [CrossRef] [PubMed]
24. Ehlerding, E.B.; Lacognata, S.; Jiang, D.; Ferreira, C.A.; Goel, S.; Hernandez, R.; Jeffery, J.J.; Theuer, C.P.; Cai, W. Targeting angiogenesis for radioimmunotherapy with a (177)Lu-labeled antibody. *Eur. J. Nucl. Med. Mol. Imaging* **2018**, *45*, 123–131. [CrossRef] [PubMed]
25. Hennrich, U.; Kopka, K. Lutathera((R)): The First FDA- and EMA-Approved Radiopharmaceutical for Peptide Receptor Radionuclide Therapy. *Pharmaceuticals* **2019**, *12*, 114. [CrossRef]
26. Fallah, J.; Agrawal, S.; Gittleman, H.; Fiero, M.H.; Subramaniam, S.; John, C.; Chen, W.; Ricks, T.K.; Niu, G.; Fotenos, A.; et al. FDA Approval Summary: Lutetium Lu 177 Vipivotide Tetraxetan for Patients with Metastatic Castration-Resistant Prostate Cancer. *Clin. Cancer Res.* **2023**, *29*, 1651–1657. [CrossRef] [PubMed]
27. Parus, J.L.; Pawlak, D.; Mikolajczak, R.; Duatti, A. Chemistry and bifunctional chelating agents for binding (177)Lu. *Curr. Radiopharm.* **2015**, *8*, 86–94. [CrossRef]
28. Suzuki, H.; Kannaka, K.; Uehara, T. Approaches to Reducing Normal Tissue Radiation from Radiolabeled Antibodies. *Pharmaceuticals* **2024**, *17*, 508. [CrossRef]
29. Wu, C.; Kobayashi, H.; Sun, B.; Yoo, T.M.; Paik, C.H.; Gansow, O.A.; Carrasquillo, J.A.; Pastan, I.; Brechbiel, M.W. Stereochemical influence on the stability of radio-metal complexes in vivo. Synthesis and evaluation of the four stereoisomers of 2-(p-nitrobenzyl)-trans-CyDTPA. *Bioorg Med. Chem.* **1997**, *5*, 1925–1934. [CrossRef]
30. Milenic, D.E.; Garmestani, K.; Chappell, L.L.; Dadachova, E.; Yordanov, A.; Ma, D.; Schlom, J.; Brechbiel, M.W. In vivo comparison of macrocyclic and acyclic ligands for radiolabeling of monoclonal antibodies with 177Lu for radioimmunotherapeutic applications. *Nucl. Med. Biol.* **2002**, *29*, 431–442. [CrossRef]
31. Bailey, G.A.; Price, E.W.; Zeglis, B.M.; Ferreira, C.L.; Boros, E.; Lacasse, M.J.; Patrick, B.O.; Lewis, J.S.; Adam, M.J.; Orvig, C. H(2)azapa: A versatile acyclic multifunctional chelator for (67)Ga, (64)Cu, (111)In, and (177)Lu. *Inorg. Chem.* **2012**, *51*, 12575–12589. [CrossRef] [PubMed]
32. Kobayashi, A.; Azuma, K.; Ikeda, K.; Inoue, S. Mechanisms Underlying the Regulation of Mitochondrial Respiratory Chain Complexes by Nuclear Steroid Receptors. *Int. J. Mol. Sci.* **2020**, *21*, 6683. [CrossRef] [PubMed]
33. Hartmann, J.T.; Haap, M.; Kopp, H.G.; Lipp, H.P. Tyrosine kinase inhibitors—A review on pharmacology, metabolism and side effects. *Curr. Drug Metab.* **2009**, *10*, 470–481. [CrossRef] [PubMed]
34. Ahmadzadehfar, H.; Eppard, E.; Kurpig, S.; Fimmers, R.; Yordanova, A.; Schlenkhoff, C.D.; Gartner, F.; Rogenhofer, S.; Essler, M. Therapeutic response and side effects of repeated radioligand therapy with 177Lu-PSMA-DKFZ-617 of castrate-resistant metastatic prostate cancer. *Oncotarget* **2016**, *7*, 12477–12488. [CrossRef]
35. Baum, R.P.; Kulkarni, H.R.; Singh, A.; Kaemmerer, D.; Mueller, D.; Prasad, V.; Hommann, M.; Robiller, F.C.; Niepsch, K.; Franz, H.; et al. Results and adverse events of personalized peptide receptor radionuclide therapy with (90)Yttrium and (177)Lutetium in 1048 patients with neuroendocrine neoplasms. *Oncotarget* **2018**, *9*, 16932–16950. [CrossRef]

antibodies

Article

NK Cytotoxicity Mediated by NK-92 Cell Lines Expressing Combinations of Two Allelic Variants for *FCGR3*

Marta Freitas Monteiro, Maria Papaserafeim, Matteo Andreani, Aline Réal, Athanasios Kouklas, Daniela Reis Galvão, Jörg D. Seebach * and Gisella L. Puga Yung *

Laboratory of Translational Immunology, Department of Medicine, Division of Immunology and Allergology, University Hospitals Geneva, Medical Faculty, CH-1211 Geneva, Switzerland
* Correspondence: joerg.seebach@hug.ch (J.D.S.); gisella.pugayung@unige.ch (G.L.P.Y.); Tel.: +41-22-372-93-72 (J.D.S.); +41-22-379-9492 (G.L.P.Y.)

Abstract: Natural killer (NK) cells play an important role in the surveillance of viral infections and cancer. NK cell antibody-dependent cellular cytotoxicity (ADCC) and direct cytotoxicity are mediated by the recognition of antibody-coated target cells through the Fc gamma receptor IIIA (FcγRIIIa/CD16) and by ligands of activating/inhibitory NK receptors, respectively. Allelic variants of the *FCGR3A* gene include the high-affinity single-nucleotide polymorphism (SNP) rs396991 (V176F), which is associated with the efficacy of monoclonal antibody (mAb) therapies, and the SNP rs10127939 (L66H/R). The contribution of *FCGR3A* SNPs to NK cell effector functions remains controversial; therefore, we generated a panel of eight NK-92 cell lines expressing specific combinations of these SNPs and tested their cytotoxicities. NK-92 cells were stably transfected with plasmids containing different combinations of *FCGR3A* SNPs. Messenger RNA and FcγRIIIa/CD16 cell surface expressions were detected using new generation sequencing (NGS) and flow cytometry, respectively. All FcγRIIIa/CD16-transfected NK-92 cell lines exhibited robust ADCC against three different target cell lines with minor differences. In addition, enhanced direct NK cytotoxicity against K562 target cells was observed, suggesting a mechanistic role of FcγRIIIa/CD16 in direct NK cytotoxicity. In conclusion, we generated eight FcγRIIIa/CD16-transfected NK-92 cell lines carrying different combinations of two of the most studied *FCGR3A* SNPs, representing the major genotypes described in the European population. The functional characterization of these cell lines revealed differences in ADCC and direct NK cytotoxicity that may have implications for the design of adoptive cancer immunotherapies using NK cells and tumor antigen-directed mAbs.

Keywords: NK-92; *FCGR3A*; CD16; ADCC; L66H/R; V176F; L43H/R; V158F

1. Introduction

Natural killer (NK) cells are a subpopulation of lymphocytes that play an important role in the immune response to viral infections and in cancer immunosurveillance [1–4]. The cardinal effector functions of NK cells are (i) the lysis of transformed (stressed) target cells triggered by the predominant expression of ligands for activating NK receptors, leading to "natural" or "direct" NK cytotoxicity; (ii) the induction of apoptosis through death receptors; and (iii) the recognition of antibody-coated target cells through their Fc gamma receptor IIIA (FcγRIIIa/CD16), leading to antibody-dependent cellular cytotoxicity (ADCC) [4]. In addition to the level of CD56 expression, the presence of FcγRIIIa/CD16 on the cell surface allows for the discrimination between NK subpopulations that are highly cytotoxic ($CD56^{dim}CD16^{pos}$) or predominantly cytokine-producing NK cells ($CD56^{bright}CD16^{dim/neg}$) [1,5].

The FcγRIIIa/CD16 receptor consists of two extracellular Ig-like domains, D1 and D2; a stalk region; a transmembrane domain; and a short cytoplasmic tail and has a low affinity for immunoglobulin G (IgG) subclasses compared to other Fc-gamma receptors [6]. The Fc fragments of IgGs bind to the FcγRIIIa/CD16 at the beginning of D2 near the hinge region between D1 and D2 [7–9]. In addition to NK cells, subsets of monocytes and macrophages and small subpopulations of dendritic cells and T cells express FcγRIIIa/CD16 on their surface [10–12]. FcγRIIIa/CD16 has different affinities for the different IgG subclasses [9] and, upon activation, is either internalized for recycling or cleaved by ADAM17 ("a disintegrin and metalloprotease 17"), which cleaves the stalk region and thereby reduces NK cytotoxicity [13,14].

Several single nucleotide polymorphisms (SNPs) have been described for *FCGR3A*. The most studied SNP is the rs396991 (V176F) located in D2, which affects the binding of IgG to the receptor [15,16]. Furthermore, the tri-allelic SNP rs10127939 (L66H/R) in D1 is relevant for NK cytotoxicity; it influences the binding of IgG subclasses to the receptor, with H having a higher binding to human IgG1, IgG3, and IgG4 than L [17]. However, the consequences of combinations of these two SNP in the context of human disease and intravenous immunoglobulin G and mAb therapies remain to be investigated [18–24].

Therapeutic mAbs are engineered to target specific molecules. EGFR is a transmembrane protein, a member of a large family of receptors involved in developmental biology and cancer. EGFR overexpression, mutation, or both contribute to the malignant phenotype of many cancer cells. Thus, the design of therapeutic mAbs, such as cetuximab, that target EGFR has led to improvements in the outcome of cancer patients [25]. Rituximab and Obinutuzumab are IgG1 subclass mAbs that bind to CD20, a transmembrane protein expressed on B cells. Although the exact biological function and physiological ligand remain elusive, CD20 plays a major role in regulating B-cell activation and proliferation by acting as a Ca^{2+} channel. There are two types of anti-CD20 mAbs based on their ability or inability to translocate CD20 to lipid rafts after binding, type-I and -II, respectively [26]. Clinically, anti-CD20 mAb therapies are largely used to treat B-cell-derived hematologic malignancies, autoimmune diseases, and humoral transplant rejection. Since the approval of rituximab [27] and several other therapeutic antibodies, it remains controversial whether or not the clinical efficacy of mAb therapy is associated with the *FCGR3A* rs396991 (V176F) SNP [28–39]. The clinical consequences of rs10127939 and rs396991 combinations in the context of human disease and intravenous immunoglobulin G and mAb therapies remain to be investigated [18,21–26]. Furthermore, *FCGR3A* SNPs are clinically relevant in infections such as malaria, severe COVID-19, bacteremia in transplant patients, periodontitis, as well as in Kaposi's sarcoma in HIV-infected patients and autoimmune diseases, including lupus and rheumatoid arthritis [21,40–42].

The $CD3^{neg}CD56^{pos}$ human NK-92 cell line, characterized by a large granular lymphocytic phenotype, was originally obtained from a patient with non-Hodgkin's lymphoma [43,44]. Cell culture of NK-92 cells is dependent on high doses of interleukin-2 (IL2), and the function and phenotype resemble activated NK cells with granzyme- and perforin-containing granules and the surface expression of CD2, CD7, CD25, CD69, NKp30, NKp46, and CD94 [43–45]. In contrast, NK-92 cells lack FcγRIIIa/CD16 expression and, consequently, do not mediate ADCC [46]. To evaluate ADCC in vitro, NK-92 cells have been transfected with murine CD16 and high-affinity CD16 CAR in the past [47,48] and transduced with low-affinity CD16 to test therapeutic mAbs [49].

The aim of the present study was to generate and compare transfectants derived from the NK-92 cell line carrying different combinations of the rs10127939 (L66H/R) and rs396991 (V176F) *FCGR3A* polymorphisms. In particular, the effect of specific SNP combinations on NK cytotoxicity, both ADCC and direct killing, was investigated.

2. Materials and Methods

2.1. Reagents

Biologicals: The human type-II anti-CD20 antibody (Obinutuzumab, non-glycoengineered) used in this study was kindly provided by Christian Klein from Roche Innovation Center, Zurich. Human anti-EGFR (anti-EGFR-IgG1) antibody was purchased from Invivogen (San

Diego, CA, USA). Human recombinant IL2 was obtained from Novartis Pharma (Proleukin, Basel, Switzerland).

Antibodies: Three different antibody clones against FcγRIIIa/CD16 (B73.1, 3G8, and MEM154), anti-HLA-ABC, and anti-CD58 were used for flow cytometry. Mouse IgG1 and IgG2a isotype controls were also included. All antibodies were conjugated to either fluorescein isothiocyanate (FITC) or to R-phycoerythrin (PE). Additionally, F(ab')$_2$ fragment goat anti-human IgG, F(ab')$_2$ fragment specific (BCR blocking), goat anti-human κ F(ab')$_2$-PE, and goat polyclonal isotype control-PE were used (Table S1).

Chemicals, culture media, and supplements: Probenecid, bovine serum albumin (BSA), ethylenediaminetetraacetic acid (EDTA), and fetal bovine serum (FBS) were purchased from Sigma-Aldrich (St. Louis, MO, USA). Buffers Dulbecco's Phosphate-Buffered Saline (PBS), Dulbecco's Modified Eagle Medium (DMEM), Roswell Park Memorial Institute 1640 (RPMI), Minimum Essential Medium Amino Acids Solution (EAA), Non-Essential Amino Acids Solution (NEAA), 4-(2-hydroxyethyl)-1-piperazineethanesulfonic acid (HEPES), penicillin/streptomycin (Pen/Strep), sodium pyruvate, and geneticin were all from Gibco (Grand Island, NY, USA). The L-Alanyl-L-Glutamine (L-Glu/L-Ala) was from Bioswisstec AG, Schaffhausen, Switzerland. AB serum was obtained from the *"Centre de Transfusion de Genève"* of our institution, (Table S1)

Cell lines, PBMC isolation, and culture conditions: The cell lines A-431, Daudi, K562, Raji, and NK-92 were acquired from ATCC. (Manassas, VA, USA) NK-92 transduced with low-affinity CD16, with F in amino acid position 176, was a kind gift from Kerry Campbell and NantKwest (noGFP-CD16 176F NK-92.05, US patent No. 7,618,817; EU patent No. 1771471). Of note, positions 66 and 176 correspond to 48 and 158, respectively, depending on whether the leader peptide is considered (Table S1).

The A431 culture medium consisted of DMEM supplemented with 20 mM HEPES, 10% FCS, 1 mM sodium pyruvate, 2 mM L-Glu/L-Ala, 100 IU/mL of 1× NEAA and 1× EAA, and 100 µg/mL Pen/Strep. The Daudi, K562, and Raji culture media consisted of RPMI-1640 containing 25 mM HEPES, 10% FCS, 2 mM L-Glu/L-Ala, and 100 µg/mL Pen/Strep. NK-92 cells and the FcγRIIIa/CD16 stable transfectants were maintained in RPMI-1640 medium containing 25 mM HEPES, 15% FCS, 5% AB serum, 2 mM L-Glu/L-Ala, 1 mM sodium pyruvate, Pen/Strep (10,000 U/mL), and 100 IU/mL of IL2. In all cases, cells were maintained in 5% CO_2 humidified incubator at 37 °C. In the case of ADCC assays, NK-92 cells were cultured without IL2 for 16 h to reduce direct cytotoxicity.

Human NK cells were isolated from the blood of healthy donors. Peripheral blood mononuclear cells (PBMCs) were separated using Ficoll-Paque PLUS density gradient centrifugation GE Healthcare (Uppsala, Sweden) centrifugation at 900× g for 20 min at room temperature without a brake. The cells were then washed twice in PBS and stored in cold FACS buffer (PBS with 0.1% BSA) before antibody staining for flow cytometry analysis.

2.2. Construction of Plasmids Carrying FCGR3A Allelic Variants and NK-92 Transfectant Cell Lines

The gene source for generating the different *FCGR3A* gene variants was the commercial vector GenScriptpUC19-Amp *FCGR3A* transcript variant 3 (NM_001127593.1) (GenScript, Piscataway, NJ, USA). *FCGR3A* was subcloned into the multiple cloning site 2 (mcs2) of the pVITRO-neo-mcs plasmid (InvivoGen, San Diego, CA, USA) through the ligation of PCR products obtained via amplification using the polymerase Q5 high-fidelity DNA polymerase (New England BioLabs, NEB, Ipswich, MA, USA). We attempted to use the mcs1 and mcs2 to generate "heterozygous" combinations of *FCGR3A*, but this proved impossible due to constant deletions, rearrangements, or aberrations after the transformation of the constructs into *E. coli*. Detailed information on the primers used can be found in Table S2. The PCR conditions consisted of 1 cycle of denaturation at 98 °C for 30 s, 30 cycles consisting of denaturation at 98 °C for 10 s, annealing at 71 °C for 30 s, and extension at 72 °C for 30 s, ending with a final extension at 72 °C for 2 min. After amplification, DNA was purified using a Monarch PCR & DNA Cleanup Kit (NEB). pVITRO-neo-mcs and

PCR products were digested in a 25 μL reaction with high-fidelity AgeI and NheI (NEB) and ligated with Instant Sticky-end Ligase master mix (NEB) in a vector-to-insert ratio of 1:3. The resulting 5 μL reaction was transformed into 50 μL of chemically competent NEB 5-alpha competent *E. coli* (high efficiency, NEB by propagation; all products details in Table S1).

To generate the *FGCR3A* variants conferred by the rs10127939 (L66H/R) and rs396991 (V176F) SNPs, site-directed mutagenesis was performed using the Q5 Site-Directed Mutagenesis kit (Promega, Madison, WI, USA) in the subcloned pVITRO-neo-CD16A_VL_{MCS2} b vector with the primers described in Table S2 and according to the manufacturer's specifications. All the PCR conditions for the site-directed mutagenesis were similar: 1 cycle at 98 °C for 30 s; 25 cycles of denaturation at 98 °C for 30 s, annealing at 66 °C for 30 s, and extension at 72 °C for 2 min; ending with 1 cycle at 72 °C for 3 min 36 s. The 5 μL of the resulting reaction kinase, ligase, and DpnI mix, which contained the vector, was transformed into 50 μL of chemically competent NEB 5-alpha Competent *E. coli* for propagation according to the manufacturer's instructions and cultured for 16 h on Luria broth agar plates with 100 μg/mL ampicillin (Sigma-Aldrich). Large amounts of plasmids were recovered using the classic phenol/chloroform method [50]. Finally, Sanger sequencing (Microsynth AG, Balgach, SG, Switzerland) was performed at all steps to confirm that the constructs were as expected.

To generate the NK-92 transfectants, 2 μg of pVITRO-neo-mcs plasmids carrying the elongation factor 1 alpha (EF-1α) promoter, which allows for the stable and long-term expression of the genes in transfected cells, were used. Subsequently, 4×10^6 NK-92 cells were transfected in the presence of 100 μL of Solution R from the Amaxa Cell Line Nucleofector Kit R via electroporation using the program A-24 of the Nucleofector I device according to the manufacturer's instructions (Lonza AG, Stein, Switzerland). Since the pVITRO-neo-mcs plasmid contains a neomycin resistance gene, 600 μg/mL of geneticin was added 48 h after electroporation and maintained along with the cultures to maintain the selection pressure. The control conditions included cells electroporated with empty pVITRO-neo-mcs and pulsed only. All cells were maintained in a 5% CO_2 humidified incubator at 37 °C. Single plasmid transfections were performed to obtain single *FCGR3A* variants (LL_VV, LL_FF, RR_VV, and HH_VV), while co-transfections were required to obtain allelic *FGCR3A* variants.

2.3. Determination of FCGR3A Variants in NK-92 Transfectants from Messenger RNA Followed by New-Generation Sequencing (NGS) Quantification

To detect the expression levels of the different *FCGR3A* variants in the NK-92 transfectants, total RNA was extracted using Trizol Reagent (Invitrogen, Carlsbad, CA, USA) according to the manufacturer's instructions. The cDNA synthesis was conducted using a ProtoScript II First Strand cDNA Synthesis Kit according to the manufacturer's instructions, starting with 0.5 μg of total RNA and 6 μM Random Primer Mix (both NEB) in a 20 μL final volume. The controls included tubes containing all RT reagents minus the RNA template or reverse transcriptase, designated "NAC-RT" and "no-RT" controls, respectively.

For cDNA synthesis, 1.25 μL of the cDNA reaction was used in a 25 μL reaction containing 0.5 U of Q5 High-Fidelity (NEB); 1.5 mM $MgCl_2$ (Promega, Madison, WI, USA); 0.2 mM Deoxynucleotide (dNTP) Solution Mix (NEB); 0.2 μM of each primer (designated Seq 48_158 fw and Seq 48_158 rev, Microsynth AG, Balgach, Switzerland, Table S2); and in 1 × Q5 reaction buffer (NEB). The cDNA amplification PCR consisted of a denaturation step at 95 °C for 30 s; followed by 30 cycles of denaturation at 98 °C for 5 s, annealing at 65 °C for 20 s, and extension at 72 °C for 15 s; ending with a final extension at 72 °C for 2 min. Negative PCR controls consisted of the following: all PCR reagents plus (i) "NAC-RT control", (ii) "no-RT control", and (iii) minus the cDNA template, designated "NAC RT-PCR", "no-RT/ PCR", and "NAC-PCR" controls, respectively.

The PCR products were run on 1.2% agarose gels at a constant 75 volts for 1 h, and bands were visualized after 30 min of staining with a solution consisting of a 1:3000 dilution in water of GelRed Nucleic Acid Gel Stain (Biotium Inc., Fremont, CA, USA) and 0.1 M NaCl for 30 min according to the manufacturer's instructions. The 100 bp ladder (ThermoScientific, Waltham, MA, USA) was used to determine DNA size.

PCR products were purified using Monarch PCR & DNA Cleanup Kit (NEB), resuspended in nuclease-free H_2O, and quantified using the Qubit 4 fluorometer (ThermoFisher Scientific, Waltham, MA, USA) coupled to the Qubit 1× dsDNA High Sensitivity (HS) kit according to the manufacturer's instructions. Next-generation sequencing (NGS) was performed at the iGE3 Genomics Platform (University of Geneva, Geneva, Switzerland). Briefly, libraries were generated using 100 ng of each PCR product and the TruSeq Nano DNA High Throughput Library Prep kit (Illumina, San Diego, CA, USA). Libraries were pooled at equimolar proportions (10 nM) before hybridization to a NovaSeq 6000 (Illumina) flow cell and paired-end 100 bp (PE100) sequencing methods. The quantification of *FCGR3A* variant expression was performed using the iGE3 Bioinformatics Platform (University of Geneva). Briefly, an average of approximately 10 million paired-end reads were obtained for each sample. Sequencing quality control (QC) was performed using the FastQC analysis tool [v0.12.1]. The reads were then aligned to the Ensembl GRCh38 genome reference using STAR v.2.7.4. The average alignment rate was 99.2%. The different genotypes at the SNP positions rs10127939 and rs396991 were then assessed using igvtools (IGV v.2.16.2). Additional alignment analysis was performed using the igvtools command "*samtools-flagstat*", and no "*QC failed*" flagged reads were identified during the alignment analysis. For all sequencing samples in the dataset, 100% of the reads were correctly paired, and there were no singletons (i.e., reads with a sequence present exactly once, which is unique among the reads). This confirmed that all paired-end reads originated from the same DNA molecule and that, therefore, the identified SNPs were correctly matched.

2.4. Surface Marker Expression Using Flow Cytometry Analysis, Cell Sorting, and Cytotoxicity Assays with NK-92 Cell Lines

Direct labeling was used to assess the cell surface expression of FcγRIIIa/CD16 on NK-92 transfectants using three different mAb clones (3G8, B73.1, and MEM154), while HLA-ABC and CD58 expressions were assessed on targets cells. Cells were washed with FACS buffer and incubated for 30 min at 4 °C with saturating amounts of antibodies or isotype-matched controls. After the labeling, excess antibody was removed by washing with FACS buffer. To test for CD20 and EGFR antigen expressions on target cells, indirect labeling was required, starting with the B-cell receptor blockade using 0.4 mg/mL $F(ab')_2$ fragment goat anti-human IgG for 30 min at 4 °C. After incubation with anti-CD20 or anti-EGFR at concentrations ranging from 0.125 to 20 μg/mL, a secondary polyclonal antibody, goat anti-human κ $F(ab')_2$-PE, was used for final labeling. Data acquisition and analysis were performed using an Attune NxT Acoustic Focusing Cytometer (ThermoFisher Scientific) and FlowJo version 10.5.3 software (RRID:SCR_008520), respectively, using the gating strategy depicted in Figure S1. The surface expression levels were compared by calculating the geometric mean fluorescence intensity ratios (MFIRs) as previously described [51].

NK-92 transfectants were sorted for FcγRIIIa/CD16 expression using a MoFlo Astrios cell sorter (Beckman Coulter Life Sciences, Brea, CA, USA) at our institution's flow cytometry core facility. Briefly, 20×10^6 cells were stained with the mAb 3G8 and the appropriate isotype control antibody for 30 min at 4 °C, washed, and resuspended in 2 mL FACS buffer plus 2.5 mM EDTA. The 3G8-positive cell fraction was then collected, centrifuged, and seeded at 0.3×10^6 cells/mL in the culture medium as previously described. Twenty-four hours after sorting, 600 μg/mL of geneticin was added to the cell cultures. The transfected NK-92 cells were cultured in the presence of 100 IU/mL of IL2.

2.5. Cytotoxicity Assays by NK-92 Cell Lines

Non-radioactive DELFIA EuTDA Cytotoxicity Reagents (PerkinElmer, Waltham, MA, USA) were used to determine the level of NK-92 cell-mediated cytotoxicity according to the manufacturer's instructions and as previously described [52]. Briefly, the target cells

(K562, Daudi, Raji, and A431) were labeled with BATDA for 20 min and washed with PBS containing 20 mM HEPES and 2.5 mM probenecid to inhibit dye leakage.

For the ADCC assays, NK-92 cell lines were deprived of IL2 overnight, washed, and maintained in a serum-free AIM-V medium. Fixed effector-to-target (E:T) ratios of 5:1 and 10:1 were used for the anti-CD20 and -EGFR dose-response assays, respectively. Direct cytotoxicity was assessed in all ADCC experiments by co-culturing the target cells and NK-92 effector cells in the absence of mAbs. The plates were incubated for 1 h when co-cultured with Daudi cells or 2 h for the other target cells at 5% CO_2 and 37 °C.

For direct cytotoxicity assays against K562 targets, NK-92 cell lines were maintained in culture medium with 200 IU/mL IL2 and used at E:T ratios of 20:1; 10:1; 5:1; and 2.5:1.

In both assays, the maximum release was obtained by the lysis of labeled target cells (Lysis Buffer, PerkinElmer). To assess cytotoxicity following co-incubation, the released TDA ligand in the supernatants was chelated with europium salt and measured in a time-resolved fluorometer (EnVision 2014 Multilabel reader, PerkinElmer). Cytotoxicity was expressed as a percentage (%) of specific lysis.

For the ADCC dose-response assays, the curves were fitted using the three-parameter dose-response curves for the agonist model. The agonist was the mAb, anti-CD20 or -EGFR, and the EC_{50} was estimated accordingly, where EC_{50} was the concentration at which the mAb produced a half-maximum response. In direct cytotoxicity assays, plots were fitted using the asymptote-modified exponential growth model of Pross et al. [53] to obtain the lytic units, defined as the number of effector cells required to lyse a given number of target cells. We defined the reference lytic level as 30% (LU_{30}) for 10^7 effector cells. The EC_{50} and LU_{30} were estimated, and all curves were plotted using Prism GraphPad (version 9.5.0, RRID:SCR_002798).

2.6. Statistical Analysis

Analysis was performed with GraphPad Version 10.2.0. Comparisons of paired determinations were performed using a ratio-paired *t*-test or mixed-effect analysis with Šídák's multiple comparisons test. For unmatched determinations, the Brown–Forsythe and Welch ANOVA with Dunnett's T3 multiple comparisons test correction or one-way ANOVA corrected by Tukey's multiple comparisons test were used to compare among cell lines. Other statistic tests are indicated in the text. The correlation between the 3G8 MFIR and EC_{50} was analyzed by calculating Spearman's Rank Correlation Coefficient. The statistical test used is indicated in each figure legend. The significance is indicated by the *p*-values in the graphs.

3. Results

3.1. Expression of mRNA of FCGR3A SNPs rs10127939 and rs396991 in Transfected NK-92 Cell Lines

The vector design for the *FCGR3A* SNP variants, rs10127939 (L66H/R) and rs396991 (V176F), started from the *FCGR3A* transcript variant 3 (NM_001127593.1). Sequence analysis confirmed that this plasmid contains the bases encoding the V and L for rs396991 and rs10127939, respectively (sequence provided in Figure S2). After initial subcloning into the multiple cloning site 2 of the pVITRO-neo-mcs, only the pVITROneoCD16A_VL_{MCS2} clone b was kept and used as the starting material for the site-directed mutagenesis. Thus, pVITROneoCD16A_VH_{MCS2} clones a/c, pVITROneoCD16A_VR_{MCS2} clones a/b/c, and pVITROneoCD16A_FL_{MCS2} clones 4/13/14/15 were generated and sequenced for quality control purposes. The constructs of all SNP variants for *FCGR3A* used in this study are shown in Table S3 and were used for the transfection of NK-92 cells.

Huang et al. previously reported that the NK-92 cell line carries the *FCGR3A* gene but fails to express it on the surface [46]. Moreover, the expression of transfected genes can be impaired either by de-novo methylation [54] or by their random integration into transcriptionally silent DNA regions [55]. Therefore, the expression of each *FCGR3A* transgene was assessed at the mRNA level using RT-PCR with primers designed to cover rs10127939 (L66H/R) and rs396991 (V176F), spanning exons 4 and 5 and excluding the <3.4 Kb intron 4. Deep sequencing of the RT-PCR amplicons allowed us to identify each variant and determine its frequency in NK-92 transfectants.

The expression of a single *FCGR3A* variant was confirmed in the cell lines generated using single-plasmid transfection, namely NK-92 $^{LL\_FF}$, NK-92 $^{LL\_VV}$, NK-92 $^{RR\_VV}$, and NK-92 $^{HH\_VV}$. The expressions of two FCGR3A allelic variants were confirmed in NK-92 $^{LL\_VF}$, NK-92 $^{LR\_VV}$, NK-92 $^{LH\_VV}$, and NK-92 $^{LR\_VF}$. In the latter three cell lines, the ratio of the two transgene variants was approximately 1:3, possibly due to the intrinsic nature of random plasmid integration in the genome (Figure 1b). The results were confirmed via the sequencing of amplicons derived from the mRNA of two healthy human donors (HD.CMU.003, HD.CMU.005), whose genotypes have been previously confirmed using Sanger sequencing Figure S3.

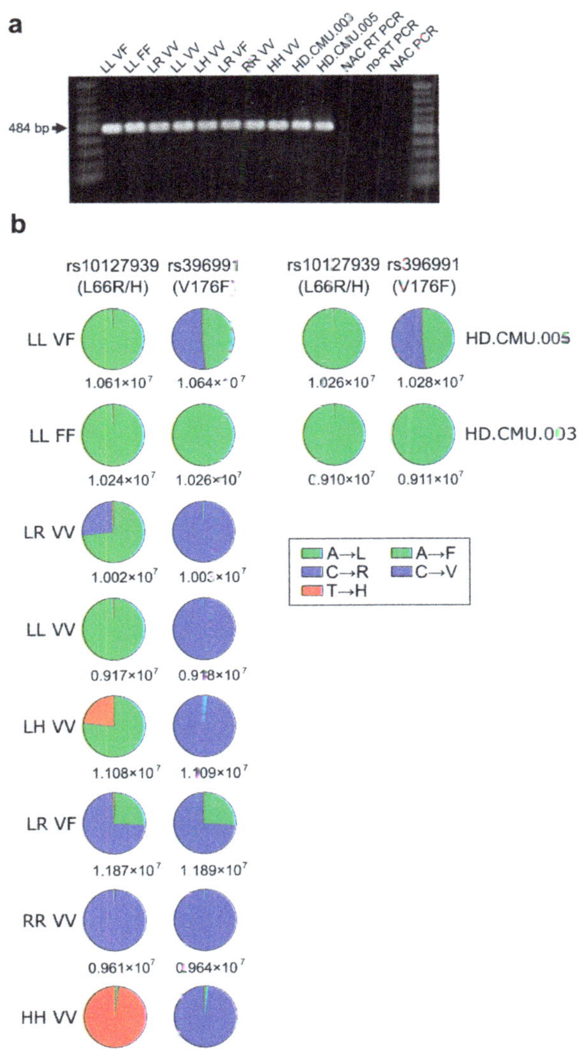

Figure 1. Messenger expression and NGS of NK-92 transfectants for the rs10127939 and rs396991 SNPs. (**a**) Total RNA was extracted from NK-92 transfectants and converted into cDNA. The cDNA

was then amplified for a 484 bp region covering both SNPs, rs10127939 and rs396991. The resulting amplicons were run in a 1.2% agarose gel and stained with RedGel. These 484 bp fragments were used for the NGS analysis. The gel includes controls for cDNA synthesis and PCR, specifically the "No amplification controls" (NAC) for NAC RT-PCR and NAC PCR, as well as a control without reverse transcriptase (no-RT PCR). (**b**) The pie charts show the NGS reads corresponding to rs10127939 (L66H/R) and rs396991 (V176F) for each of the NK-92 transfectants. The total number of reads is shown below each pie chart, with bases represented by color codes. For each SNP, a specific base read corresponds to amino acids: leucine (L), arginine (R), histidine (H), phenylalanine (P), and valine (V).

In summary, the expression of either one or two sets of SNPs, rs10127939 (L66H/R) and rs396991 (V176F), in the eight different NK-92 cell lines generated in this study was confirmed through the NGS of amplicons derived from reverse-transcribed mRNA.

3.2. *Stable FcγRIIIa/CD16 Cell Surface Expression of Different Combinations of the FCGR3A SNPs rs10127939 and rs396991*

FcγRIIIa/CD16 cell surface expression of the eight NK-92 cell lines (Table S4), successfully transfected with different combinations of the SNP variants rs10127939 and rs396991 was determined using flow cytometry. Figure 2 shows the level of expression using three different mAb clones. Clone B73.1 binds to domain 1 (D1) of the FcγRIIIa/CD16 molecule, recognizing predominantly the L, weakly the R, but not the H variant; clone 3G8 binds to D2; and clone MEM154 has been reported to discriminate V from F by strong binding to V (Figure 2a) [15]. Figure 2b shows representative histograms for the relative expression of FcγRIIIa/CD16 compared to matched isotype controls when the transfectant cells were cultured in the presence of 100 IU/mL of IL2. Clone B73.1 showed high expression for transfectants with only L, i.e., NK-92 $^{LL\_VF}$, NK-92 $^{LL\_FF}$, and NK-92 $^{LL\_VV}$ (MFIRs of 14.3 ± 7.9, 27.1 ± 14.0, and 25.4 ± 9.0, respectively), lower for the other combinations (MFIRs ranging from 3.2 ± 1.6 to 7.9 ± 1.9), and no staining for NK-92 $^{HH\_VV}$ (MFIR of 0.8 ± 0.1; Table S5, left column). In our hands, MEM154 stained the NK-92 transfectants poorly (right column), with MFIRs ranging from 1.4 ± 0.3 to 5.5 ± 1.2 (Table S5, right column). In the case of clone 3G8, the MFIR varied between 13.0 ± 4.2 (NK-92 $^{LR\_VF}$) and 91.9 ± 38.4 (NK-92 $^{LL\_VV}$) (Table S5, middle column). Nonetheless, FcγRIIIa/CD16 expression was higher on freshly isolated NK cells compared to transfected NK-92 cells. Here, the clone MEM154 clearly discriminated between the donors with different V176F variants (Figure 2b,c). Our NK-92 $^{LL\_FF}$ cell line showed 14- to 12-fold higher levels of FcγRIIIa/CD16 expression for B73 and 3G8, respectively, compared to the transduced noGFP-CD16 176F NK-92.05 cell line with the same combination of *FCGR3A* SNPs previously described by Binyamin et al. [49] (Figure 2b,c and Table S5). The parental NK-92 cell line, as well as NK-92 ^{pVITRO}, had no detectable levels of FcγRIIIa/CD16 on its surface (Figure 2e,f). At last, FcγRIIIa/CD16 expression on NK-92 transfectants was stable after selection and cell sorting, as it did not change for more than five months or 80 cell culture passages (Figure 2g).

Thus, the eight transfected NK-92 cell lines not only express the different combinations of *FCGR3A* SNPs rs10127939 (L66H/R) and rs396991 (V176F) at the mRNA level, but also show stable surface expression.

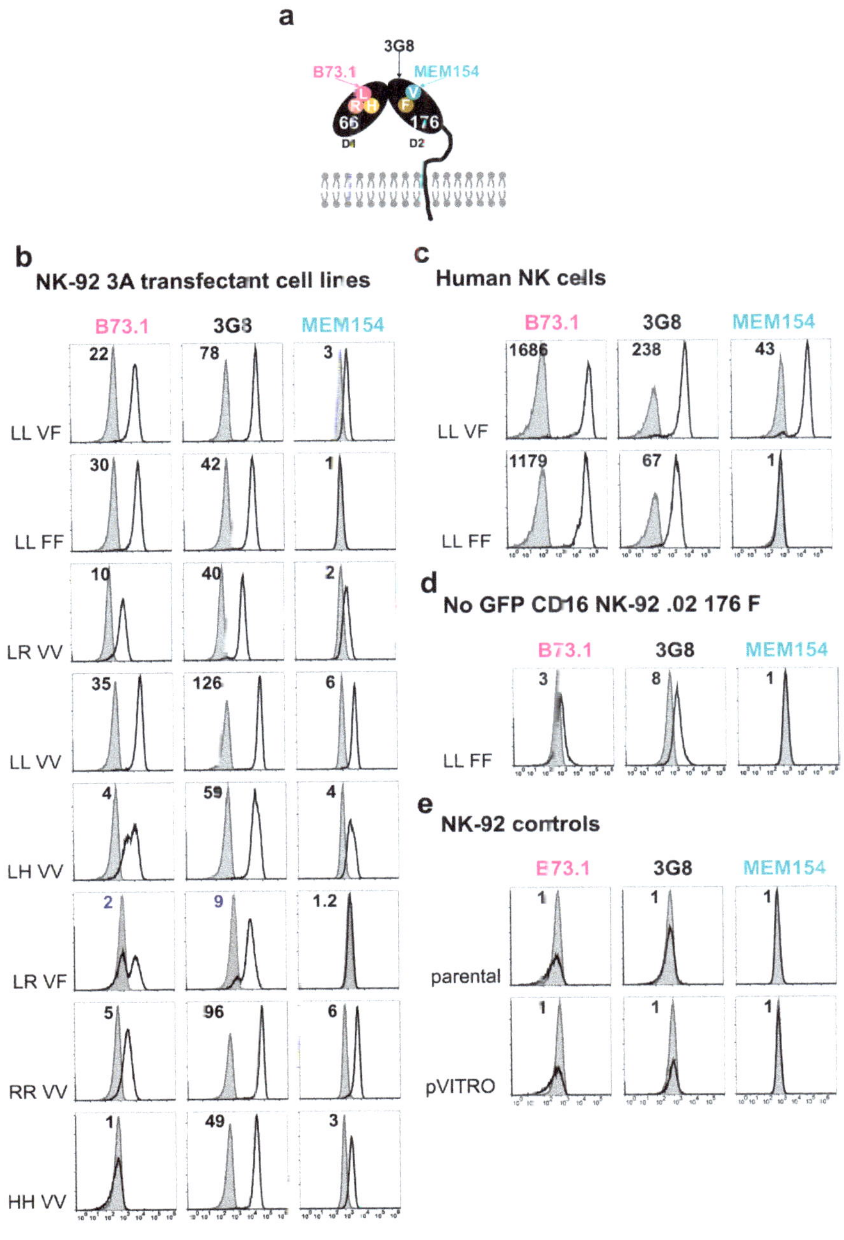

Figure 2. *Cont.*

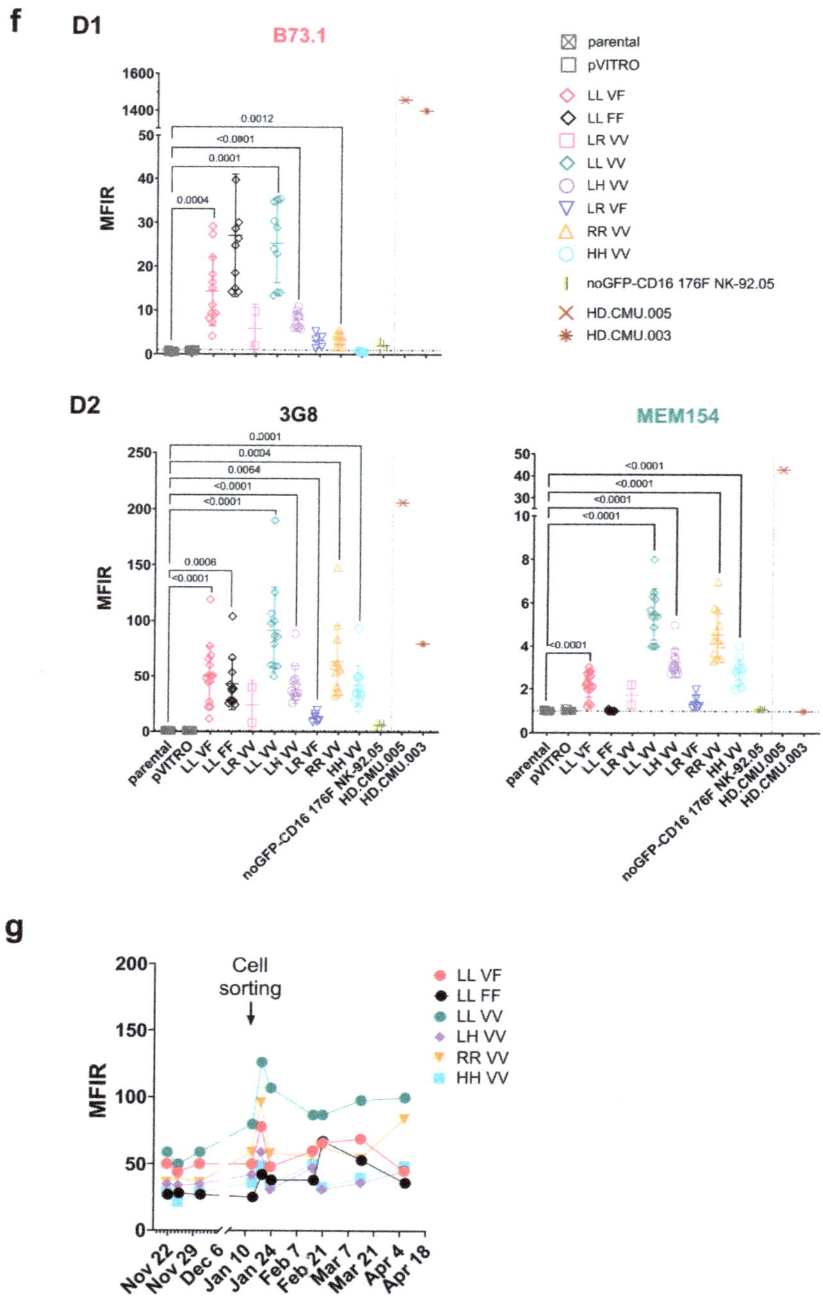

Figure 2. Surface expression of FcγRIIIa/CD16 on NK-92 transfectants using different anti-CD16 monoclonal antibody clones. Transfected NK-92 cell lines were tested for the surface expression of FcγRIIIa/CD16 using the mAb clones B73.1, 3G8, and MEM154 and matching isotype controls (grey), followed by flow cytometry analysis. (**a**) Schematic overview of the epitopes recognized by the anti-CD16 monoclonal antibody clones. These cover the different L66H/R and V176F variants of the FcγRIIIa/CD16 receptor located in domain 1 (D1) and domain 2 (D2), respectively. MEM154 only

recognizes valine (V) at position 176 but not phenylalanine (F), whereas the clone B73.1 recognizes leucine (L) at position 66, binds poorly to arginine (R), but does not recognize histidine (H). In the case of clone 3G8, it recognizes a different epitope of D1, independent of the polymorphisms studied. (**b–e**) Representative histograms showing the recognition of FcγRIIIa/CD16 by the three clones. The L66H/R and V176F polymorphisms present in each cell line/donor are indicated next to the histograms, and the mean intensity ratio (MFIR) for each histogram is described in the upper-left corner. (**b**) The eight transfected cell lines ($n = 10$, except for NK-92 $^{LL\_VF}$, NK-92 $^{LR\_VV}$, and NK-92 $^{LR\_VF}$, where n = 13, 2, and 5). (**c**) Single determination for the level of expression of FcγRIIIa/CD16 for two of the most common variants in fresh PBMCs gated on $CD3^{-}CD56^{+}$ NK cells are shown for donors HD.CMU.003 and HD.CMU.005. (**d**) Evaluation of FcγRIIIa/CD16 in the transduced cell line noGFP-CD16 176F NK-92.05 used as reference [49] (n = 3). (**e**) Transfection controls, including parental NK-92 and NK-92 ^{pVITRO} (n = 12 and 15, respectively). (**f**) Summary plots of FcγRIIIa/CD16 expression shown as MFIRs for the staining with anti-CD16 clones B731, 3G8, and MEM154 for all NK-92 cell lines (transfectants and transduced), including expression on fresh NK cells. One-way ANOVA (Brown–Forsythe and Welch) analysis for all cell lines except for human NK cells. (**g**) Time course of FcγRIIIa/CD16 expression as measured using 3G8 mAb FACS staining in six NK-92 transfectants over six months (80 passages), including a freeze/thaw cycle and cell sorting of the transfectants in January 2022 (arrow).

The culture, proliferation, and cytotoxicity of the NK-92 cell line are IL2 dependent [43,44]. Nevertheless, to better distinguish between direct cytotoxicity and ADCC, we deprived NK-92 cells of IL2 prior to the assays [56,57]. It is also known that the activation of human NK cells leads to the membrane shedding of FcγRIIIa/CD16 by ADAM-17 (13, 14). Thus, it was of interest to determine whether overnight IL2 withdrawal from the cultures affected FcγRIIIa/CD16 surface expression in six of our NK-92 transfectants. Flow cytometric analysis using the 3G8 clone demonstrated that four out of the six transfectants tested (NK-92 $^{LL\_VF}$, NK-92 $^{LL\_VV}$, NK-92 $^{LH\_VV}$, and NK-92 $^{RR\_VV}$) exhibited increased FcγRIIIa/CD16 expressions when deprived of IL2 overnight, with adjusted *p*-values of 0.0015, 0.0141, 0.0397, and 0.0060, respectively (Figure 3a). In contrast, the two other cell lines, NK-92 $^{LL\_FF}$ and NK-92 $^{HH\_VV}$, did not significantly change their FcγRIIIa/CD16 expressions. NK-92 $^{LL\_VV}$ showed the highest expression (3G8 MFIR values of 83 ± 24.0 and 135.2 ± 46.7, with and without IL2, respectively) and NK-92 $^{HH\_VV}$ the lowest expression (37.1 ± 9.68 and 48.4 ± 17.6, with and without IL2, respectively). The pooled mean FcγRIIIa/CD16 expression of the six NK-92 transfectants increased upon IL2 deprivation ($p = 0.0010$), with variations from 1.3- to 2-fold higher MFIRs in the case of NK-92 $^{HH\_VV}$, and NK-92 $^{RR\_VV}$, or NK-92 $^{LL\_VF}$ (Figure 3b). The same analysis, using clone MEM154, also revealed a significant increase in FcγRIIIa/CD16 expression upon IL2 deprivation for most of the cell lines, with the exception of NK-92 $^{LL\_FF}$ and NK-92 $^{HH\_VV}$. For the clone B73.1, only NK-92 $^{LL\_VF}$ showed an increase in FcγRIIIa/CD16 ($p = 0.0213$) after IL2 deprivation (Figure S4). The overnight removal of IL2 resulted in a slight reduction in the viability of NK-92 cells, from 91.6 ± 4.4% to 83.7 ± 2.4%, which was nevertheless statistically significant ($p = 0.0007$). In all but one instance, the viability of the cells remained above 70% (Figure 3c).

In conclusion, overnight deprivation significantly increased the surface expression of FcγRIIIa/CD16 in the NK-92 transfectants.

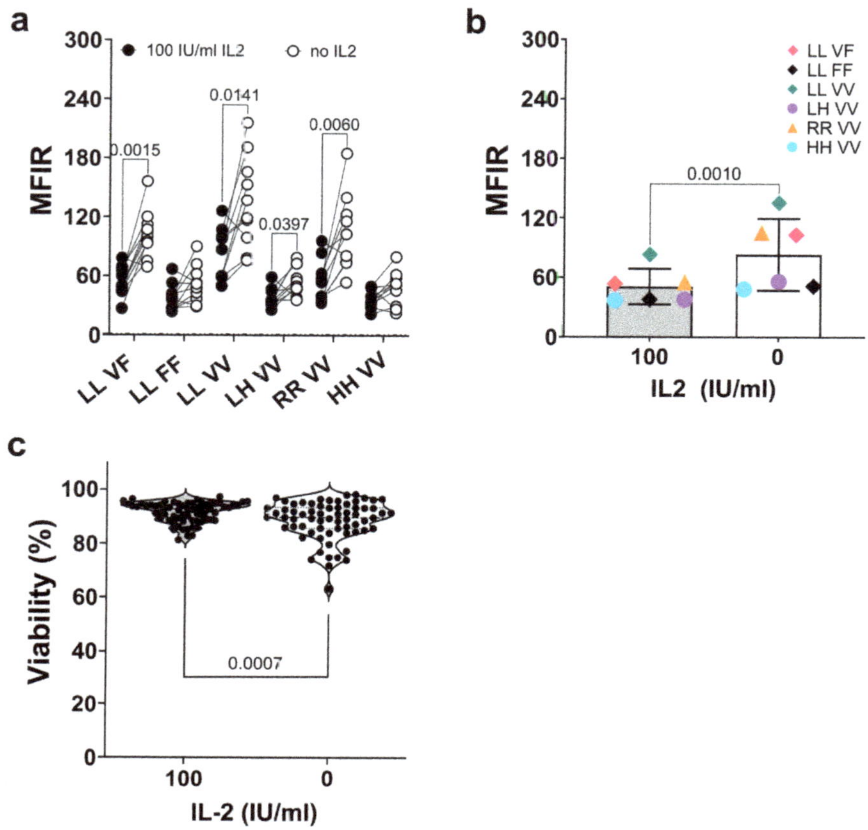

Figure 3. Interleukin-2 deprivation increased FcγRIIIa/CD16 expression on NK-92 transfectants. (**a**) Analysis of FcγRIIIa/CD16 expression using flow cytometry of the NK-92 cell lines stained with clone 3G8 after three days in culture with 100 IU/mL of IL2 (black circles) or IL2-deprived overnight (open circles) for six NK-92 transfectants. (**b**) Summary plot of the average of the mean fluorescence intensity ratio (MFIR) for each cell line analyzed under the two conditions previously described. In all cases, comparisons were made using a ratio-paired *t*-test, with n = 10. No values are shown if $p \geq 0.05$. (**c**) Cell viability of NK-92 cell lines after short-term IL-2 deprivation. Different NK-92 cell lines, including both transfected and non-transfected cells, were cultured in the presence of IL-2 (100 IU/mL) or in the absence of IL-2 (0) for 16 h. Cell viability was analyzed using flow cytometry using the dead cell exclusion dye 7-AAD, and comparisons were made using a paired *t*-test, with n = 67.

3.3. Characterization of Cell Surface Antigen Expression on Target Cells

The target cells used to study cytotoxicity were characterized to avoid the misinterpretation of the results due to different levels of antigen expression. Figure 4a shows that the level of CD20 expression on Raji cells was higher than on Daudi cells, with MFIR values at 1250 ng/mL of 59.2 ± 5.4 and 20.2 ± 2.3, respectively ($p = 0.0003$, unpaired *t*-test). On the other hand, EGFR expression on A431 cells, as measured using cetuximab staining, reached saturation at 2000 ng/mL, with MFIR values of 153.0 ± 15.9 and 165.1 ± 18.8 at 2000 ng/mL and 20 μg/mL, respectively (Figure 4b). Additionally, the target cells were tested for two antigens that may affect NK cytotoxicity, HLA-ABC and CD58. We confirmed that the prototypic cell line K562 used in direct NK-92 cytotoxicity assays, as well as Daudi cells, lacked HLA-ABC expression (MFIRs of 1.09 ± 0.04 and 1.13 ± 0.04, respectively). In contrast, Raji and A431 expressed high (MFIR 30.2 ± 7.4) and moderate (MFIR 17.7 ± 4.5) HLA-ABC levels, respectively (Figure 4c, left). Finally, CD58, which is the ligand for the co-stimulatory receptor CD2 on NK cells, was expressed at higher levels on Raji compared to the other cell lines with MFIRs of 55.4 ± 1.1, 26.7 ± 2.2, 102.7 ± 21.7, and 37.6 ± 4.7 for K562, Daudi, Raji, and A431, respectively (Figure 4c, right).

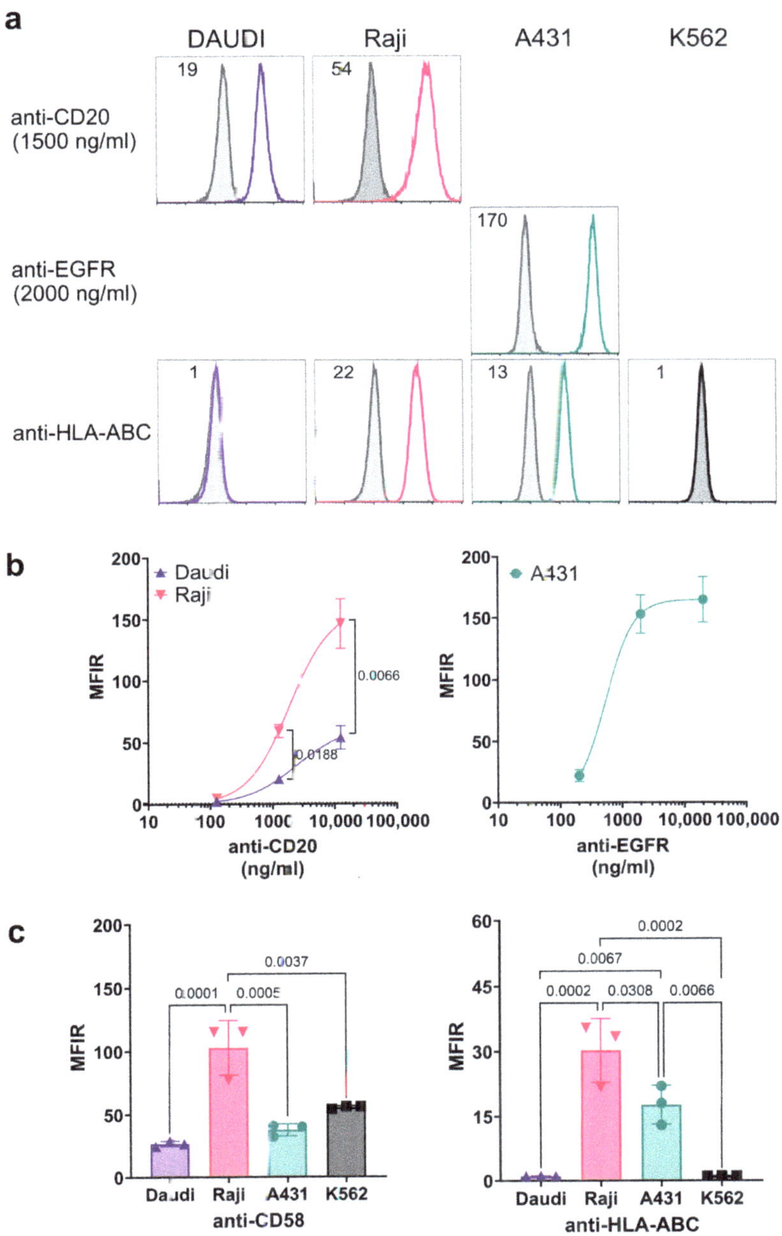

Figure 4. Expression of surface antigen on target cells. (a) Representative histogram overlays for the expressions of CD20, EGFR, and MHC-I (HLA-ABC), conducted by indirect and direct antibody staining

using the humanized anti-CD20 and -EGFR at 1.25 and 2 µg/mL, respectively, and the mouse anti-HLA-ABC at a saturation condition. The antigen expression in each cell line is indicated next to the histogram, and the mean intensity ratio (MFIR) for each histogram is described in the upper-left corner. Gray histograms correspond to isotype controls. (**b**) Pooled data (n = 3) of MFIRs with respect to target cell lines at three different primary antibody concentrations for anti-CD20 and -EGFR. (**c**) Pooled data (n = 3) for HLA-ABC and CD58 expressions in target cells used in the study. Each symbol represents an independent experiment.

3.4. FcγRIIIa/CD16-Transfected NK-92 Cell Lines Induce Antibody-Dependent Cell-Mediated Cytotoxicity (ADCC)

The parental NK-92 cell line has been extensively used to dissect the mechanisms of direct NK cytotoxicity. However, the lack of FcγRIIIa/CD16 expression makes them unsuitable for ADCC studies. The generation of a panel of NK-92 cell lines expressing different combinations of *FCGR3A* SNP variants, rs10127939 (L66H/R) and rs396991 (V176F), allowed us to test for differences in ADCC depending on a given SNP combination. First, one of the most common *FCGR3A* genotypes in the population, i.e., homozygosity for L and F, was analyzed. ADCC mediated by a transfected NK-92 $^{LL\_FF}$ cell line against Daudi target cells was induced by an anti-CD20 mAb as compared to the ADCC induced by matching-genotype NK cells isolated from either fresh blood or buffy coat. As little as 125 ng/mL of anti-CD20 induced the maximal lysis of Daudi cells (~80%, Figure 5a) with no apparent differences between NK-92 $^{LL\_FF}$ and primary NK cells despite the fact that primary NK cells have higher levels of FcγRIIIa/CD16 on their surface (Figure 1b,c).

ADCC was demonstrated for all NK-92 cell lines transfected with the allelic variants of FcγRIIIa/CD16, as shown in the summary plots (Figure 5b,c). The direct cytotoxicity of NK-92 transfectants against Daudi cells in the absence of anti-CD20, a control included in all experiments, was significantly lower than ADCC, with minor variations within the six NK-92 transfectants (Figure 5b, open circles). The FcγRIIIa/CD16 deficient controls, parental NK-92 and empty vector NK-92 ^{pVITRO} transfectant, induced direct cytotoxicity of the Daudi target cells with 12.2 ± 7.2% and 11.7 ± 6.4% of specific lysis, respectively. The addition of 500 ng/mL anti-CD20 antibody to the NK-92 control cells increased specific lysis by 11% compared to direct cytotoxicity (23.5 ± 9.2% and 21.5 ± 4.0%, respectively; Figure 5b, star symbols). Since the only difference between the two conditions was the presence of the anti-CD20 antibody, this additional killing was related to the direct induction of programmed cell death, a mechanism of target cell destruction that can be observed with type-II anti-CD20 antibodies such as Obinutuzumab [58–60]. Importantly, all eight NK-92 transfectants induced ADCC, significantly different from parental NK-92 and the NK-92 ^{pVITRO} cell lines, using Daudi cells as target ($p \leq 0.0001$).

Overall, ADCC mediated by the transfected NK-92 cell lines against both CD20-expressing target cells was antibody dose-dependent when assayed at a fixed E:T ratio of 5:1 (Figure 5c: top Daudi; middle Raji). Maximum lysis of Daudi cells was achieved at 12.5 ng/mL of anti-CD20, with 56.3% and 55.9% specific lysis for NK-92 $^{LR\_VV}$ and NK-92 $^{LR\_VF}$, respectively, compared to an average of 79.4 ± 6.6% specific lysis for the other FcγRIIIa/CD16-expressing NK-92 transfectants. Of note, the low-affinity transfectant NK-92 $^{LL\_FF}$ showed 79.1% specific lysis. In terms of the EC_{50}, NK-92 $^{LL\_FF}$ and NK-92 $^{HH\_VV}$ had the highest (2.13 and 2.23 ng/mL), while NK-92 $^{LR\_VV}$ had the lowest (0.02 ng/mL) value; all other transfectants had an EC_{50} ranging from 0.62 to 1.64 ng/mL. Nevertheless, using Raji target cells, which express more than twice as much CD20 as Daudi cells (Figure 4), the average maximum specific lysis (72.6 ± 5.6%) for the six NK-92 transfectants tested was observed at an anti-CD20 mAb concentration of 125 ng/mL. ADCC against Raji target cells mediated by NK-92 $^{LL\_FF}$ showed an EC_{50} of 6.22 ng/mL in contrast to the average of 1.1 ng/mL observed for the other five NK-92 transfectants.

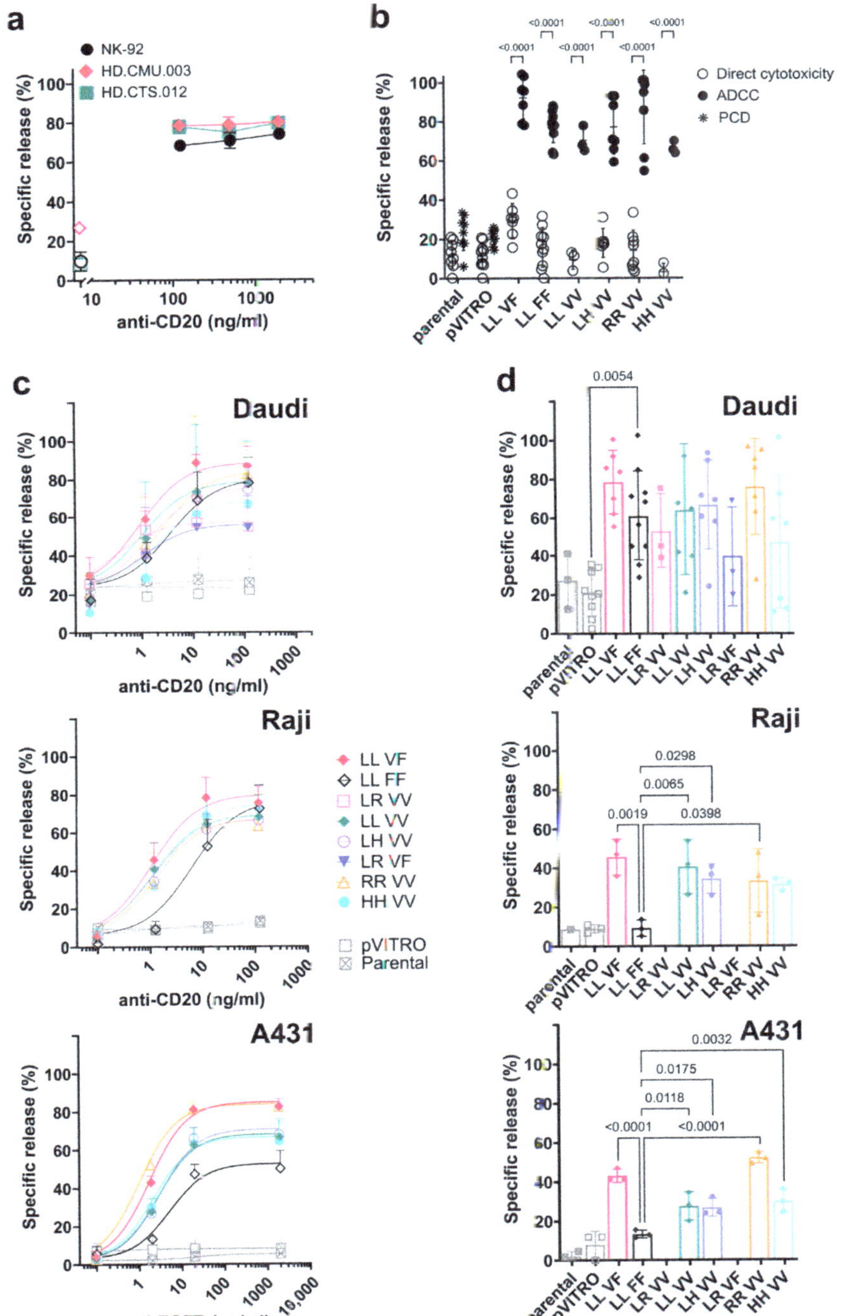

Figure 5. All NK-92 transfectants expressing FcγRIIIa/CD16 variants perform ADCC against antibody-coated cells. ADCC assays were performed using either Daudi or Raji cells as targets opsonized with the anti-CD20 antibody at an effector-to-target ratio (E:T) of 5:1 or A431 cells with anti-EGFR antibody at an E:T ratio of 10:1. The non-radioactive cytotoxic assays lasted for one and

two hours for anti-CD20 and anti-EGFR, respectively. The graphs show the percentage of specific release (%) mean ± SD versus anti-CD20 or anti-EGFR. (**a**) Comparison of ADCC between the NK-92 $^{LL\_VF}$ transfectant cells (black circle), NK cells from healthy donors isolated from blood (HD.CMU.003, magenta diamond), and a buffy coat (HD.CTS.012, green square) used as effectors, and Daudi cells as targets opsonized with different concentrations of anti-CD20 antibody (125, 500, and 2000 ng/mL) at an E:T of 5:1 (one experiment per donor). Direct cytotoxicity controls (i.e., no antibody, open symbols) are included. (**b**) This panel compares the six NK-92 cell lines and NK-92 controls co-cultured with Daudi cells alone (direct cytotoxicity, white circle), together with 500 ng/mL of anti-CD20 (ADCC, black circles), or NK-92 controls (parental and pVITRO) with targets and anti-CD20 (programmed cell death, PCD, stars). (**c**) The summary plots display the mean ± SD of the specific release (%) for ADCC by different transfected NK-92 cell lines, shown for Daudi (top, n = 5), Raji (middle, n = 3), and A431 (bottom, n = 3). The ADCC dose-response curves were fitted using the dose-response curves for the agonist model, where the agonist was the mAb, anti-CD20 or -EGFR, and the EC_{50} was estimated accordingly. (**d**) ADCC was measured at non-saturating concentrations of anti-CD20 and -EGFR mAbs, corresponding to 1.25 ng/mL and 2 ng/mL, respectively. The targets used were Daudi (n = 3 to 10, upper), Raji (n = 3, middle), and A431 (n = 3, bottom). Data were analyzed using one-way ANOVA, and multiple comparisons were made against NK-92 $^{LL\_FF}$.

Finally, ADCC against the A431, a cell line expressing the EGFR antigen, was demonstrated for all NK-92 transfectants, with a maximum of lysis reached at 20 ng/mL. In the presence of an anti-EGFR mAb, the variants NK-92 $^{LL\_VF}$ and NK-92 $^{RR\_VV}$ showed the highest ADCC (85.3% and 84.1% specific lysis, respectively), followed by NK-92 $^{LL\_VV}$, NK-92 $^{LH\_VV}$, and NK-92 $^{HH\_VV}$ (68.5, 71.1, and 67.1%, respectively), whereas the low-affinity variant NK-92 $^{LL\_FF}$ showed only 52.8% of the maximum specific lysis (Figure 5c, bottom) and the highest EC_{50} (5.02 ng/mL, Table 1).

Table 1. EC_{50} obtained from ADCC dose-dependent curves using different NK-92 cell lines against target cells.

	Daudi			Raji			A431		
Transfectant	**n**	*** R^2**	**‡ EC_{50} (ng/mL)**	**n**	**R^2**	**EC_{50} (ng/mL)**	**n**	**R^2**	**EC_{50} (ng/mL)**
NK-92 $^{LL\_VF}$	2, 5	0.9139	0.99	3	0.9268	1.1	3	0.9715	1.8
NK-92 $^{LL\_FF}$	4, 8	0.8504	2.13	3	0.9039	6.2	3	0.9011	5.0
NK-92 $^{LR\_VV}$	3	0.5515	0.02	ND			ND		
NK-92 $^{LL\_VV}$	4	0.5578	1.02	3	0.8626	1.2	3	0.9284	2.8
NK-92 $^{LH\_VV}$	2, 4	0.7898	1.58	3	0.9106	1.1	3	0.9472	3.1
NK-92 $^{LR\_VF}$	3	0.5181	0.62	ND			ND		
NK-92 $^{RR\_VV}$	2, 5	0.7526	1.64	3	0.9749	1.4	3	0.9689	1.1
NK-92 $^{HH\_VV}$	2, 4	0.5679	2.23	3	0.8003	1.8	3	0.9379	2.2
Parental	3, 6	0.0476	-;	3		-	3	0.0097	-
NK-92 ^{pVITRO}	7, 9	0.0738	-	3		-	3	0	-

* The goodness of fit (R^2) for the curves was calculated. For $R^2 < 0.5$, poor fitting to asymptotic curves. ‡ Estimate of EC_{50} for calculated from the fitted curves. *Abbreviations:* EC_{50}, effective concentration to reach 50% of cytotoxicity; ND, not determined.

To better describe the differences between cell lines, particularly the NK-92 $^{LL\_FF}$ variant, we analyzed the specific release of each target cell line at non-saturating concentrations of mAbs, which were 1.25 ng/mL and 2 ng/mL for anti-CD20 and -EGFR, respectively. The comparisons showed statistical differences in ADCC mediated by NK-92 $^{LL\_FF}$ when the Raji and A421 cells were used as targets but not when Daudi cells were used (Figure 5d).

Finally, we tested whether there was a correlation between ADCC and FcγRIIIa/CD16 surface expression as detected using the 3G8 clone in overnight IL2-deprived NK-92 transfectants. No correlation was found for any of the three different target cell lines with R squared values of 0.1564, 0.2671, and 0.2110 and a two-tailed *p* of 0.4377, 0.938, and 0.3594 for Daudi, Raji, and A431 cells, respectively (Figure S5).

In conclusion, all NK-92 cells transfected with different combinations of SNPs for *FCGR3A*, rs10127939 (L66H/R) and rs396991 (V176F), mediated ADCC, which could be clearly distinguished from direct cytotoxicity.

There were minor differences in the lytic activity induced by anti-CD20 antibody binding to FcγRIIIa/CD16, independent of the genetic variant involved. Nevertheless, more pronounced differences were observed for anti-EGFR and against CD20-positive Raji target cells at non-saturating concentrations of anti-CD20 mAb. For example, ADCC against EGFR-positive A431 target cells in the presence of cetuximab was lower with the low-affinity variant expressed on NK-92 $^{LL\_FF}$ compared to the high-affinity variants.

3.5. Direct Cytotoxicity against K562 Target Cells Mediated by FcγRIIIa/CD16-Transfected NK-92 Cell Lines

Direct cytotoxicity of IL2-deprived NK-92 transfectants against Daudi target cells showed little variation among the six cell lines tested (open circles, Figure 5b). To further test the direct NK cytotoxicity mediated by the FcγRIIIa/CD16-expressing NK-92 transfectants, we performed cytotoxicity assays against the prototypic HLA-ABC-negative K562 target cells in the presence of 200 IU/mL of IL2 (Figure 6). Several transfected NK-92 lines showed increased direct cytotoxicity compared to the FcγRIIIa/CD16-negative NK-92 control effector cells, parental NK-92, and NK-92 ^{pVITRO} lines (Figure 6a). To show all the pooled data, we plotted the results for all the different cell lines as described in the Materials section to obtain the LU_{30}, i.e., the number of effector cells required to kill 30% of the K562 expressed per 10^7 effectors (Table 2). Thus, the NK-92 $^{LL\_VV}$ and NK-92 $^{LR\_VF}$ transfectants showed the same LU_{30} values as the control cell lines (parental and pVITRO), followed by the NK-92 $^{LL\_FF}$ and NK-92 $^{LH\_VV}$ transfectants. NK-92 $^{LR\_VV}$ and NK-92 $^{HH\_VV}$ required three times less effectors than controls to lyse 30% of the targets. Finally, the NK-92 $^{LL\_VF}$ transfectant showed enhanced direct cytotoxicity (Figure 6a).

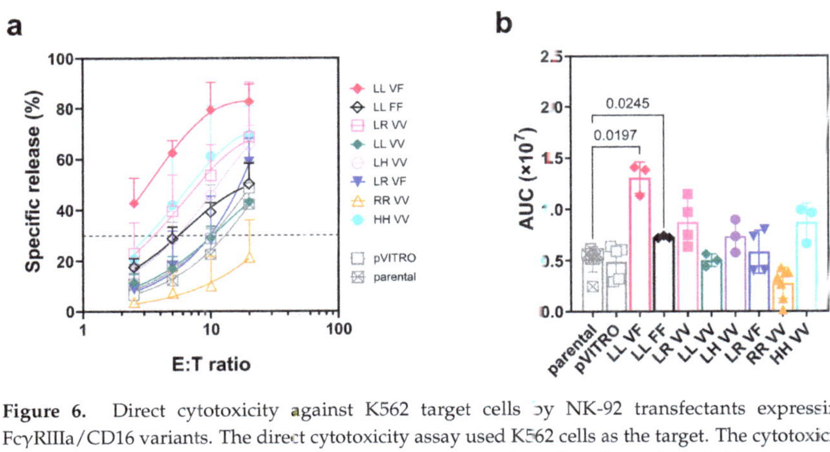

Figure 6. Direct cytotoxicity against K562 target cells by NK-92 transfectants expressing FcγRIIIa/CD16 variants. The direct cytotoxicity assay used K562 cells as the target. The cytotoxicity assay quantified the release of a non-radioactive compound resulting from the lysis/destruction of target cells over a two-hour period. NK-92 transfectants were cultured with 200 IU/mL of IL2 for two days and challenged with K562 at different effector-to-target (E:T) ratios. (**a**) Summary plots of specific release (%) are shown as the specific release mean ± SD of four independent experiments for each target cell line. Lytic units at 30% specific lysis (LU_{30}) for 10^7 effector cells were obtained by fitting the plot of pooled data using an asymptote-modified exponential growth model. (**b**) This is a summary plot that compares the mean ± SD of the area under the curve (AUC) calculated from individual cytotoxicity assays for the eight transfectants, including the parental and NK-92 ^{pVITRO} control cell lines. A comparison was made between the mean of NK-92 parental and the remaining NK-92 transfectants using Dunnett's T3 multiple comparisons test. Values are not shown if $p \geq 0.05$.

Table 2. Comparison of the direct cytotoxicity of NK-92 transfectants against K562 cells by using the lytic units at 30% specific lysis (LU_{30}) for 10^7 effector cells.

Name	Goodness of Fit, R^2 *	Maximum Specific Lysis (%)	$LU_{30}/10^7$ Effectors
NK-92 $^{LL\_VF}$	0.8095	83	63
NK-92 $^{LL\_FF}$	0.8588	52	18
NK-92 $^{LR\_VV}$	0.6181	71	28
NK-92 $^{LL\_VV}$	0.9251	55	10
NK-92 $^{LH\_VV}$	0.8174	80	18
NK-92 $^{LR\_VF}$	0.7148	100 *	11
NK-92 $^{RR\_VV}$	0.3276	77	n/a †
NK-92 $^{HH\_VV}$	0.6209	73	31
parental	0.6521	91	8
NK-92 ^{pVITRO}	0.5487	83	10

* The goodness of fit (R^2) for the curves was calculated. For $R^2 < 0.5$, poor fitting curves. Constriction, specific release < 100%. † n/a, not applicable, as the cells never reached 30% of specific lysis.

For the two most common *FCGR3A* allelic variant combinations in the European population, NK-92 $^{LL\_VF}$ and NK-92 $^{LL\_FF}$, the increase in direct NK cytotoxicity, measured as the area under the curve (AUC), reached statistical significance, with *p*-values of 0.0197 and 0.0245, respectively. In contrast, the NK-92 $^{RR\_VV}$ transfectant showed a slightly lower cytotoxicity compared to the controls, which did not reach statistical significance (Figure 6b).

Finally, direct cytotoxicity mediated by IL2-deprived NK-92 transfectants against all four target cell lines was analyzed. The absence of HLA-ABC in K562 and Daudi, as compared to Raji and A431, was associated with higher direct cytotoxicity, when tested at an E:T ratio of 5:1, with specific lysis of 12.6 ± 12.8% (n = 11), 21.9 ± 7.8% (n = 11), 5.1 ± 2.7% (n = 8), and 4.1 ± 1.4% (n = 8), respectively, and with small differences between the different NK-92 transfectants. In contrast, no differences were observed against target cells expressing HLA-ABC (Figure S6).

In summary, this study demonstrated that all NK-92 cells transfected with different combinations of SNP for *FCGR3A* can effectively mediate ADCC, which is distinct from direct cytotoxicity. The lytic activity observed in ADCC varied depending on the genetic variant and the type of target cells. The expression of FcγRIIIa/CD16 resulted in an increase in direct cytotoxicity on some NK-92 transfectants as compared to parental NK-92 cells. These findings open new avenues of investigation.

4. Discussion

Immortal cell lines derived from leukemia or lymphoma are powerful tools for the study of NK cell biology (44). The NK-92 cell line was first reported by Klingemann and colleagues in 1994 [43–45]. This cell line lacks FcγRIIIa/CD16 surface expression for largely unexplained reasons and, therefore, does not perform ADCC. In recent years, attempts have been made to generate NK cell lines expressing FcγRIIIa/CD16 to study ADCC and for adoptive cancer immunotherapy using tumor antigen-directed mAbs [61]. Several publications describe the generation of NK-92 cell lines expressing FcγRIIIa/CD16 by viral transduction [48,49] or electroporation [47,62,63]. More recently, the modification of the promoter region of *FCGR3A* in the parental NK-92 cell line using CRISPR genome editing technology resulted in the endogenous expression of FcγRIIIa/CD16 [46]. Other attempts include tools to test the ADCC capacity of murine monoclonal antibodies using NK-92 cells transduced with murine CD16 [47]. Moreover, CAR-modified NK-92 cell lines against various tumor types and an NK-92 cell line expressing the high-affinity 176V variant of *FCGR3A*, also called high-affinity NK (haNK), have been reported. A robust antitumor activity in vitro and in vivo has been demonstrated using irradiated haNK cells in combination with IgG1 anti-tumor mAbs [64–66].

The focus of all these studies was on the SNP rs396991 (V176F), and to our knowledge, only Rataj et al. evaluated the impacts of both *FCGR3A* SNPs on ADCC. Nonetheless, the latter study analyzed the effect of FcγRIIIa/CD16 expression on T-cell function by transducing human T cells with three homozygous combinations of *FCGR3A* SNPs (HH_VV, LL_VV, and LL_FF) [67]. NK-92 cells were not tested in the latter study. A correlation between the two SNPs has been reported in European and Colombian populations [15,68]. This finding in European and Colombian populations was confirmed using an 'LDpair Tool' analysis [https://ldlink.nih.gov/, accessed on 27 July 2023], which only reports the L and H alleles for the tri-allelic variant rs10127939 (L66H/LR) without considering the R allele Table S6. In contrast, a linkage disequilibrium for rs10127939 and rs396991 was not found when 'all populations' (5008 individuals) were tested [69].

The present study is the first to successfully generate and characterize eight different NK-92 cell lines carrying these two most studied SNPs of *FCGR3A*. Despite the fact that there are 18 theoretical combinations, our panel of eight transfected NK-92 cell lines represents all combinations of the rs396991 (V176F) and rs10127939 (L66H/R) SNPs present in a human cohort of 87 European individuals, according to the study of Koene et al. The two most common combinations were LL_FF (32.2%) and LL_VF (39.1%) [15].

4.1. *Expression of FcγRIIIa/CD16*

The surface expression of the FcγRIIIa/CD16 depends on the association with signaling adaptor molecules such as TCRζ-chain (CD247) and/or FcγR1 [70]. Since Maki et al. demonstrated that NK-92 expresses CD247 at the mRNA level [45], we did not consider co-transfecting adaptor molecules to our NK-92 transfectants.

The analysis of FcγRIIIa/CD16 protein expression using different anti-CD16 mAb clones demonstrated levels within the same order of magnitude for all NK-92 transfectants. Moreover, FcγRIIIa/CD16 expression was higher compared to the previously reported transduced NK-92 cells [49] and, most importantly, was stable over time. In general, the expression levels of FcγRIIIa/CD16 on normal human NK cells were one to two orders of magnitude higher than on our NK-92 transfectants. Possible explanations for this finding include the (i) weaker activity of the plasmid promoter compared to the endogenous promoter, (ii) the downregulation of FcγRIIIa/CD16 in highly activated NK-92 cells due to the presence of high concentrations of IL2 in the culture media, (iii) the shedding of FcγRIIIa/CD16 by ADAM-17 [13,14], and (iv) elevated levels of miR218 compared to blood NK cells [71]. Indeed, we showed that overnight IL2 deprivation doubled the surface expression of FcγRIIIa/CD16 in the transfected NK-92 cell lines. However, without further experiments, we cannot draw firm conclusions to explain the overall lower FcγRIIIa/CD16 expression levels observed in the transfected NK-92 cell lines.

In our hands, the clone MEM154 discriminated poorly between V and F for the V176F SNP in transfectants, contrary to what we and others have previously described in NK cells isolated from peripheral blood [15] probably due to the lower expression of FcγRIIIa/CD16 in the transfected NK-92 cell lines. Indeed, the MEM154 signal was the weakest of all antibody clones tested. A problem with the reagent was excluded because MEM154 easily identified the SNP rs396991 (V176F) in primary NK cells. On the other hand, we are aware that clone 3G8 does not bind equally to F and V allotypes (rs396991), showing stronger binding to V, which is consistent with our flow cytometry results. Unfortunately, we did not have access to the reagent reported by Li et al. that binds equally to F and V allotypes [72]. In conclusion, our results indicate differences in the overall FcγRIIIa/CD16 expression levels on the NK-92 transfectants as compared to primary NK cells.

4.2. *ADCC Mediated by NK-92 Expressing FcγRIIIa/CD16*

Regardless of the difference in the expression level between the NK-92 transfectants and primary NK cells carrying the same SNP combination (LL_FF), ADCC was equivalent for CD20-expressing target cells. This indicates that the FcγRIIIa/CD16 expression levels on the NK-92 transfectants are sufficient for ADCC under our experimental conditions. Previously, it was reported that the combination of rs396991 (V176F) and rs10127939 (L66H/R) affected the binding of IgG, dividing the individuals in the study population into high (66RR or HH with 176VF or 176VV), intermediate (66LH or LR with 176VF or 176VV), and low binders (66LL with 176VF or 176FF; 66RR or 66HH with 176FF; 66LH with 176FF; 66RH with 176FF). However, ADCC assays were not performed in this study to analyze the effect of different SNPs combinations on NK cell function [73]. Interestingly, the IgG low binders, grouped by the combinations of rs396991 (V176F) and rs10127939 (L66H/R), conferred an increased risk for lupus nephritis in African Americans. The authors speculate that less efficient immune complex handling by low binders may contribute to lupus pathogenesis and that targeting FcγRIIIa/CD16 to improve receptor expression and function may be a potential therapeutic approach. To our knowledge, there are no other studies reporting the clinical relevance of different combinations of rs396991 and rs10127939 *FCGR3A* SNPs. However, rs396991 has been extensively studied and reported to be relevant in infection, autoimmunity, and inflammation [21,40–42].

Although IL2, added to maintain NK-92 cultures, has a major impact on the overall NK killing machinery as recently highlighted by transcriptomic analysis for activating and inhibitory receptors [74], ADCC mediated by NK-92 transfectants was clearly depended on IgG1 monoclonal antibody concentrations, as demonstrated by the dose-response curves. No major differences were observed in the lytic activity against two different CD20-positive target cell lines at saturating antibody concentrations, regardless of the genetic variant tested. However, compared to NK-92 $^{LL\_VF}$ and other variants, ADCC induced by the low-affinity variant NK-92 $^{LL\_FF}$ was indeed lower against EGFR-positive A431 target cells in the presence of cetuximab and against CD20-positive Raji target cells at non-saturating concentrations of anti-CD20 mAb. In addition to the role of the two SNPs, rs396991 (V176F) and rs10127939 (L66H/R), other factors may be considered. Several other molecular interactions between the antigen, the mAb, and the FcγRIIIa/CD16 could be involved to explain these findings. These include (i) the antigen density on the target cells; (ii) the epitope recognized by the antibody paratope (which may induce or block conformational changes leading to oligo- or dimerization of the antigen); or (iii) the fluidity of the cell membrane (given by fatty acid composition and raft formation), which may favor or impede the binding of the Fc portion to the FcγRIIIa/CD16. For example, the anti-EGFR mAbs cetuximab and panitumumab both interact with the extracellular domain of EGFR. Bárta et al. found that the binding of these mAbs to the EGFR depends on the cell line used to measure the antibody affinity. The anti-EGFR binding affinity to the keratinocyte cell line HaCaT was higher and more heterogeneous as compared to A431. These differences, detected using interaction map analysis, suggest that it is not the target but the supramolecular context in which the antigen is located that determines the binding of the mAb [75]. Thus, the conflicting data stemming from therapeutic mAb studies could be the result of focusing only on the genetic variations in *FCGR3A* and not considering variations in the antigen levels on the targeted cells.

4.3. Direct or Natural Cytotoxicity of NK-92 Transfectants

Unexpectedly, several FcγRIIIa/CD16-transfected NK-92 lines showed increased direct cytotoxicity against HLA-ABC class-negative K562 target cells compared to FcγRIIIa/CD16-negative NK-92 control cells. A role for FcγRIIIa/CD16 in direct cytotoxicity independent of antibody ligation has been reported, but the mechanisms remain elusive [76]. Target cells may express putative ligands for FcγRIIIa/CD16; alternatively, FcγRIIIa/CD16 may interact with other activating receptors expressed on NK cells. Supporting this hypothesis, physiological interactions between FcγRIIIa/CD16 and the co-stimulatory receptor CD2 were shown to be necessary to induce direct NK cytotoxicity but not ADCC in a patient with EBV-driven Castleman's disease homozygous for H in position 66 (rs10127939) [19]. Interactions between CD58 and CD2 are essential for immune synapse formation and target cell elimination by NK cells [77,78]; however, the observed differences in ADCC and direct NK cytotoxicity could not be explained by differences in the expression levels of CD58 on our target cells. Furthermore, CD2 is expressed by NK-92 cells [43–45,77,78] and by our NK-92 transfectants, including the parental line and NK-92 ^{pVITRO} transfected NK-92 cells Figure S7).

The NK-92 $^{RR\_VV}$ cell line showed lower direct cytotoxicity against K562 cells compared to NK-92 $^{LL\_VF}$. Here, we could speculate that single or multiple random plasmid integrations into the genomic DNA led to a dysregulation of pathways involved in direct cytotoxicity [79]. It is worth mentioning that the parental NK-92 cell line has numerical and structural cytogenic alterations, making it difficult to predict which chromosomes these alterations will be inserted into [43]. At last, the NK-92 $^{HH\_VV}$ transfectant showed similar direct NK cytotoxicity against K562 targets compared to the parental NK-92 line. This finding contrasts with some previous reports of patients homozygous for H at position 66 of FcγRIIIa/CD16 (rs10127939) with recurrent infections and deficient NK cytotoxicity [18–20]. However, our results are consistent with findings in a patient with DiGeorge syndrome, homozygous for H, and with normal NK cytotoxicity, leading the authors to conclude that the L66H variant is unlikely to be a direct genetic cause of NK cell dysfunction [80].

5. Conclusions

In summary, we have generated eight NK-92 transfectants carrying different combinations of two important SNPs of *FCGR3A*, covering all major genotypes described in the human population. The functional characterization of these cell lines revealed differences in ADCC and direct NK cytotoxicity that may have implications for immunotherapy. The use of either patient-derived NK cells or "off-the-shelf" NK-92 cells offers the potential to modify or include the *FCGR3A* SNP combination in a more effective manner, depending on the desired activity. For instance, the LL_VF SNP combination demonstrated potent ADCC and direct cytotoxicity. In addition, other mAb engineering approaches such as afucosylation alter the affinity of IgG to the FcγRIIIa/CD16.

Supplementary Materials: The following supporting information can be downloaded at: https://www.mdpi.com/article/10.3390/antib13030055/s1, Figure S1: Gating strategy used in flow cytometry analysis of NK-92 cell lines; Figure S2: GenScriptpUC19-Amp *FCGR3A* transcript variant 3; Figure S3: Sanger sequencing of the messenger RNA of *FCGR3A* coming from human NK cells; Figure S4: Clones B73.1 and MEM154 confirmed the findings of 3G8 for the increase in FcγRIIIa/CD16 due to the lack of interleukin-2 during overnight cultures; Figure S5: Correlation in NK-92 transfectants between the expression of FcγRIIIa/CD16 and antibody-dependent cell-mediated cytotoxicity (ADCC); Figure S6: Comparison of direct cytotoxicity between target cells expressing, or not, HLA-ABC; Figure S7: Expression of CD2 on NK-92 cell lines. Table S1: Detailed list of reagents, antibodies, culture media, supplements, and cell lines used; Table S2: List of primer pairs used for the generation of *FCGR3A* allelic variants for SNPs rs396991 and rs10127939, including verification steps; Table S3: Names of constructs and allelic variants used in the current study; Table S4: Nomenclature of NK-92 cell lines carrying, or not, FCGR3A allelic variants, and plasmid used for their generation; Table S5: Expression level of FcγRIIIa/CD16 on NK-92 cells and NK cells using three different monoclonal antibody clones. Table S6: Linkage disequilibrium (LD) analysis for rs10127939 and rs396991 using LDpair Tool.

Author Contributions: Conceptualization, J.D.S. and G.L.P.Y.; methodology, M.F.M., M.P. and G.L.P.Y.; validation, M.F.M., J.D.S. and G.L.P.Y.; formal analysis, M.F.M., M.P., M.A., A.R. and G.L.P.Y.; investigation, M.F.M., M.P., M.A., A.R., A.K., D.R.G. and G.L.P.Y; resources, J.D.S.; writing—original draft preparation, M.F.M.; writing—review and editing, M.F.M., M.P., M.A., A.R., A.K., J.D.S. and G.L.P.Y.; visualization, G.L.P.Y.; supervision, J.D.S. and G.L.P.Y.; project administration, J.D.S. and G.L.P.Y.; funding acquisition, J.D.S. All authors have read and agreed to the published version of the manuscript.

Funding: This research was funded by a Private Foundation and the Swiss National Science Foundation, SNSF # 10030_201231 and CRSII5_198577.

Institutional Review Board Statement: The study was conducted in accordance with the Declaration of Helsinki and approved by the Cantonal Research Ethics Committee (CCER, *Commission cantonale d'éthique de la recherche*) (CER18-00552, 11 July 2018).

Informed Consent Statement: Informed consent was obtained from all subjects involved in the study.

Data Availability Statement: The data presented in this study are available on request from the corresponding authors.

Acknowledgments: The authors would like to thank the University of Geneva and CSL Behring for their support; Christian Klein at Roche Innovation Center Zurich for providing the non-glycoengineered Obinutuzumab anti-CD20 monoclonal antibody; and Kerry Campbell at Fox Chase Cancer Center, Philadelphia, PA, and NantKwest for sharing the CD16-transduced NK-92 cell line.

Conflicts of Interest: The authors declare no conflicts of interest except for M.P. and G.L.P.Y., whose salaries were partially financed by CSL Behring. The funder had no role in the design of the study; in the collection, analyses, or interpretation of data; in the writing of the manuscript; or in the decision to publish the results.

References

1. Del Zotto, G.; Marcenaro, E.; Vacca, P.; Sivori, S.; Pende, D.; Della Chiesa, M.; Moretta, F.; Ingegnere, T.; Mingari, M.C.; Moretta, A.; et al. Markers and Function of Human Nk cells in Normal and Pathological Conditions. *Cytom. Part B Clin. Cytom.* **2017**, *92*, 100–114. [CrossRef] [PubMed]
2. Krabbendam, L.; Bernink, J.H.; Spits, H. Innate Lymphoid Cells: From Helper to Killer. *Curr. Opin. Immunol.* **2020**, *68*, 28–33. [CrossRef] [PubMed]
3. Seillet, C.; Brossay, L.; Vivier, E. Natural Killers or Ilc1s? That is the Question. *Curr. Opin. Immunol.* **2020**, *68*, 48–53. [CrossRef] [PubMed]
4. Vivier, E.; Raulet, D.H.; Moretta, A.; Caligiuri, M.A.; Zitvogel, L.; Lanier, L.L.; Yokoyama, W.M.; Ugolini, S. Innate or Adaptive Immunity? The Example of Natural Killer Cells. *Science* **2011**, *331*, 44–49. [CrossRef] [PubMed]
5. Del Zotto, G.; Antonini, F.; Pesce, S.; Moretta, F.; Moretta, L.; Marcenaro, E. Comprehensive Phenotyping of Human Pb Nk Cells by Flow Cytometry. *Cytom. A* **2020**, *97*, 891–899. [CrossRef] [PubMed]
6. Nimmerjahn, F.; Ravetch, J.V. Fcgamma Receptors as Regulators of Immune Responses. *Nat. Rev. Immunol.* **2008**, *8*, 34–47. [CrossRef] [PubMed]
7. Caaveiro, J.M.; Kiyoshi, M.; Tsumoto, K. Structural Analysis of Fc/Fcgammar Complexes: A Blueprint for Antibody Design. *Immunol. Rev.* **2015**, *268*, 201–221. [CrossRef] [PubMed]
8. Canfield, S.M.; Morrison, S.L. The Binding Affinity of Human Igg for Its High Affinity Fc Receptor Is Determined by Multiple Amino Acids in the Ch2 Domain and Is Modulated by the Hinge Region. *J. Exp. Med.* **1991**, *173*, 1483–1491. [CrossRef] [PubMed]
9. Hogarth, P.M.; Pietersz, G.A. Fc Receptor-Targeted Therapies for the Treatment of Inflammation, Cancer and Beyond. *Nat. Rev. Drug Discov.* **2012**, *11*, 311–331. [CrossRef] [PubMed]
10. Bournazos, S.; Wang, T.T.; Ravetch, J.V. The Role and Function of Fcgamma Receptors on Myeloid Cells. *Microbiol. Spectr.* **2016**, *4*, MCHD-0045-2016. [CrossRef]
11. Bournazos, S.; Woof, J.M.; Hart, S.P.; Dransfield, I. Functional and Clinical Consequences of Fc Receptor Polymorphic and Copy Number Variants. *Clin. Exp. Immunol.* **2009**, *157*, 244–254. [CrossRef] [PubMed]
12. Shamir, J.; Slomp, A.; Boross, P.; Leussen, J.H.W. Fc Receptors. *eLS*. 2013. Available online: www.els.net (accessed on 1 March 2013).
13. Lajoie, L.; Congy-Jolivet, N.; Bolzec, A.; Gouilleux-Gruart, V.; Sicard, E.; Sung, H.C.; Peiretti, F.; Moreau, T.; Vie, H.; Clemenceau, B.; et al. Adam17-Mediated Shedding of Fcgammariiia on Human Nk Cells: Identification of the Cleavage Site and Relationship with Activation. *J. Immunol.* **2014**, *192*, 741–751. [CrossRef]
14. Romee, R.; Foley, B.; Lenvik, T.; Wang, Y.; Zhang, B.; Ankarlo, D.; Luo, X.; Cooley, S.; Verneris, M.; Walcheck, B.; et al. Nk Cell Cd16 Surface Expression and Function Is Regulated by a Disintegrin and Metalloprotease-17 (Adam17). *Blood* **2013**, *121*, 3599–3608. [CrossRef] [PubMed]
15. Koene, H.R.; Kleijer, M.; Algra, J.; Roos, D.; von dem Borne, A.E.; de Haas, M. Fc Gammariiia-158v/F Polymorphism In-fluences the Binding of Igg by Natural Killer Cell Fc Gammariiia, Independently of the Fc Gammariiia-48l/R/H Phenotype. *Blood* **1997**, *90*, 1109–1114. [CrossRef] [PubMed]
16. Ravetch, J.V.; Perussia, B. Alternative Membrane Forms of Fc Gamma Riii(CD16) on Human Natural Killer Cells and Neutrophils. Cell Type-Specific Expression of Two Genes That Differ in Single Nucleotide Substitutions. *J. Exp. Med.* **1989**, *170*, 481–497. [CrossRef] [PubMed]
17. de Haas, M.; Koene, H.R.; Kleijer, M.; de Vries, E.; Simsek, S.; van Tol, M.J.; Roos, D.; Borne, A.E.v.D. A Triallelic Fc Gamma Receptor Type Iiia Polymorphism Influences the Binding of Human IgG by Nk Cell Fc Gamma Riiia. *J. Immunol.* **1996**, *156*, 2948–2955. [CrossRef]

18. de Vries, E.; Koene, H.R.; Vossen, J.M.; Gratama, J.W.; von dem Borne, A.E.; Waaijer, J.L.; Haraldsson, A.; de Haas, M.; van Tol, M.J. Identification of an Unusual Fc Gamma Receptor Iiia (Cd16) on Natural Killer Cells in a Patient with Recurrent Infections. *Blood* **1996**, *88*, 3022–3027. [CrossRef] [PubMed]
19. Grier, J.T.; Forbes, L.R.; Monaco-Shawver, L.; Oshinsky, J.; Atkinson, T.P.; Moody, C.; Pandey, R.; Campbell, K.S.; Orange, J.S. Human Immunodeficiency-Causing Mutation Defines Cd16 in Spontaneous Nk Cell Cytotoxicity. *J. Clin. Investig.* **2012**, *122*, 3769–3780. [CrossRef] [PubMed]
20. Jawahar, S.; Moody, C.; Chan, M.; Finberg, R.; Geha, R.; Chatila, T. Natural Killer (Nk) Cell Deficiency Associated with An Epitope-Deficient Fc Receptor Type Iiia (Cd16-II). *Clin. Exp. Immunol.* **1996**, *103*, 408–413. [CrossRef] [PubMed]
21. Li, X.; Gibson, A.W.; Kimberly, R.P. Human Fcr Polymorphism and Disease. *Curr. Top. Microbiol. Immunol.* **2014**, *382*, 275–302.
22. Orange, J.S. Natural Killer Cell Deficiency. *J. Allergy Clin. Immunol.* **2013**, *132*, 515–525. [CrossRef] [PubMed]
23. Papaserafeim, M.; Jandus, P.; Ferfoglia, R.I.; Nieke, J.F.; Vonarburg, C.; Spirig, R.; Yung, G.L.P.; Seebach, J.D. Effect of Intravenous Igg Therapy on Natural Killer Cell Function Related to Fc Gamma Receptor Gene Expression. *J. Allergy Clin. Immunol.* **2020**, *146*, 667–670. [CrossRef]
24. Pradier, A.; Papaserafeim, M.; Li, N.; Rietveld, A.; Kaestel, C.; Gruaz, L.; Vonarburg, C.; Spirig, R.; Yung, G.L.P.; Seebach, J.D. Small-Molecule Immunosuppressive Drugs and Therapeutic Immunoglobulins Differentially Inhibit Nk Cell Effector Functions in vitro. *Front. Immunol.* **2019**, *10*, 556. [CrossRef] [PubMed]
25. Mitchell, R.A.; Luwor, R.B.; Burgess, A.W. Epidermal Growth Factor Receptor: Structure-function Informing the Design of Anticancer Therapeutics. *Exp. Cell Res.* **2018**, *371*, 1–19. [CrossRef] [PubMed]
26. Pierpont, T.M.; Limper, C.B.; Richards, K.L. Past, Present, and Future of Rituximab-the World's First Oncology Monoclonal Antibody Therapy. *Front. Oncol.* **2018**, *8*, 163. [CrossRef]
27. Grillo-Lopez, A.; White, C.; Dallaire, B.; Varns, C.; Shen, C.; Wei, A.; Leonard, J.; McClure, A.; Weaver, R.; Cairelli, S.; et al. Rituximab the First Monoclonal Antibody Approved for the Treatment of Lymphoma. *Curr. Pharm. Biotechnol.* **2000**, *1*, 1–9. [CrossRef] [PubMed]
28. Boero, S.; Morabito, A.; Banelli, B.; Cardinali, B.; Dozin, B.; Lunardi, G.; Piccioli, P.; Lastraioli, S.; Carosio, R.; Salvi, S.; et al. Analysis of in Vitro Adcc and Clinical Response to Trastuzumab: Possible Relevance of Fcgammariiia/Fcgammariia Gene Polymorphisms and Her-2 Expression Levels on Breast Cancer Cell Lines. *J. Transl. Med.* **2015**, *13*, 324. [CrossRef] [PubMed]
29. Calemma, R.; Ottaiano, A.; Trotta, A.M.; Nasti, G.; Romano, C.; Napolitano, M.; Galati, D.; Borrelli, P.; Zanotta, S.; Cassata, A.; et al. Fc Gamma Receptor Iiia Polymorphisms in Advanced Colorectal Cancer Patients Correlated with Response to Anti-Egfr Antibodies and Clinical Outcome. *J. Transl. Med.* **2012**, *10*, 232. [CrossRef]
30. Gavin, P.G.; Song, N.; Kim, S.R.; Lipchik, C.; Johnson, N.L.; Bandos, H.; Finnigan, M.; Rastogi, P.; Fehrenbacher, L.; Mamounas, E.P.; et al. Association of Polymorphisms in Fcgr2a and Fcgr3a with Degree of Trastuzumab Benefit in the Adjuvant Treatment of Erbb2/Her2-Positive Breast Cancer: Analysis of the Nsabp B-31 Trial. *JAMA Oncol.* **2017**, *3*, 335–341. [CrossRef]
31. Goulet, D.R.; Atkins, W.M. Considerations for the Design of Antibody-Based Therapeutics. *J. Pharm. Sci.* **2020**, *109*, 74–103. [CrossRef]
32. Guillerey, C.; Huntington, N.D.; Smyth, M.J. Targeting Natural Killer Cells in Cancer Immunotherapy. *Nat. Immunol.* **2016**, *17*, 1025–1036. [CrossRef] [PubMed]
33. Lin, T.S.; Flinn, I.W.; Modali, R.; Lehman, T.A.; Webb, J.; Waymer, S.; Moran, M.E.; Lucas, M.S.; Farag, S.S.; Byrd, J.C. Fcgr3a and Fcgr2a Polymorphisms May Not Correlate with Response to Alemtuzumab in Chronic Lymphocytic Leukemia. *Blood* **2005**, *105*, 289–291. [CrossRef] [PubMed]
34. Mellor, J.D.; Brown, M.P.; Irving, H.R.; Zalcberg, J.R.; Dobrovic, A. A Critical Review of the Role of Fc Gamma Receptor Polymorphisms in the Response to Monoclonal Antibodies in Cancer. *J. Hematol. Oncol.* **2013**, *6*, 1. [CrossRef] [PubMed]
35. Negri, F.V.; Musolino, A.; Naldi, N.; Bortesi, B.; Missale, G.; Laccabue, D.; Zerbini, A.; Camisa, R.; Chernyschova, N.; Bisagni, G.; et al. Role of Immunoglobulin G Fragment C Receptor Polymorphism-Mediated Antibody-Dependant Cellular Cytotoxicity In Colorectal Cancer Treated With Cetuximab Therapy. *Pharmacogenomics J.* **2014**, *14*, 14–19. [CrossRef] [PubMed]
36. Park, S.J.; Hong, Y.S.; Lee, J.L.; Ryu, M.H.; Chang, H.M.; Kim, K.P.; Ahn, Y.C.; Na, Y.S.; Jin, D.H.; Yu, C.S.; et al. Genetic Polymorphisms of Fcgammariia and Fcgammariiia Are Not Predictive of Clinical Outcomes after Ce-tuximab Plus Irinotecan Chemotherapy in Patients with Metastatic Colorectal Cancer. *Oncology* **2012**, *82*, 83–89. [CrossRef] [PubMed]
37. Seidel, U.J.; Schlegel, P.; Lang, P. Natural Killer Cell Mediated Antibody-Dependent Cellular Cytotoxicity in Tumor Immunotherapy with Therapeutic Antibodies. *Front. Immunol.* **2013**, *4*, 76. [CrossRef] [PubMed]
38. Stork, A.C.; Notermans, N.C.; van den Berg, L.H.; Schellevis, R.D.; Niermeijer, J.M.; Nederend, M.; Leusen, J.H.; van der Pol, W.L. Fcgamma Receptor Iiia Genotype Is Associated with Rituximab Response in Antimyelin-Associated Glycoprotein Neuropathy. *J. Neurol. Neurosurg. Psychiatry* **2014**, *85*, 918–920. [CrossRef] [PubMed]
39. Taylor, R.J.; Chan, S.L.; Wood, A.; Voskens, C.J.; Wolf, J.S.; Lin, W.; Chapoval, A.; Schulze, D.H.; Tian, G.; Strome, S.E. Fcgammariiia Polymorphisms and Cetuximab Induced Cytotoxicity in Squamous Cell Carcinoma of the Head and Neck. *Cancer Immunol. Immunother.* **2009**, *58*, 997–1006. [CrossRef]
40. Munde, E.O.; Okeyo, W.A.; Raballah, E.; Anyona, S.B.; Were, T.; Ong'echa, J.M.; Perkins, D.J.; Ouma, C. Association between Fcγ Receptor Iia, Iiia and Iiib Genetic Polymorphisms and Susceptibility to Severe Malaria Anemia in Children in Western Kenya. *BMC Infect. Dis.* **2017**, *17*, 289. [CrossRef] [PubMed]

41. Shimizu, S.; Tanaka, Y.; Tazawa, H.; Verma, S.; Onoe, T.; Ishiyama, K.; Ohira, M.; Ide, K.; Ohdan, H. Fc-Gamma Receptor Polymorphisms Predispose Patients to Infectious Complications after Liver Transplantation. *Am. J. Transplant.* **2016**, *16*, 625–633. [CrossRef]
42. Vietzen, H.; Danklmaier, V.; Zoufaly, A.; Puchhammer-Stockl, E. High-Affinity Fcgammariiia Genetic Variants and Potent Nk Cell-Mediated Antibody-Dependent Cellular Cytotoxicity (Adcc) Responses Contributing to Severe COVID-19. *Genet. Med.* **2022**, *24*, 1449–1458. [CrossRef]
43. Drexler, H.G.; Matsuo, Y. Malignant Hematopoietic Cell Lines: In Vitro Models for the Study of Natural Killer Cell Leukemia-Lymphoma. *Leukemia* **2000**, *14*, 777–782. [CrossRef]
44. Gong, J.H.; Maki, G.; Klingemann, H.G. Characterization of a Human Cell Line (Nk-92) with Phenotypical and Functional Characteristics of Activated Natural Killer Cells. *Leukemia* **1994**, *8*, 652–658. [PubMed]
45. Maki, G.; Klingemann, H.-G.; Martinson, J.A.; Tam, Y.K. Factors Regulating the Cytotoxic Activity of the Human Natural Killer Cell Line, Nk-92. *J. Hematotherapy Stem Cell Res.* **2001**, *10*, 369–383. [CrossRef]
46. Huang, R.-S.; Shih, H.-A.; Lai, M.-C.; Chang, Y.-J.; Lin, S. Enhanced Nk-92 Cytotoxicity by CRISPR Genome Engineering Using Cas9 Ribonucleoproteins. *Front. Immunol.* **2020**, *11*, 1008. [CrossRef] [PubMed]
47. Clémenceau, B.; Vivien, R.; Pellat, C.; Foss, M.; Thibault, G.; Vié, H. The human natural killer cytotoxic cell line Nk-92, Once Armed with A Murine CD16 Receptor, Represents a Convenient Cellular Tool for the Screening of Mouse mAbs According to their ADCC Potential. *mAbs* **2013**, *5*, 587–594. [CrossRef] [PubMed]
48. Zhao, H.; Zhou, Z.; Li, G.; Liu, G.; Lin, S.; Chen, W.; Xiong, S. An Nk Cell Line (Nk92-41bb) Expressing High Levels of Granzyme Is Engineered to Express the High Affinity Chimeric Genes Cd16/Car. *Cytotechnology* **2021**, *73*, 539–553. [CrossRef] [PubMed]
49. Binyamin, L.; Alpaugh, R.K.; Hughes, T.L.; Lutz, C.T.; Campbell, K.S.; Weiner, L.M. Blocking Nk Cell Inhibitory Self-Recognition Promotes Antibody-Dependent Cellular Cytotoxicity in a Model of Anti-Lymphoma Therapy. *J. Immunol.* **2008**, *180*, 6392–6401. [CrossRef] [PubMed]
50. Joseph, S.; Maniatis, T.; Fritsch, E.F. *Molecular Cloning: A Laboratory Manual*, 2nd ed.; Cold Spring Harbor Laboratory: Cold Spring Harbor, NY, USA, 1989.
51. Millard, A.-L.; Valli, P.V.; Stussi, G.; Mueller, N.J.; Yung, G.P.; Seebach, J.D. Brief Exercise Increases Peripheral Blood Nk Cell Counts without Immediate Functional Changes, but Impairs their Responses to Ex Vivo Stimulation. *Front. Immunol.* **2013**, *4*, 125. [CrossRef] [PubMed]
52. Freitas Monteiro, M.; Papaserafeim, M.; Real, A.; Yung, G.L.P.; Seebach, J.D. Anti-Cd20 Rituximab Igg1, Igg3, and Igg4 but Not Igg2 Subclass Trigger Ca^{2+} Mobilization and Cytotoxicity in Human Nk Cells. *J. Leukoc. Biol.* **2020**, *108*, 1409–1423. [CrossRef]
53. Pross, H.F.; Baines, M.G.; Rubin, P.; Shragge, P.; Patterson, M.S. Spontaneous Human Lymphocyte-Mediated Cytotoxicity against Tumor Target Cells. Ix. The Quantitation of Natural Killer Cell Activity. *J. Clin. Immunol.* **1981**, *1*, 51–63. [CrossRef]
54. Garrick, D.; Fiering, S.; Martin, D.I.; Whitelaw, E. Repeat-Induced Gene Silencing in Mammals. *Nat. Genet.* **1998**, *18*, 56–59. [CrossRef] [PubMed]
55. Festenstein, R.; Tolaini, M.; Corbella, P.; Mamalaki, C.; Parrington, J.; Fox, M.; Miliou, A.; Jones, M.; Kioussis, D. Locus Control Region Function and Heterochromatin-Induced Position Effect Variegation. *Science* **1996**, *271*, 1123–1125. [CrossRef]
56. Berg, M.; Lundqvist, A.; McCoy, P.; Samsel, L.; Fan, Y.; Tawab, A.; Childs, R. Clinical-Grade Ex Vivo-Expanded Human Natural Killer Cells Up-Regulate Activating Receptors and Death Receptor Ligands and Have Enhanced Cytolytic Activity against Tumor Cells. *Cytotherapy* **2009**, *11*, 341–355. [CrossRef]
57. Törnroos, H.; Hägerstrand, H.; Lindqvist, C. Culturing the Human Natural Killer Cell Line Nk-92 in Interleukin-2 and Interleukin-15—Implications for Clinical Trials. *Anticancer. Res.* **2019**, *39*, 107–112. [CrossRef]
58. Herter, S.; Herting, F.; Mundigl, O.; Waldhauer, I.; Weinzierl, T.; Fauti, T.; Muth, G.; Ziegler-Landesberger, D.; Van Puijenbroek, E.; Lang, S.; et al. Preclinical Activity of the Type Ii Cd20 Antibody Ga101 (Obinutuzumab) Compared with Rituximab and Ofatumumab in Vitro and in Xenograft Models. *Mol. Cancer Ther.* **2013**, *12*, 2031–2042. [CrossRef]
59. Ivanov, A.; Beers, S.A.; Walshe, C.A.; Honeychurch, J.; Alduaij, W.; Cox, K.L.; Potter, K.N.; Murray, S.; Chan, C.H.; Klymenko, T.; et al. Monoclonal Antibodies Directed to CD20 and Hla-Dr can Elicit Homotypic Adhesion Followed by Lysosome-Mediated Cell Death in Human Lymphoma and Leukemia Cells. *J. Clin. Investig.* **2009**, *119*, 2143–2159. [CrossRef]
60. Meyer, S.; Evers, M.; Jansen, J.H.M.; Buijs, J.; Broek, B.; Reitsma, S.E.; Moerer, P.; Amini, M.; Kretschmer, A.; Broeke, T.T.; et al. New Insights in Type I and II CD20 Antibody Mechanisms-of-Action with a Panel of Novel CD20 Antibodies. *Br. J. Haematol.* **2018**, *180*, 808–820. [CrossRef]
61. Klingemann, H. The Nk-92 cell line—30 years later: Its Impact on Natural Killer Cell Research and Treatment of Cancer. *Cytotherapy* **2023**, *25*, 451–457. [CrossRef]
62. Schnueriger, A.; Grau, R.; Sondermann, P.; Schreitmueller, T.; Marti, S.; Zocher, M. Development of a Quantitative, Cell-Line Based Assay to Measure Adcc Activity Mediated by Therapeutic Antibodies. *Mol. Immunol.* **2011**, *48*, 1512–1517. [CrossRef]
63. Yamashita-Kashima, Y.; Iijima, S.; Yorozu, K.; Furugaki, K.; Kurasawa, M.; Ohta, M.; Fujimoto-Ouchi, K. Pertuzumab in Combination with Trastuzumab Shows Significantly Enhanced Antitumor Activity in Her2-Positive Human Gastric Cancer Xenograft Models. *Clin. Cancer Res.* **2011**, *17*, 5060–5070. [CrossRef] [PubMed]
64. Fabian, K.P.; Hodge, J.W. The Emerging Role of Off-the-Shelf Engineered Natural Killer Cells in Targeted Cancer Immunotherapy. *Mol. Ther. Oncolytics* **2021**, *23*, 266–276. [CrossRef] [PubMed]

65. Fujii, R.; Schlom, J.; Hodge, J.W. A potential Therapy for Chordoma via Antibody-Dependent Cell-Mediated Cytotoxicity Employing Nk or High-Affinity Nk Cells in Combination with Cetuximab. *J. Neurosurg.* **2018**, *128*, 1419–1427. [CrossRef] [PubMed]
66. Jochems, C.; Hodge, J.W.; Fantini, M.; Fujii, R.; Morillon, Y.M.; Greiner, J.W., 2nd; Padget, M.R.; Tritsch, S.R.; Tsang, K.Y.; Campbell, K.S.; et al. An Nk Cell Line (Hank) Expressing High Levels of Granzyme and Engineered to Express the High Affinity Cd16 Allele. *Oncotarget* **2016**, *7*, 86359–86373. [CrossRef] [PubMed]
67. Rataj, F.; Jacobi, S.J.; Stoiber, S.; Asang, F.; Ogonek, J.; Tokarew, N.; Cadilha, B.L.; van Puijenbroek, E.; Heise, C.; Duewell, P.; et al. High-Affinity CD16-Polymorphism and Fc-Engineered Antibodies Enable Activity of CD16-Chimeric Antigen Receptor-Modified T Cells for Cancer Therapy. *Br. J. Cancer* **2019**, *120*, 79–87. [CrossRef]
68. Perez-Romero, C.A.; Sanchez, I.P.; Naranjo-Piedrahita, L.; Orrego-Arango, J.C.; Muskus-Lopez, C.E.; Rojas-Montoya, W.; Restrepo, J.L.F.; Trujillo-Vargas, C.M. Frequency Analysis of the G.7081t>G/a and G.10872t>G Polymorphisms in the Fcgr3a Gene (Cd16a) Using Nested Pcr and Their Functional Specific Effects. *Genes Immun.* **2019**, *20*, 39–45. [CrossRef] [PubMed]
69. Machiela, M.J.; Chanock, S.J. LDlink: A Web-Based Application for Exploring Population-Specific Haplotype Structure and Linking Correlated Alleles of Possible Functional Variants. *Bioinformatics* **2015**, *31*, 3555–3557. [CrossRef] [PubMed]
70. Blázquez-Moreno, A.; Park, S.; Im, W.; Call, M.J.; Call, M.E.; Reyburn, H.T. Transmembrane Features Governing Fc Receptor CD16A Assembly with CD16A Signaling Adaptor Molecules. *Proc. Natl. Acad. Sci. USA* **2017**, *114*, E5645–E5654. [CrossRef] [PubMed]
71. Victor, A.R.; Weigel, C.; Scoville, S.D.; Chan, W.K.; Chatman, K.; Nemer, M.M.; Mao, C.; Young, K.A.; Zhang, J.; Yu, J.; et al. Epigenetic and Posttranscriptional Regulation of CD16 Expression during Human Nk Cell Development. *J. Immunol.* **2018**, *200*, 565–572. [CrossRef]
72. Li, W.; Yang, H.; Dimitrov, D.S. Identification of High-Affinity Anti-CD16A allotype-Independent Human Antibody Domains. *Exp. Mol. Pathol.* **2016**, *101*, 281–289. [CrossRef]
73. Dong, C.; Ptacek, T.S.; Redden, D.T.; Zhang, K.; Brown, E.E.; Edberg, J.C.; McGwin, G., Jr.; Alarcon, G.S.; Ramsey-Goldman, R.; Reveille, J.D.; et al. Fcgamma Receptor Iiia Single-Nucleotide Polymorphisms and Haplotypes Affect Human Igg Binding and Are Associated with Lupus Nephritis in African Americans. *Arthritis Rheumatol.* **2014**, *66*, 1291–1299. [CrossRef]
74. Deuse, T.; Hu, X.; Agbor-Enoh, S.; Jang, M.K.; Alawi, M.; Saygi, C.; Gravina, A.; Tediashvili, G.; Nguyen, V.Q.; Liu, Y.; et al. The Sirpalpha-Cd47 Immune Checkpoint in Nk Cells. *J. Exp. Med.* **2021**, *218*, e20200839. [CrossRef]
75. Barta, P.; Andersson, K.; Trejtnar, F.; Buijs, J. Exploring Time-Resolved Characterization of the Heterogeneity and Dynamics of Ligand-Receptor Interactions on Living Cells. *J. Anal. Oncol.* **2014**, *3*, 94–104. [CrossRef]
76. Mandelboim, O.; Malik, P.; Davis, D.M.; Jo, C.H.; Boyson, J.E.; Strominger, J.L. Human CD16 as a Lysis Receptor Mediating Direct Natural Killer Cell Cytotoxicity. *Proc. Natl. Acad. Sci. USA* **1999**, *96*, 5640–5644. [CrossRef]
77. Binder, C.; Cvetkovski, F.; Sellberg, F.; Berg, S.; Visbal, H.P.; Sachs, D.H.; Berglund, E.; Berglund, D. Cd2 Immuno-biology. *Front. Immunol.* **2020**, *11*, 1090. [CrossRef]
78. Zhang, Y.; Liu, Q.; Yang, S.; Liao, Q. Cd58 Immunobiology at a Glance. *Front. Immunol.* **2021**, *12*, 705260. [CrossRef]
79. Carlsten, M.; Childs, R.W. Genetic Manipulation of Nk Cells for Cancer Immunotherapy: Techniques and Clinical Implications. *Front. Immunol.* **2015**, *6*, 266. [CrossRef]
80. Izadi, N.; Sun, M.; Mace, E.M.; O'gorman, M.R.G.; Church, J.A. Undetectable Nk Cells Due to the Fcgr3a Variant, 166h, Which May Not Be Directly Disease-Causing. *J. Clin. Immunol.* **2021**, *41*, 1957–1959. [CrossRef]

antibodies

MDPI

Article

A Novel Tetravalent Bispecific Immune Cell Engager Activates Natural Killer Cells to Kill Cancer Cells without Mediating Fratricide

Ge Yang [1,†], Shahryar Khoshtinat Nikkhoi [1,†], Hajar Owji [1], Geng Li [1], Mohammad Massumi [1], Jessica Cervelli [2], Venu Gopal Vandavasi [3] and Arash Hatefi [1,4,*]

1 Department of Pharmaceutics, Rutgers University, Piscataway, NJ 08854, USA
2 Environmental and Occupational Health Science Institute, Flow Cytometry Core Facility, Rutgers University, Piscataway, NJ 08854, USA
3 Department of Chemistry, Biophysics Core Facility, Princeton University, Princeton, NJ 08544, USA
4 Cancer Pharmacology Program, Cancer Institute of New Jersey, New Brunswick, NJ 08901, USA
* Correspondence: ahatefi@pharmacy.rutgers.edu; Tel.: +1-848-445-6366; Fax: +1-732-445-3134
† These authors contributed equally to this work.

Abstract: We previously reported the structure, affinity, and anticancer activity of a bivalent bispecific natural killer cell engager (BiKE) composed of one anti-CD16a VHH and one anti-HER2 VHH fused via a linker. In this study, we explored the engineering of a tetravalent BiKE by fusing two anti-CD16a and two anti-HER2 VHHs in tandem, using bivalent BiKE as a template. The tetravalent BiKE was genetically engineered, and its tertiary structure was predicted using in silico modeling. The antigen binding and affinity of the tetravalent BiKE were assessed using ELISA, flow cytometry, and biolayer interferometry. The ability of the BiKEs to kill cancer cells was evaluated through classical and residual antibody-dependent cellular cytotoxicity (ADCC) assays. Additionally, we investigated the potential for NK cell fratricide via CD16a-CD16a crosslinking. Our results revealed that the tetravalent BiKE exhibited at least 100-fold higher affinity toward its target antigens compared to its bivalent counterpart. The residual ADCC assay indicated that the tetravalent BiKE was more effective in killing cancer cells than the bivalent BiKE, attributable to its lower K_{off} value, which prolonged its binding to NK cell surfaces. Fratricide assays demonstrated that neither the bivalent nor the tetravalent BiKE mediated fratricide. Notably, our findings showed that daratumumab-induced NK fratricide was restricted to CD38-CD38 crosslinking and was not related to ADCC via CD16a-CD38 crosslinking. This study is the first in the literature to show the successful engineering of a tetravalent immune cell engager composed of tandem VHH units, which achieves high affinity and anticancer activity without mediating fratricide.

Keywords: multi-paratope antibody; BiKE; immune cell engager; VHH; tetravalent bispecific antibody; fratricide; daratumumab; CD16a; laNK92

1. Introduction

Variable domains of heavy chain-only antibodies (VHH) are single-domain antibodies derived from camelid heavy chain-only antibodies [1]. VHHs were first discovered in 1993 and are commonly referred to as nanobodies, a term originally coined by the company Ablynx in 2003 [2]. The monomeric, highly structured scaffold of VHHs imparts antigen binding abilities comparable to conventional monoclonal antibodies (mAbs) but with ten times smaller size and greater thermostability. These characteristics make VHHs suitable for the modular design of multivalent antibody drugs. The small size and stable architecture of the VHHs facilitate improved avidity through the creation of bi-, tri, or multi-paratope forms. Multivalent VHHs, composed of repeating units of the same VHH, often exhibit significantly higher overall functional affinity and activity due to increased avidity compared to their monovalent counterpart [3]. In this context, the objective of this study was to examine the physicochemical properties and biological activities of a tetravalent bispecific immune cell engager and demonstrate its superior activity to its bivalent counterpart.

Previously, we reported the sequence of a bispecific killer cell engager (BiKE) that targets the CD16a activating receptor on natural killer (NK) cells and the human epidermal growth factor receptor 2 (HER2) on cancer cells [4]. This BiKE, composed of one anti-CD16 VHH (clone C1) and one anti-HER2 VHH (clone E5), is recombinantly fused via a semi-flexible human muscle aldolase (HMA) linker [4]. For simplicity, we refer to this bivalent bispecific antibody as BiKE-E5:C1. To create a tetravalent BiKE, we employed protein engineering techniques to develop BiKE-2E5:2C1, which consists of 2 anti-CD16a VHHs fused in tandem with two anti-HER2 VHHs. Both constructs were characterized for their binding affinity to target antigens using ELISA, biolayer interferometry (BLI), and flow cytometry. The ability of these constructs to activate NK cells and induce cytotoxicity against HER2+ cancer cells was evaluated using an antibody-dependent cell-mediated cytotoxicity (ADCC) assay. NK cell activation by BiKEs was further validated by measuring the release of cytokines IFN-γ and TNF-α via ELISA. Finally, we assessed the potential of both bivalent and tetravalent BiKEs to induce fratricide due to the crosslinking of two or more NK cells.

2. Experimental Methods

2.1. Cell Culture and Handling

SKOV-3 ovarian cancer cells (Cat# HTB-77), NK92 cells (Cat# CRL-2407), and laNK92 (CD16a^{F176}, Cat#PTA-8837) cells were purchased from American Type Cell Culture (Manassas, VA, USA). LaNK92 cells are not currently available at ATCC but can be purchased directly from Brink Biologics through an MTA. The laNK92 cells are genetically engineered NK92 cells that constitutively express the low-affinity CD16a receptor, making them a suitable CD16+ NK cell line for evaluation of the activity of NK cell engagers using ADCC assay [5]. Peripheral blood NK cells (PB-NK; Cat# 2W-501) were purchased from Lonza Bioscience and cultured as recommended by the vendor. LaNK92 cells were cultured in α-MEM (ThermoFisher, Somerset, NJ, USA, Cat#12561056), supplemented with 12.5% FBS (ThermoFisher, 10437028), 12.5% horse serum (ThermoFisher, 16050122), 0.2 mM myo-inositol (Sigma Aldrich, Burlington, VT, USA, Cat# I7508-100G), 0.02 mM folic acid (Sigma Aldrich, F8758-5G), 100 U/mL penicillin-streptomycin (ThermoFisher, 15140122), 0.1 mM β-mercaptoethanol (ThermoFisher, 21985023), and 100 U/mL IL-2 (Peprotech, Cranbury, NJ, USA, Cat# 200-02) NK92 and LaNK92 cells were passaged every two days and seeded at a density of 0.2 million/mL in a non-treated flask. SKOV-3 cells were cultured in McCoy'5A media (ThermoFisher, 16600108) supplemented with 10% FBS and 100 U/mL penicillin-streptomycin. SKOV-3 cells were passaged every 5–7 days and seeded in a treated flask.

2.2. Expression and Purification of BiKEs

The bivalent BiKE-E5:C1 was expressed and purified as we described previously [4,6]. To express the tetravalent BiKE-2E5:2C1, the gene encoding BiKE-2E5:2C1 was synthesized and cloned into a pET28b vector by Genscript (Piscataway, NJ, USA). The vector was then transformed into a SHuffle DE3 E. coli strain (New England Biolabs, Ipswich, MA, USA, Cat# C3026J) for BiKE expression. Following inoculation of overnight culture at 30 °C in LB medium (IBI Scientific, Dubuque, IA, USA, Cat# IB49020) supplemented with 50 µg/mL kanamycin (ThermoFisher, Cat# 11815024), the culture was induced at OD_{600} of 0.5 and with 0.5 mM IPTG (Teknova, Hollister, CA, USA, Cat# I3325) and continued at 25 °C for 18 h. Cells were pelleted and resuspended in lysis buffer composed of 0.5 mg/mL lysozyme (ThermoFisher, Cat# 89833), 10 µg/mL RNase A (Sigma, Cat# R4875), 5 µg/mL DNase I (Sigma, Cat# 10104159001) in pH 7.0 tris-phosphate-saline (TBS) buffer. The suspension was then incubated at room temperature for 40 min while stirring using a plate stirrer. Subsequently, it was sonicated for 5 min on ice (3 s on, 5 s off, amplitude 75%, 120 Watt, 20 kHz). Following centrifugation at 30,000× *g* at 4 °C for 45 min, the supernatant was recovered and incubated with 1 mL slurry of Ni-NTA beads (Qiagen, Germantown, TN, USA, Cat# 30210) pre-equilibrated with 10 mL of pH 7.0 TBS, at 4 °C for 1 h. The mixture was washed once with 20 mL pH 7.0 TBS buffer containing 0.05% Tween 20 (TBS-0.05%T) and once with 20 mL DPBS (ThermoFisher, Cat# 14190144). Then, 1 mL of DPBS containing 100 U WELQut protease (ThermoFisher, Cat# EO0861) was added for overnight digestion to remove the histag. Next, BiKE-2E5:2C1 underwent a second round of purification using resin conjugated with an anti-myc tag antibody (ThermoFisher, Cat# 20169) using the purification method recommended by the manufacturer. Elution was accomplished using 1 mL of 3M sodium thiocyanate (Spectrum Chemical MFG, New Brunswick, TN, USA, Cat# S1495). Eluted BiKE was concentrated and exchanged into DPBS buffer using a Centriprep®30K centrifugal filter column (Sigma). The purity was assessed using SDS-PAGE gel stained with colloidal Coomassie blue (Biorad, San Francisco, CA, USA, Cat# 1610803), and the concentration was determined using the BCA assay following the manufacturer's protocol (ThermoFisher, Cat# 23225).

2.3. In Silico Modeling of Bivalent BiKE-E5:C1 and Tetravalent BiKE-2E5:2C1

The amino acid sequences of BiKEs were utilized as input for structure prediction. Specifically, ChimeraX software (Version 1.8) was employed, leveraging the AlphaFold algorithm for 3D structure prediction. The ImmunoGenetics (IMGT)/V-Quest database was used to identify the complementarity-determining regions (CDRs) in each binder. Subsequently, the predicted 3D models were refined using GalaxyRefine (http://galaxy.seoklab.org/cgi-bin/submit.cgi?type=REFINE) (Access date: 8 January 2024), which performs atomic-level energy minimization and optimizes hydrogen bond networks. Finally, the quality of the refined tertiary structures was validated through Ramachandran plot analysis using PROCHECK (SAVES) (https://saves.mbi.ucla.edu/) (Access date: 8 January 2024).

2.4. Evaluation of the Ability of BiKE to Bind to CD16a and HER2 Antigens Using ELISA

To assess the binding activity, 0.1 µg of recombinant human CD16a (AcroBiosystem, Newark, NJ, USA, Cat# CDA-H82E8) or HER2 (AcroBiosystem, HE2-H82E2) protein in coating buffer (Biorad, Cat# EUF030C) was applied onto a 96-well plate and incubated at 4 °C overnight. After one wash with 250 µL DPBS containing 0.05% Tween 20 (DPBS-T), the plate was blocked with 3% skim milk (Merck, Cat# 1.15363.0500) in DPBS for 1.5 h. Next, 100 µL of BiKEs at equimolar concentrations, i.e., 1 µg/mL of BiKE-E5:C1 and 2 µg/mL of BiKE-2E5:2C1, was added to each well. 3G8 (anti-CD16a antibody, Invitrogen, MA1-10112) and trastuzumab (anti-HER2 antibody, Novus, NBP2-75895) were used as controls. Following five washes, 100 µL of secondary HRP-conjugated anti-myc tag antibody (Abcam, Waltham, MA, USA, Cat# ab1326), or HRP-conjugated goat anti-mouse IgG antibody (Abcam, Waltham, MA, USA, Ab6789), or HRP-conjugated goat anti-human IgG antibody (Abcam, Waltham, MA, USA, Ab97165) diluted to 0.1 ug/mL was added. Each well was then washed five times before the addition of TMB substrate (ThermoFisher, Cat# 34029). The reaction was stopped after 2–5 min with 1 N H_2SO_4, and measurements were taken using a Tecan Infinity® PRO plate reader ((Tecan, Morrisville, NC, USA) at 450 nm with a reference wavelength of 650 nm.

2.5. Evaluation of the Ability of BiKE to Recognize HER2 and CD16a Antigens on Cells Using Flow Cytometry

To analyze binding activity to HER2 and CD16a, 0.5 million SKOV-3 cells (HER2+) were resuspended in 100 µL of staining buffer (DPBS containing 3% FBS). Equimolar amounts of BiKE-E5:C1 (0.5 µg) or BiKE-2E5:2C1 (1 µg) were added to the resuspended cells and incubated for 45 min at 4 °C. After incubation, the cells were washed and stained with secondary anti-myc tag antibody (Abcam, Cat#ab223896). Following three additional washes, the cells were analyzed using a flow cytometer.

Similarly, for laNK92 cells (CD16+), 0.5 million cells were resuspended in 100 μL of staining buffer and stained with either BiKE-E5:C1 (0.5 μg) or BiKE-2E5:2C1 (1 μg). After washing, cells were stained with 1 μg of rabbit anti-VHH antibody (Genscript, Piscataway, NJ, USA, Cat# A01861). The cells were then washed again and stained with goat anti-rabbit antibody conjugated with Alexa Fluor 647 (Abcam, Cat# ab150079). The stained cells were analyzed using a CytoFlex® flow cytometer (Beckman Coulter, Jersey City, NJ, USA) at the Flow Cytometry Core Facility of Rutgers University.

2.6. Measurement of the Binding Affinities of BiKEs Using Biolayer Interferometer

Using a previously published method [4], the binding affinity (K_D), rate constant of association (k_{on}), and rate constant of dissociation (k_{off}) were measured using a biolayer interferometer (Octet RED96e, Sartorius, Bohemia, NY, USA) at the Biophysics Core Facility at Princeton University. In brief, streptavidin biosensors (Octet®) were soaked inside DPBS containing 0.1% casein, with either biotinylated HER2 or biotinylated CD16a, until a 1 nm shift was noted by the instrument. Subsequently, the sensor was kept in a wash buffer for baselining, followed by loading HER2 or CD16a at concentrations of 60, 30, 15, 7.5, 3.75, 1.875, and 0 nM in the association step for 5 min. For the acquisition of the data, in the dissociation step, the sensors were dipped in the wash buffer (DPBS containing 0.1% Casein). The data were analyzed with Octet Data Analysis HT 11.1 software at the core facility. For fitting the 1:1 binding model in analysis, the sensorgrams were subtracted from the reference. The program "association and dissociation" was used to analyze affinity and kinetics.

2.7. Measurement of Antibody-Dependent Cellular Cytotoxicity

For the antibody-dependent cellular cytotoxicity (ADCC) assay, we adhered to our previously published methodology [7]. Briefly, SKOV-3 cells were seeded in a 96-well plate at a density of 10,000 cells per well one day before the experiment. On the following day, the supernatants were removed, and the antibodies, diluted to the indicated concentrations, were added to each well containing SKOV-3 cells for opsonization for 30 min. Concurrently, laNK92 cells were resuspended at the specified densities to achieve different effector:target (E:T) ratios and then incubated with the SKOV-3 cells. After the incubation period, the plate was washed three times with DPBS, followed by the addition of the AlamarBlue™ (ThermoFisher, Cat#A50101) viability assay reagent and further incubation for 2 h at 37 °C. Measurements were taken using an excitation wavelength of 560 nm and an emission wavelength of 590 nm with a bandwidth of 5.0 nm.

2.8. Residual Antibody-Dependent Cellular Cytotoxicity Assay

For the residual antibody-dependent cellular cytotoxicity assay, we followed procedures adapted from a previously published method [8]. Briefly, SKOV-3 cells were seeded in a 96-well plate at a density of 10,000 cells per well one day before the experiment. On the day of the experiment, laNK92 cells were resuspended in media supplemented with BiKEs at the indicated concentrations and incubated at 37 °C for 30 min for opsonization. The laNK92 cells were then washed twice by laNK92 media at 37 °C to remove any unbound BiKEs. Subsequently, the laNK92 cells were resuspended in media for an additional 30 min to allow for the dissociation of the BiKEs. The laNK92 cells were pelleted, resuspended in media, and added to the 96-well plate containing the pre-seeded SKOV-3 cells at different E:T ratios. The plate was then incubated for 4 h at 37 °C. Following the incubation, AlamarBlue™ reagent was added, and cell viability was measured as previously described.

2.9. Measurement of Fratricide

To study fratricide, 50,000 laNK92 cells were first resuspended in fresh media and then added to a flat-bottom low-attachment 96-well plate (Corning, NY, USA, Cat#3370)

Daratumumab (ThermoFisher, Cat# MA5-41886), 3G8 (ThermoFisher, Cat# 16-0166-65), trastuzumab (Novus, Centennial, CO, USA, Cat# NBP2-75895), BiKE-E5-C1, or BiKE-2E5:2C1 were added to each well at specified concentrations and incubated for 2 or 24 h at 37 °C. In experiments that lasted 24 h, after an initial 2 h incubation with the antibody, 50 μL of 40% AlamarBlue™ reagent diluted in NK cell culture media was added to each well, followed by an additional 22 h incubation (total 24 h). NK cell viability was measured using the AlamarBlue™ kit and protocol.

To quantify fratricide after a 2 h incubation with antibodies, we employed the Trypan blue assay as cell viability assays that rely on the metabolic activity of the cells were not suitable. The reason is that NK cells have low metabolic activity, and incubation of NK cells with reagents such as AlamarBlue™ does not provide sufficient sensitivity (change in color) to detect dead cells. In the Trypan blue assay, 50,000 NK cells were first incubated with either daratumumab alone for 2 h or with Fc blocker or 3G8 for 30 min, followed by incubation with daratumumab at indicated concentrations for an additional 1.5 h. Aggregated cells were then dispersed into single cells by pipetting, diluted in trypan blue (0.4%, ThermoFisher Cat# 15250-061) in a 1:1 ratio, loaded onto cell counter slides, and analyzed using a Countless 3 cell counter (ThermoFisher) (n = 12). The gating method for viability was as follows: living cells (size = 70; brightness = 255; circularity = 100); dead cells (size = 0; brightness = 0; circularity = 0). A 100% viability corresponded to the total number of viable cells in the control group (treated with culture media), and relative viability was determined by the ratio of the total number of living cells in the treatment groups compared to that of the control groups.

2.10. Measurement of the NK Cell Degranulation and Release of Cytokines and Effector Proteins

Measurement of the surfaced CD107a (LAMP-1) on NK cells was adapted from a previously published method [4]. Briefly, 10 nM of anti-CD107a antibody was mixed with 40,000 laNK92 cells per well in a non-treated 96-well plate and incubated for two hours at 37 °C. Following incubation, the plate was centrifuged, and the cells were washed three times with staining buffer (DPBS supplemented with 3% FBS). Subsequently, the cells were stained with APC-conjugated anti-CD107a antibody (R&D Systems, Minneapolis, MN, USA, Cat#IC4800A), supplemented with Fc blocker solution (ThermoFisher, Cat#14-9161-73), and incubated for 1 h at 4 °C. Finally, the cells were centrifuged and washed three times, resuspended, and analyzed using a Beckman Coulter CytoFLEX Cytometer at the Rutgers University Flow Cytometry Core Facility.

Cytokine concentrations before and after NK cell activation were measured using our previously published method [4]. In brief, SKOV-3 cells were seeded at a density of 10,000 cells per well in a 96-well plate one day prior to the experiment. The next day, 10 nM of BiKE or control antibodies were added to the plate and incubated at 37 °C for 30 min. Then, NK cells were added to the plate and incubated for the specified duration. The plate was centrifuged at 2000× g for 10 min to pellet the cells, and the supernatant was transferred to a non-treated 96-well plate. The concentrations of TNF-α (R&D Systems, Cat# DY210-05), IFN-γ (R&D Systems, Cat# DY285B-05), and granzyme B (R&D Systems, Cat# DGZB00) were measured using ELISA kits following the manufacturers' protocols.

3. Results and Discussion

3.1. Prediction of 3D Molecular Structure of BiKE-2E5:2C1

To predict the tertiary structure of the tetravalent BiKE-2E5:2C1, we utilized the AlphaFold 2 algorithm. The AlphaFold 2 is a high-precision software that uses artificial intelligence algorithms to model protein structures [9]. The predicted structures for BiKEs reveal that both antibodies are composed of nine anti-parallel β-sheets, which are typical of immunoglobulin structures (Figure 1). The structure of tetravalent BiKE-2E5:2C1 resembles a horseshoe, with all the CDRs accessible for binding, oriented toward the surface. The CDR3, shown in green, is the most variable and the longest CDR, playing a critical role in antigen binding, and is also accessible in all four binders. One advantage of the horseshoe structure, compared to a linear one, is that it creates a tighter immunological synapse, which contributes to enhanced cytotoxicity [10]. The Ramachandran analysis of both structures revealed that over 96% of the amino acid residues were in the most favored region, and less than 6% were in the allowed region, with an ERRAT score of 100%, indicating no errors in the modeled residues.

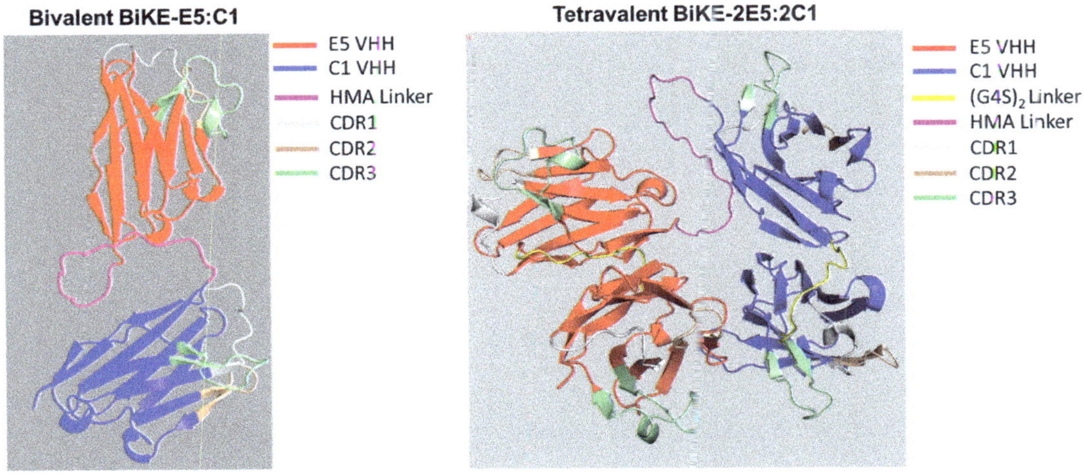

Figure 1. The predicted model of tetravalent and bivalent BiKEs' tertiary structures using AlphaFold software. While the bivalent BiKE appears to have assumed a linear structure, the tetravalent BiKE appears U-shaped.

3.2. Tetravalent BiKE-2E5:2C1 Binds to Its Target Antigens Similar to Bivalent BiKE-E5:C1

In the next step, we investigated whether the tetravalent BiKE retained its activity and could recognize its target antigens, addressing a common concern with multi-domain fusion proteins where one domain might interfere with another, rendering the fusion protein inactive. To test this, we first expressed and purified both BiKE constructs to high purity. The SDS-PAGE analysis confirmed that both BiKEs were over 95% pure, making them suitable for binding activity studies (Figure 2A). We then performed an ELISA to assess the ability of tetravalent BiKE to bind to HER2 and CD16a antigens, using the bivalent BiKE as a positive control due to its previously validated activity [4]. The ELISA results demonstrated that BiKE-2E5:2C1 could recognize both HER2 and CD16a antigens similarly to BiKE-E5:C1 (Figure 2B).

Next, we evaluated the ability of the tetravalent BiKE-2E5:2C1 to recognize HER2 and CD16a antigens on the surface of SKOV-3 (HER2+) and laNK92 (CD16a+) cells. Flow cytometry, with BiKE-E5:C1 as the positive control, showed that the tetravalent BiKE-2E5:2C1 was as effective as the bivalent BiKE-E5:C1 in binding to its target antigens on the cell surface. Overall, the ELISA and flow cytometry data indicated that both constructs were equally active, with no significant differences in antigen targeting, thereby validating that the paratopes on the tetravalent BiKE-2E5:2C1 were effectively exposed.

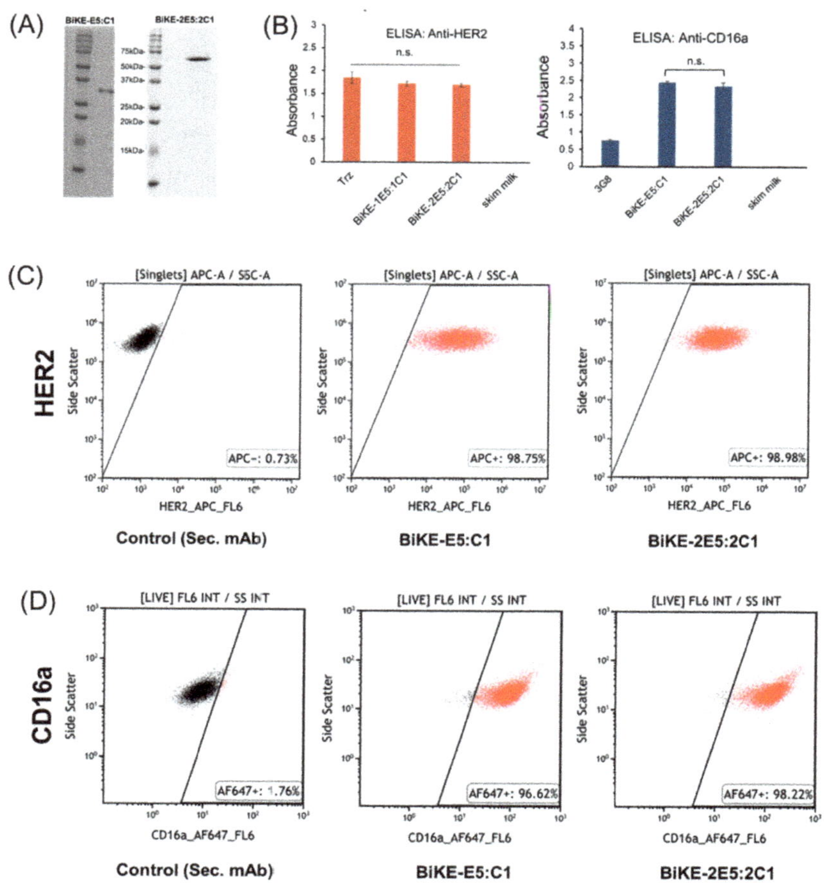

Figure 2. Characterization of the tetravalent BiKE-2E5:2C1 and its comparison to the bivalent BiKE-E5:C1. (**A**) SDS-PAGE analysis of the purified BiKEs. (**B**) ELISA for recognition of HER2 and CD16a antigens by the tetravalent and the bivalent BiKEs. 3G8 and trastuzumab (Trz) are used as positive controls. n.s.: not significant (*t*-test, $p > 0.05$). (**C**) Flow cytometry dot plots showing recognition of HER2 antigen on the surface of HER2+ SKOV-3 cells. (**D**) Flow cytometry dot plots showing recognition of CD16a antigen on the surface of CD16a+ laNK92 cells.

3.3. *Tetravalent BiKE-2E5:2C1 Exhibits Picomolar Affinity toward Its Target Antigens*

We have previously demonstrated that bivalent BiKE-E5:C1, due to its high-affinity interaction with the CD16a receptor, could activate the NK cells to kill cancer cells more efficiently than the best-in-class anti-HER2 antibody, trastuzumab [4]. Given the importance of antibody affinity toward their target antigens in determining their activity, we investigated whether tetravalent BiKE-2E5:2C1 could bind to its target antigens with higher avidity than bivalent BiKE-E5C1. To this end, we employed BLI to measure the affinity of both BiKEs toward the HER2 antigen. The BLI sensograms revealed K_D values of 56 pM for tetravalent BiKE-2E5:2C1 and 3.8 nM for the bivalent BiKE-E5:C1 (Figure 3A–C). Next, we assessed the affinities of BiKEs toward the CD16a antigen. The BLI sensograms showed the affinities of 106 pM for the tetravalent BiKE-2E5:2C1 and 11 nM for bivalent BiKE-E5:C1 (Figure 3B,C). While most antibodies have K_D values in the range of 10^{-7} to 10^{-9} M, high-affinity antibodies have K_D values in the sub-nanomolar range (10^{-9} M). Therefore, the tetravalent BiKE-2E5:2C1 can be considered a very high-affinity antibody due to its picomolar affinity toward its target antigens. Overall, the BLI data demonstrated a significant enhancement of binding affinity when the bivalent BiKE-E5:C1 was engineered as the tetravalent BiKE-2E5:2C1, conferring enhanced affinity through avidity. These findings prompted us to conduct an ADCC assay to measure and compare the anticancer activity of the tetravalent and the bivalent BiKEs.

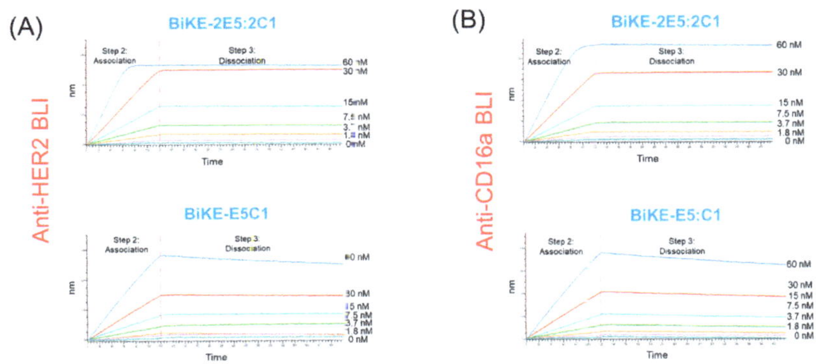

(C)

		K_{on}	K_{off}	K_D
HER2 BLI	BiKE-E5:C1	7.09 x 10^4	2.67 x 10^{-4}	3.77 x 10^{-9}
	BiKE-2E5:2C1	3.03 x 10^5	1.70 x 10^{-5}	5.62 x 10^{-11}
CD16a BLI	BiKE-E5:C1	4.42 x 10^4	4.84 x 10^{-4}	1.10 x 10^{-8}
	BiKE-2E5:2C1	1.85 x 10^5	1.96 x 10^{-5}	1.06 x 10^{-10}

Figure 3. Measurement of K_D (binding constant), K_{on} (association constant), and K_{off} (dissociation constant) of the tetravalent BiKE-2E5:2C1 and the bivalent BiKE:E5C1 using BLI. (**A**,**B**) BLI sensograms showing the binding kinetics of BiKEs for HER2 and CD16a antigens. (**C**) The measured K_D, K_{off}, and K_{on} values of BiKEs for HER2 and CD16a antigens.

3.4. Tetravalent BiKE-2E5:2C1 Has Higher Anticancer Activity than Bivalent BiKE-E5:C1

Multivalent antibodies contain multiple paratopes, allowing them to target selected antigens of interest with enhanced avidity and reduced dissociation rates, ultimately resulting in higher efficacy. Clinical data have shown a direct correlation between the affinity of antibodies for CD16a receptors and therapeutic outcomes [11,12]. Given that the tetravalent BiKE-2E5:2C1 exhibited approximately 100-fold higher affinity toward its target antigens compared to the bivalent BiKE-E5:C1, we sought to determine if this increased affinity would translate into enhanced anticancer activity. To test this, we first conducted a classical ADCC assay using laNK92 (CD16a+) cells as effector cells and SKOV-3 cells as targets. We have previously demonstrated that the bivalent BiKE could activate CD16-expressing immune cells, including PB-NK cells, laNK92 cells, and THP-1-CD16a macrophages [7]. For the ADCC assay in this study, we utilized both laNK92 and PB-NK cells. The results indicated no significant difference in anticancer activity between the tetravalent and bivalent BiKEs regardless of the NK cell type used (Figure 4A,B). This observation was unexpected because measurement of the concentrations of released cytokines (i.e., TNF-α and IFN-γ) from the PB-NK cells illustrated that at almost all E:T ratios, the tetravalent BiKE-2E5:2C1 activated NK cells more efficiently than the bivalent BiKE-E5:C1 (t-test, $p < 0.05$) (Figure 4C,D). To resolve this ambiguity, we performed a residual ADCC assay, which serves as a proxy for the in vivo persistence of an antibody. We removed excess unbound BiKE in each well and placed BiKE-coated NK cells under sink conditions to accelerate the dissociation of lower affinity BiKE molecules. This approach allowed high affinity BiKE molecules that remained bound to the surface of the NK cells to interact with SKOV-3 cancer cells. The rationale was that the higher avidity and the resultant lower k_{off} for the CD16a receptor of the tetravalent BiKE would lead to greater antibody retention on the pre-opsonized NK cells. The results of the residual ADCC assay demonstrated that the tetravalent BiKE-2E5:2C1 was significantly more effective in killing cancer cells than the bivalent BiKE-E5:C1 (Figure 4E). The prior literature on potent high-affinity antibodies targeting CD16a receptors also supports the fact that residual ADCC is a more effective method for capturing the differences between tetravalent and bivalent antibodies, especially when both are highly potent [8]. Our earlier work has shown that the bivalent BiKE could activate NK cells to kill not only HER2-positive cancer cells but also those with low HER2 expression [7]. Given that the tetravalent BiKE exhibits higher affinity with lower K_{off} for both CD16a and HER2 antigens and is more efficient at killing cancer cells under sink conditions than its bivalent counterpart, it may offer the potential to target and eradicate low HER2-expressing cancer cells with greater specificity and efficiency. This could ultimately lead to improved therapeutic responses with fewer side effects in low HER2-expressing cancer patients (e.g., breast, ovarian, and lung) due to prolonged recruitment and activation of immune effector cells.

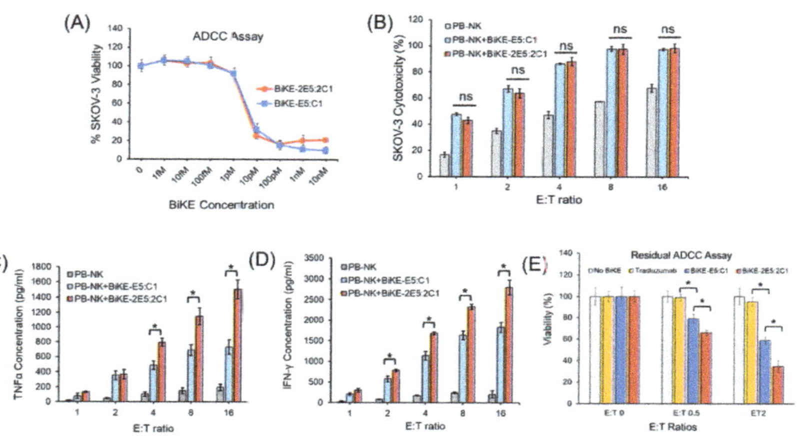

Figure 4. Measurement of the difference in anticancer activity of the tetravalent BiKE-2E5:2C1 and the bivalent BiKE-E5:C1. (**A**) Classical ADCC assay using laNK92 cells and SKOV-3 cells at fixed E:T ratio of 4 but variable BiKE concentrations. (**B**) Classical ADCC assay using PB-NK cells and SKOV-3 cells at various E:T ratios but fixed BiKE concentration (100 nM). (**C**,**D**) Measurement of the IFN-γ and TNF-α after incubation of SKOV-3 cells with PB-NK cells in presence of BiKE using ELISA. (**E**) Residual ADCC assay using laNK92 cells and SKOV-3 cells at different E:T ratios. * *t*-test, $p < 0.05$, ns: not significant (n = 3).

3.5. Tetravalent BiKE-2E5:2C1 Targeting CD16a Receptors Does Not Mediate NK Fratricide

In this study, we evaluated the safety of our BiKE platform by assessing the risk of intercellular CD16a-CD16a crosslinking and the mediation of NK fratricide—a phenomenon where NK cells kill each other, often seen as an unwanted side effect. Fratricide has been documented with daratumumab, which is a bivalent anti-CD38 antibody [13,14]. Clinical data indicate that treatment with daratumumab leads to the rapid depletion of approximately 85% of NK cells, lasting up to six months post-treatment discontinuation [15,16].

Given that BiKE-2E5:2C1 is a tetravalent antibody capable of crosslinking NK cells via CD16a activating receptors, we investigated whether it could mediate NK cell fratricide. For antibody controls, we utilized daratumumab, an anti-CD38 antibody with a human IgG Fc domain, and 3G8, an anti-CD16a antibody with a mouse IgG Fc domain (Figure 5A,B). The mouse Fc domain of the 3G8 antibody is unable to engage CD16a receptors on human

NK cells, thereby restricting its interaction with NK cells solely through the Fab regions. Bivalent BiKE-E5:C1, which is incapable of crosslinking CD16a receptors, was used as a negative BiKE control. For clarity, we illustrated the molecular structures of the tetravalent and bivalent BiKEs using a ball-and-stick model (Figure 5C,D). As cell controls, we employed NK92 cells (CD16−, CD38+) and laNK92 cells (CD16+, CD38+) [17]. While NK-92 cells cannot mediate ADCC [18], laNK92 cells are highly efficient in mediating ADCC [4,7]. Therefore, the NK92/laNK92 pair were selected as effector cells for the fratricide studies.

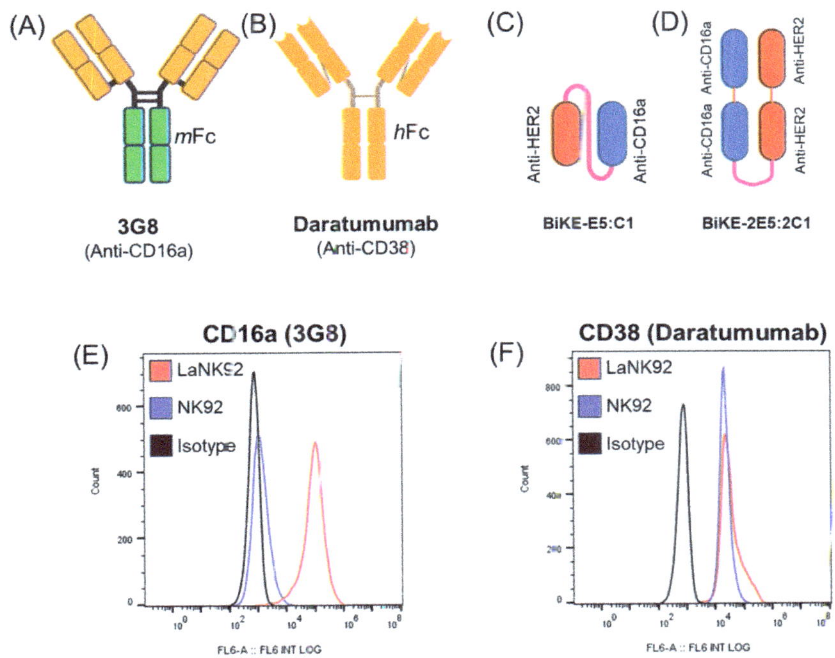

Figure 5. Specifications of the antibodies and cells used in examining fratricide. (**A**) Bivalent anti-CD16a antibody (3G8) with mouse IgG Fc domain. (**B**) Bivalent anti-CD38 antibody (daratumumab) with human IgG Fc domain. (**C**,**D**) schematic representation of tetravalent and bivalent BiKEs. (**E**) flow cytometry histogram demonstrating the expression of CD16a receptor on laNK92 cells but not NK92 cells using 3G8 antibody. (**F**) flow cytometry histogram showing expression of CD38 receptors on both laNK92 and NK92 cells using daratumumab.

Initially, NK92 cells, laNK92 cells, daratumumab, and 3G8 were used to validate the expression of CD38 receptors on both cell lines, as well as the presence of CD16a receptors on laNK92 cells but not NK92 cells (Figure 5E,F). While the ability of laNK92 cells to induce ADCC has previously been demonstrated by our group [4], the inability of NK92 cells to induce ADCC, due to the lack of CD16a expression, was confirmed using SKOV-3 cells (HER2+) as target cells (Supplementary Figure S1).

Using the aforementioned cells and antibodies, we conducted fratricide assays by incubating NK92 and laNK92 cells with daratumumab, 3G8, or a combination of daratumumab and 38 antibodies. The objective was to determine whether fratricide occurs due to CD38-CD38 crosslinking or if it could be initiated through CD38-CD16a or CD16a-CD16a crosslinking. To test this hypothesis, we performed three independent fratricide assays. The results revealed that daratumumab was equally toxic to both NK92 and laNK92 cells regardless of incubation time, whereas the 3G8 antibody did not mediate fratricide (Figure 6A–C). These observations indicated that daratumumab's toxicity was due to the crosslinking of

CD38-CD38 and not CD16a-C38 crosslinking, as NK92 cells mediated fratricide similar to laNK92 cells. This suggests that CD38-CD38 crosslinking, rather than CD16a-CD16a, is the major player in fratricide. To validate this observation, we incubated laNK92 cells with daratumumab alone or with daratumumab plus 3G8 antibodies. The results revealed that blocking of CD16a with the 3G8 antibody or Fc blocker did not significantly impact daratumumab's fratricide effect, ruling out CD16a-CD38 crosslinking as the culprit (Figure 6D) (Supplementary Figure S2). Overall, our data indicate that daratumumab is an agonistic CD38 antibody that mediates fratricide through NK cells' natural cytotoxicity capability, which is the innate ability of NK cells to lyse tumor or virally infected target cells. While some reports suggest that daratumumab mediates fratricide through crosslinking of CD16a-CD38 receptors, which is an antibody-dependent mechanism (ADCC) [13], our data do not support this idea and are more in agreement with early report by Sconocchia et al. [19].

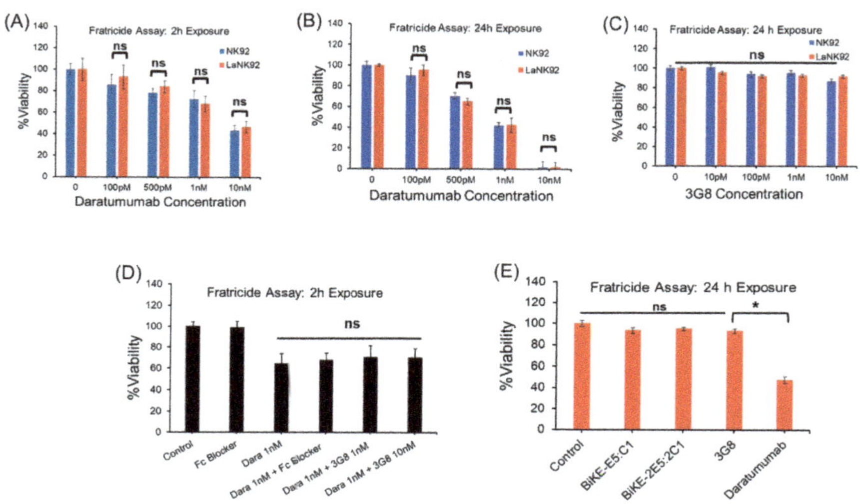

Figure 6. Evaluation of the ability of tetravalent BiKE to induce fratricide by measuring the cell viability after incubation of NK cells with antibodies. (**A**,**B**) Fratricide (ADCC) assay using NK92 and laNK92 cells in combination with daratumumab at different concentrations. (**C**) Fratricide assay using NK92 and laNK92 cells in combination with 3G8 antibody at different concentrations. (**D**) Fratricide assay using daratumumab in combination with 3G8 or Fc blocker at different concentrations. (**E**) Fratricide assay using laNK92 cells and tetravalent and bivalent BiKEs along with control antibodies. * *t*-test, $p < 0.05$. ns: not significant.

Finally, we examined whether the tetravalent BiKE-2E5:2C1 could mediate fratricide. The results of the fratricide assay showed that similar to the 3G8 antibody, the tetravalent BiKE-2E5:2C1 did not mediate fratricide (Figure 6E).

To elucidate the mechanism of action, we measured the degranulation of laNK92 cells and the release of granzyme B from activated NK cells. Flow cytometry histograms and ELISA results illustrated that only daratumumab was capable of degranulating NK cells, resulting in the release of granzyme B (Figure 7A–C) (Supplementary Figure S3). Given that CD38, through its enzymatic functions, can regulate granule polarization and degranulation of the NK cells [20,21], this explains daratumumab's ability to activate NK cells, leading to fratricide. Overall, unlike daratumumab, the tetravalent BiKE-2E5:2C1 did not compromise NK cell viability, indicating its potential to serve as a safe therapeutic platform.

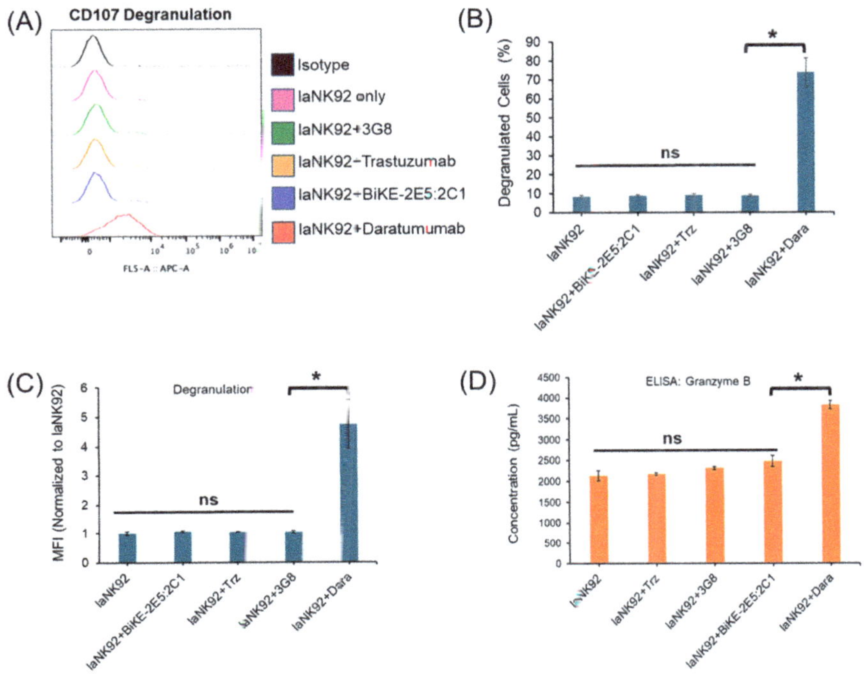

Figure 7. Determination of NK cell activation. (**A–C**) Measurement of the degranulation of laNK92 cells after incubation with tetravalent BiKE, trastuzumab (Trz.), 3G8, and daratumumab at 10 nM concentration (2 h exposure time) using flow cytometry. (**D**) Measurement of the concentration of granzyme B from antibody treated laNK92 cells using ELISA. * *t*-test, $p < 0.05$. ns: not significant.

4. Conclusions

The data presented in this manuscript demonstrate that a tetravalent BiKE can be engineered from its bivalent counterpart by fusing VHH units in tandem. The resulting tetravalent BiKE exhibits high affinity toward its target antigens due to avidity. This study is the first in the literature to show the successful engineering of a tetravalent immune cell engager composed of tandem VHH units, which achieves high affinity and anticancer activity without mediating fratricide.

Supplementary Materials: The following supporting information can be downloaded at https://www.mdpi.com/article/10.3390/antib13030075/s1, Figure S1: ADCC assay using NK92 as effector cells and SKOV-3 as target cells at E:T ratio of 4. BiKE-E5:C1 and trastuzumab (Trz) were used at 10 nM concentration. n.s.: not significant, *t*-test $p > 0.05$.; Figure S2: Fratricide assay using daratumumab alone or in combination with 3G8 or Fc blocker at different concentrations (24 h incubation time); Figure S3: Flow cytometry data for degranulation of laNK92 cells after incubation with antibodies.

Author Contributions: G.Y. and S.K.N. contributed to methodology, data acquisition, validation, formal analysis, investigation, writing, and visualization; H.O., G.L., M.M. and J.C. contributed to methodology, data acquisition, and analysis; V.G.V. contributed to methodology; A.H. contributed to conceptualization, methodology, validation, formal analysis, resources, writing, review, editing, visualization, supervision, project administration, and funding acquisition. All authors have read and agreed to the published version of the manuscript.

Funding: This work was supported by grants from the NIH/NCI (R01CA251438) and NIH/NHLBI/ Rutgers HealthAdvance joint program (U01HL150852).

Institutional Review Board Statement: This study does not involve human participants.

Informed Consent Statement: Not applicable.

Data Availability Statement: All data relevant to this study are included in the article or uploaded as Supplementary Information. The original contributions presented in this study are included in the article/Supplementary Materials; further inquiries can be directed to the corresponding author.

Acknowledgments: We would like to thank the NIH/NCI flow cytometry (P30CA072720-5924) shared resources of the Rutgers-Cancer Institute of New Jersey (NCI-designated Comprehensive Cancer Center).

Conflicts of Interest: A.H. and S.K. are the inventors named on a patent (WO2023129819) that describes the BiKE technology. The intellectual property (IP) rights to the patented technology belong to Rutgers University. The BiKE technology for oncology indications has been licensed to Sauvie Inc.

References

1. Steeland, S.; Vandenbroucke, R.E.; Libert, C. Nanobodies as therapeutics: Big opportunities for small antibodies. *Drug Discov. Today* **2016**, *21*, 1076–1113. [CrossRef] [PubMed]
2. Hamers-Casterman, C.; Atarhouch, T.; Muyldermans, S.; Robinson, G.; Hamers, C.; Songa, E.B.; Bendahman, N.; Hamers, R. Naturally occurring antibodies devoid of light chains. *Nature* **1993**, *363*, 446–448. [CrossRef] [PubMed]
3. Huet, H.A.; Growney, J.D.; Johnson, J.A.; Li, J.; Bilic, S.; Ostrom, L.; Zafari, M.; Kowal, C.; Yang, G.; Royo, A.; et al. Multivalent nanobodies targeting death receptor 5 elicit superior tumor cell killing through efficient caspase induction. *mAbs* **2014**, *6*, 1560–1570. [CrossRef] [PubMed]
4. Nikkhoi, S.K.; Li, G.; Eleya, S.; Yang, G.; Vandavasi, V.G.; Hatefi, A. Bispecific killer cell engager with high affinity and specificity toward CD16a on NK cells for cancer immunotherapy. *Front. Immunol.* **2022**, *13*, 1039969. [CrossRef] [PubMed] [PubMed Central]
5. Gopal, S.; Kwon, S.J.; Ku, B.; Lee, D.W.; Kim, J.; Dordick, J.S. 3D tumor spheroid microarray for high-throughput, high-content natural killer cell-mediated cytotoxicity. *Commun. Biol.* **2021**, *4*, 893. [CrossRef] [PubMed] [PubMed Central]
6. Nikkhoi, S.K.; Heydarzadeh, H.; Vandavasi, V.G.; Yang, G.; Louro, P.; Polunas, M.; Owji, H.; Hatefi, A. A high affinity and specificity anti-HER2 single-domain antibody (VHH) that targets trastuzumab's epitope with versatile biochemical, biological, and medical applications. *Immunol. Res.* **2024**, *72*, 103–118. [CrossRef]
7. Khoshtinat Nikkhoi, S.; Yang, G.; Owji, H.; Grizotte-Lake, M.; Cohen, R.I.; Gil Gonzalez, L.; Massumi, M.; Hatefi, A. Bispecific immune cell engager enhances the anticancer activity of CD16+ NK cells and macrophages in vitro, and eliminates cancer metastasis in NK humanized NOG mice. *J. Immunother. Cancer* **2024**, *12*, e008295. [CrossRef] [PubMed] [PubMed Central]
8. Reusch, U.; Burkhardt, C.; Fucek, I.; Le Gall, F.; Le Gall, M.; Hoffmann, K.; Knackmuss, S.H.; Kiprijanov, S.; Little, M.; Zhukovsky, E.A. A novel tetravalent bispecific TandAb (CD30/CD16A) efficiently recruits NK cells for the lysis of CD30+ tumor cells. *mAbs* **2014**, *6*, 728–739. [CrossRef] [PubMed] [PubMed Central]
9. Jumper, J.; Evans, R.; Pritzel, A.; Green, T.; Figurnov, M.; Ronneberger, O.; Tunyasuvunakool, K.; Bates, R.; Žídek, A.; Potapenko, A. Highly accurate protein structure prediction with AlphaFold. *Nature* **2021**, *596*, 583–589. [CrossRef] [PubMed]
10. Bluemel, C.; Hausmann, S.; Fluhr, P.; Sriskandarajah, M.; Stallcup, W.B.; Baeuerle, P.A.; Kufer, P. Epitope distance to the target cell membrane and antigen size determine the potency of T cell-mediated lysis by BiTE antibodies specific for a large melanoma surface antigen. *Cancer Immunol. Immunother.* **2010**, *59*, 1197–1209. [CrossRef] [PubMed] [PubMed Central]
11. Musolino, A.; Naldi, N.; Bortesi, B.; Pezzuolo, D.; Capelletti, M.; Missale, G.; Laccabue, D.; Zerbini, A.; Camisa, R.; Bisagni, G.; et al. Immunoglobulin G fragment C receptor polymorphisms and clinical efficacy of trastuzumab-based therapy in patients with HER-2/neu-positive metastatic breast cancer. *J. Clin. Oncol.* **2008**, *26*, 1789–1796. [CrossRef] [PubMed]
12. Weng, W.K.; Levy, R. Two immunoglobulin G fragment C receptor polymorphisms independently predict response to rituximab in patients with follicular lymphoma. *J. Clin. Oncol.* **2003**, *21*, 3940–3947. [CrossRef] [PubMed]
13. Wang, Y.; Zhang, Y.; Hughes, T.; Zhang, J.; Caligiuri, M.A.; Benson, D.M.; Yu, J. Fratricide of NK Cells in Daratumumab Therapy for Multiple Myeloma Overcome by Ex Vivo-Expanded Autologous NK Cells. *Clin. Cancer Res.* **2018**, *24*, 4006–4017. [CrossRef] [PubMed] [PubMed Central]
14. Lejeune, M.; Duray, E.; Peipp, M.; Clemenceau, B.; Baron, F.; Beguin, Y.; Caers, J. Balancing the CD38 Expression on Effector and Target Cells in Daratumumab-Mediated NK Cell ADCC against Multiple Myeloma. *Cancers* **2021**, *13*, 3072. [CrossRef] [PubMed] [PubMed Central]
15. Casneuf, T.; Xu, X.S.; Adams, H.C., 3rd; Axel, A.E.; Chiu, C.; Khan, I.; Ahmadi, T.; Yan, X.; Lonial, S.; Plesner, T.; et al. Effects of daratumumab on natural killer cells and impact on clinical outcomes in relapsed or refractory multiple myeloma. *Blood Adv.* **2017**, *1*, 2105–2114. [CrossRef] [PubMed] [PubMed Central]
16. Nijhof, I.S.; Casneuf, T.; van Velzen, J.; van Kessel, B.; Axel, A.E.; Syed, K.; Groen, R.W.; van Duin, M.; Sonneveld, P.; Minnema, M.C.; et al. CD38 expression and complement inhibitors affect response and resistance to daratumumab therapy in myeloma. *Blood* **2016**, *128*, 959–970. [CrossRef] [PubMed]

17. Gurney, M.; Stikvoort, A.; Nolan, E.; Kirkham-McCarthy, L.; Khoruzhenko, S.; Shivakumar, R.; Zweegman, S.; Van de Donk, N.; Mutis, T.; Szegezdi, E.; et al. CD38 knockout natural killer cells expressing an affinity optimized CD38 chimeric antigen receptor successfully target acute myeloid leukemia with reduced effector cell fratricide. *Haematologica* **2022**, *107*, 437–445. [CrossRef] [PubMed] [PubMed Central]
18. Klingemann, H.; Boissel, L.; Toneguzzo, F. Natural Killer Cells for Immunotherapy—Advantages of the NK-92 Cell Line over Blood NK Cells. *Front. Immunol.* **2016**, *7*, 91. [CrossRef] [PubMed] [PubMed Central]
19. Sconocchia, G.; Titus, J.A.; Mazzoni, A.; Visintin, A.; Pericle, F.; Hicks, S.W.; Malavasi, F.; Segal, D.M. CD38 triggers cytotoxic responses in activated human natural killer cells. *Blood* **1999**, *94*, 3864–3871. [CrossRef] [PubMed]
20. Paul, S.; Lal, G. The Molecular Mechanism of Natural Killer Cells Function and Its Importance in Cancer Immunotherapy. *Front. Immunol.* **2017**, *8*, 1124. [CrossRef] [PubMed] [PubMed Central]
21. Mallone, R.; Funaro, A.; Zubiaur, M.; Baj, G.; Ausiello, C.M.; Tacchetti, C.; Sancho, J.; Grossi, C.; Malavasi, F. Signaling through CD38 induces NK cell activation. *Int. Immunol.* **2001**, *13*, 397–409. [CrossRef] [PubMed]

antibodies

MDPI

Article

Targeting CD44 and EpCAM with Antibody Dye Conjugates for the Photoimmunotherapy of Prostate Cancer

Isis Wolf [1,2,3], Susanne Schultze-Seemann [1,2], Christian Gratzke [1,2] and Philipp Wolf [1,2,*]

[1] Department of Urology, Medical Center—University of Freiburg, 79106 Freiburg im Breisgau, Germany; isis.wolf@uniklinik-freiburg.de (I.W.); susanne.schultze-seemann@uniklinik-freiburg.de (S.S.-S.); christian.gratzke@uniklinik-freiburg.de (C.G.)
[2] Faculty of Medicine, University of Freiburg, 79110 Freiburg im Breisgau, Germany
[3] Faculty of Biology, University of Freiburg, 79104 Freiburg im Breisgau, Germany
* Correspondence: philipp.wolf@uniklinik-freiburg.de

Abstract: Background/Objectives: Photoimmunotherapy (PIT) is an innovative approach for the targeted therapy of cancer. In PIT, photosensitizer dyes are conjugated to tumor-specific antibodies for targeted delivery into cancer cells. Upon irradiation with visible light, the photosensitizer dye is activated and induces cancer-specific cell death. In the present article, we describe the PIT of prostate cancer (PC) as a therapeutic option for the targeted treatment of localized prostate cancer. Methods: We conjugated the silicon phthalocyanine dye WB692-CB2 to recombinant cysteine-modified anti-CD44 and anti-EpCAM antibodies via a maleimide linker and tested the antibody dye conjugates for PIT on PC cells and prostate cancer stem cell (PCSC)-like cells. Results: The anti-CD44 and anti-EpCAM antibody dye conjugates showed specific binding and high cytotoxicity against PC and PCSC-like cells following irradiation with red light. Combined treatment with both conjugates led to enhanced cytotoxic effects. Conclusions: PIT with our dye WB692-CB2 can serve as an effective focal therapy against prostate cancer, preserving the prostate gland and minimizing side effects. It can be employed during radical prostatectomy (RP) to treat residual tumor cells or lymph node metastases in areas where further surgical intervention is not feasible. Utilizing multiple conjugates against antigens expressed on differentiated PC and PCSC-like cells, such as CD44 and EpCAM, could be an effective method to eradicate residual cancer cells in heterogeneous tumors. This approach could reduce the risk of local recurrence after RP and thus increase the therapeutic outcome of PC patients.

Keywords: prostate cancer; photoimmunotherapy; CD44; EpCAM; antibody dye conjugates

1. Introduction

Prostate cancer (PC) is the second most common cancer in men from industrial countries with over 1.46 million new cases and 396,000 deaths reported annually [1]. Demographic changes and increasing life expectancy indicate that annual new PC cases will increase to 2.9 million in 2040 [2].

Treatment options for localized PC include radical prostatectomy (RP), active surveillance, watchful waiting, external beam radiation and brachytherapy [3]. The prerequisite for a successful treatment is the complete eradication of the tumor. In the case that residual tumor cells persist in the surgical margins, the tumor may recur locally and begin to metastasize. Biochemical recurrence affects approximately 20–40% of patients within ten years after RP [4]. Androgen deprivation therapy (ADT), either as monotherapy or

 https://doi.org/10.3390/antib14010005

in combination with radiotherapy and/or chemotherapy, has become the standard treatment for patients with relapsed castration-sensitive PC [3]. Unfortunately, the efficacy of ADT is limited. Over time, the tumor becomes hormone refractory and progresses to metastatic castration-resistant PC (mCRPC). The standard treatment for mCRPC involves ADT in combination with chemotherapy and prednisolone. However, this approach is primarily palliative and only marginally prolongs the patient's overall survival, which is on average 16 to 18 months [5]. Depending on the treatment and the progression of the disease, patients suffer from a range of symptoms and adverse effects such as incontinence, erectile dysfunction, decreased libido, reduction in muscle mass and bone density, edema and psychosomatic symptoms, which negatively affect their quality of life [3]. Given the limitations of current therapeutic strategies, there is a need for the development of novel and innovative therapeutic approaches for PC.

Despite the promising results of conventional and PSMA-targeted therapies, tumor heterogeneity remains a major cause of treatment failure in PC [6]. Patients with primary PC often harbor multiple distinct tumor foci within the prostate gland, exhibiting high levels of intra-tumoral heterogeneity [7]. Prostate tumor heterogeneity refers to cancer cells with differences in morphology, transcriptional profiles, metabolism and metastatic potential. Factors that contribute to tumor heterogeneity are genetic mutations, epigenetic alterations, influence of the tumor microenvironment, evolutionary pressure and tumor cell plasticity [7–9]. Tumor heterogeneity can also be explained by the existence of the prostate cancer stem cell (PCSC) hypothesis, which postulates that PCSCs form a small subset of cells within the tumor capable of self-renewal and differentiate into various cell types. From a therapeutic point of view, this means that all tumor cells, i.e., PCSCs and differentiated PC cells, have to be eliminated to achieve a cure. Since PCSCs are characterized by unique tumor antigens, either exclusive to them or shared with differentiated PC cells, they can be targeted using RNA knockdown, inhibitors or antibodies directed against these antigens [10].

In the present study, we introduce photoimmunotherapy (PIT) as an innovative supportive treatment during RP to target PC cells and PCSC-like cells. PIT employs tumor-specific antibodies conjugated with photoactivatable dyes (photosensitizers) that can be activated by visible light to selectively induce cancer cell death [11]. In our approach, we used antibodies against the cluster of differentiation 44 antigen (CD44) and the epithelial cell adhesion molecule (EpCAM) to target both PC cells and PCSC-like cells. CD44 is a hyaluronic acid transmembrane receptor, which serves as a cell adhesion and signaling molecule. It is associated with the self-renewal, proliferation, differentiation, invasion and metastasis of PC. EpCAM is a transmembrane glycoprotein with various cellular functions, contributing to cancer stem cell maintenance, proliferation, invasion, metastasis and therapeutic resistance [10]. We conjugated the anti-CD44 and anti-EpCAM antibodies to our silicon phthalocyanine photosensitizer dye, WB692-CB2, and evaluated these antibody dye conjugates for PIT of PC and PCSC-like cells.

2. Materials and Methods

2.1. Cell Lines and Dye

The suspension cell line EXPI293F (Thermo Fisher Scientific, Waltham, MA, USA) was propagated in serum-free FreeStyle F17 medium (Thermo Fisher Scientific) with 8 mM L-glutamine (PAN-Biotech, Aidenbach, Germany) and 0.1% pluronic F68 (PAN-Biotech). The PC cell line PC3-PSMA was a kind gift from P. Giangrande (University of Iowa, Iowa City, IA, USA). The cell line PC3-MM2 was kindly provided by the MD Anderson Cancer Center (Huston, TY, USA). This cell line is marked by different stem cell characteristics, e.g., expression of the stem cell markers CD44, CD133, ß1 integrin or ALDH1, sphere formation, high tumorigenic and metastatic potential, and therapeutic resistance, and can therefore serve as a PCSC-like cell line [12–14]. Both cell lines were cultivated in RMPI 1640 medium supplemented with 10% fetal bovine serum (FBS, Merck, Darmstadt, Germany) and 1% penicillin/streptomycin (Pen/Strep, Merck). The control cell line CHO (Thermo Fisher Scientific) was cultivated in Ham's F-12 Nutrient Mixture medium (Thermo Fisher Scientific) with 10% FBS and 1% Pen/Strep. All cell lines were propagated under sterile conditions at 37 °C in a 5% CO_2 atmosphere. Cell line identities were confirmed using short tandem repeat (STR) analysis (CLS GmbH, Eppelheim, Germany). Lysates were generated through lysis of cells in 50 mM Tris-HCl, 150 mM NaCl, 1 mM EDTA, 0.5% sodium deoxycholate, 0.05% SDS, 1% Igepal supplemented with cOmplete™ Protease Inhibitor Cocktail for 30 min on ice, and centrifugation at 18,000× *g* and 4 °C for 30 min. Protein concentration of the supernatant was determined by using the Quick Start™ Bradford Protein Assay according to manufacturer's instructions (Bio-Rad Laboratories Inc., Hercules, CA, USA). The phthalocyanine photosensitizer dye WB692-CB2 [15] was kindly provided by Jonas Storz and Reinhard Brückner, Institute for Organic Chemistry, University of Freiburg, Germany.

2.2. Cloning, Expression and Purification of the Anti-CD44 and Anti-EpCAM Antibodies

The genes encoding the human IgG1 heavy chains of the anti-CD44 antibody [16] and the anti-EpCAM antibody [17] each containing the cysteine mutations T120C and D265C (EU numbering) were synthesized (GeneArt, Invitrogen, Regensburg, Germany) and cloned into the expression vector pCSEH1c [18]. In our previous studies, both mutations were shown not to influence the specificity of an antibody targeting the prostate specific antigen by protein modeling [15]. Synthesized variable domains of the light chains (V_L) of both antibodies were cloned into the expression vector pCSL3k, containing a constant domain (CL) of a human IgG1 kappa light chain. Vectors were transformed into XL1-Blue MRF' supercompetent *E. coli* cells (Agilent Technologies, Waldbronn, Germany) and purified with help of the NucleoBond® Xtra Maxi Kit (Macherey-Nagel, Düren, Germany). The sequences of the antibody chains were then verified for accuracy through sequencing (Microsynth Seqlab, Göttingen, Germany). Antibody expression in EXPI293F cells was performed as published previously [15]. After purification by protein G affinity chromatography (Cytiva, Marlborough, USA) and dialysis against PBS, concentrations of the antibodies, called $CD44^{T120C/D265C}$ and $EpCAM^{T120C/D265C}$, were determined by Pierce BCA Protein Assay Kit (Thermo Fisher Scientific).

2.3. Conjugation of the Photosensitizer Dye WB692-CB2

For dye conjugation, 1 mg of the cysteine-modified antibodies $CD44^{T120C/D265C}$ or $EpCAM^{T120C/D265C}$ were diluted in PBS containing 1 mM EDTA (pH 7.4) and were reduced using 40 molar equivalents of Tris-(2-Carboxyethyl)phosphine Hydrochloride (TCEP, Carl Roth, Karlsruhe, Germany) at 37 °C for 3 h. The antibodies were then dialyzed overnight at 4 °C against PBS containing 1 mM EDTA (pH 7.4). Subsequently, re-oxidation was performed with 30 molar equivalents of dehydroascorbic acid (dHAA, Sigma-Aldrich, St. Louis, MI, USA) at room temperature for 4 h. For conjugation, the antibodies were incubated with 10 molar equivalents of WB692-CB2 at room temperature for 1 h. The reaction was quenched by adding 25 molar equivalents of N-acetyl-L-cysteine (Sigma-Aldrich) for 15 min, followed by overnight dialysis against PBS (pH 7.4) [15]. Unbound dye was removed by protein G affinity chromatography and the antibody dye conjugates were stored a 4 °C protected from light. The dye-to-antibody ratios of the conjugates were determined as published [15].

2.4. SDS-PAGE and Western Blots

Cell lysates, antibodies and conjugates were analyzed by SDS-PAGE using the NuPAGE® Bis-Tris Electrophoresis System (Thermo Fisher Scientific). For SDS-PAGE under reducing conditions, 3 μg of antibody or conjugate was combined with NuPAGE® LDS Sample Buffer and NuPAGE® Reducing Agent, followed by incubation for 10 min at 70 °C. Then, the samples were loaded onto a pre-cast NuPAGE® Novex® Bis-Tris Gel (Thermo Fisher Scientific). Electrophoresis was performed at 200 V for 35 min using the XCell SureLockTM Mini-Cell system (Thermo Fisher Scientific), with NuPAGE® MES Buffer supplemented with NuPAGE® Antioxidant as running buffer. For SDS-PAGE under non-reducing conditions, the same protocol using Laemmli sample buffer without NuPAGE® Reducing Agent was used. Protein detection was achieved by Coomassie staining or fluorescence-based imaging (λ_{max} = 680 nm, IVIS Spectrum, In Vivo Imaging System, PerkinElmer, Waltham, MA, USA).

For Western blot analysis, the proteins were blotted onto nitrocellulose membranes at 30 V for 1 h in NuPAGE® Transfer Buffer supplemented with 10% methanol and 1% NuPAGE® antioxidant. The membrane was blocked with 5% non-fat milk in 0.1% Tween-20/PBS (M-PBST) for 1 h, followed by incubation with mouse anti-human CD44 (#ab82529, Abcam, Cambridge, UK) or mouse anti-human EpCAM (#sc-25308, Santa Cruz Biotechnology, Dallas, TX, USA) as primary antibodies overnight at 4 °C in M-PBST. After washing with 0.1% Tween-20/PBS, the membrane was incubated with the horseradish peroxidase (HRP)-conjugated secondary antibody rabbit anti-mouse Ig/HRP (#P0161, DAKO, Hamburg, Germany) for 1 h, followed by another washing step. ß-actin as loading control was detected using the antibody mouse α-human β-actin-HRP (Proteintech Group Inc., Rosemont, IL, USA). Detection of the target proteins was conducted using an enhanced chemiluminescence solution and the INTAS Chemo Star Imager (INTAS Science Imaging Instruments, Göttingen, Germany).

2.5. Flow Cytometry

Specific binding of the antibodies or antibody dye conjugates to the PC cells was evaluated by flow cytometry. An amount of 2×10^5 cells per well was seeded in flow cytometry buffer consisting of PBS, 3% FBS, 0.1% sodium azide. Cells were then incubated with different concentrations of antibody or conjugate for one hour on ice. Following this, the cells were washed with PBS and incubated using a goat-anti-human Ig(H+L)-RPE detection antibody (Southern Biotech, Birmingham, AL, USA) for 30 min at 4 °C. After washing, cells were resuspended in flow cytometry buffer supplemented with 1 μg/mL propidium iodide. Mean fluorescence intensities of the stained cells were measured with a FACS Calibur Flow Cytometer and the CellQuest Pro software 4.0.2 (BD Biosciences, Heidelberg, Germany). Binding affinities of the antibodies or conjugates were determined using GraphPad Prism software 8.01. Dissociation constants (K_d) were determined by identifying the concentration of the antibody or conjugate required to achieve half-maximal specific binding.

2.6. Photoimmunotherapy

For PIT, 2.5×10^5 cells were seeded in 35 mm cell culture dishes and cultivated overnight at 37 °C, 5% CO_2. On the next day, fresh medium containing the appropriate treatment (medium, free dye, 10 μg/mL antibody, 10 μg/mL conjugate or 10 μg/mL of two conjugates each in the combination experiment) was administered to the cells, followed by incubation for 24 h. After washing with PBS, cells were incubated in fresh medium and irradiated with various mean irradiation light doses (λ_{max} = 690 nm) of a light-emitting diode (LED, L690-66-60, Marubeni America Corporation, Santa Clara, CA, USA). At 24 h after irradiation, cells were trypsinized and stained with Erythrosine B (Logos, Biosystems, Anyang-si, South Korea), followed by cell viability analysis with a Neubauer Counting Chamber. Numbers of living cells were normalized relative to the untreated control.

3. Results

The recombinant cysteine-engineered antibodies, CD44$^{T120C/D265C}$ and EpCAM$^{T120C/D265C}$, were successfully cloned and expressed in EXPI293F cells. Antibody yields were obtained, with 44.6 mg/L supernatant for CD44$^{T120C/D265C}$ and of 39.4 mg/L supernatant for EpCAM$^{T120C/D265C}$, in high purity after Protein G affinity chromatography (Figure 1a). SDS-PAGE under non-reducing conditions confirmed the self-assembly of the antibodies with two heavy and two light chains each, forming a whole hIgG1 molecule with a predicted size of ~150 kDa (Figure 1b).

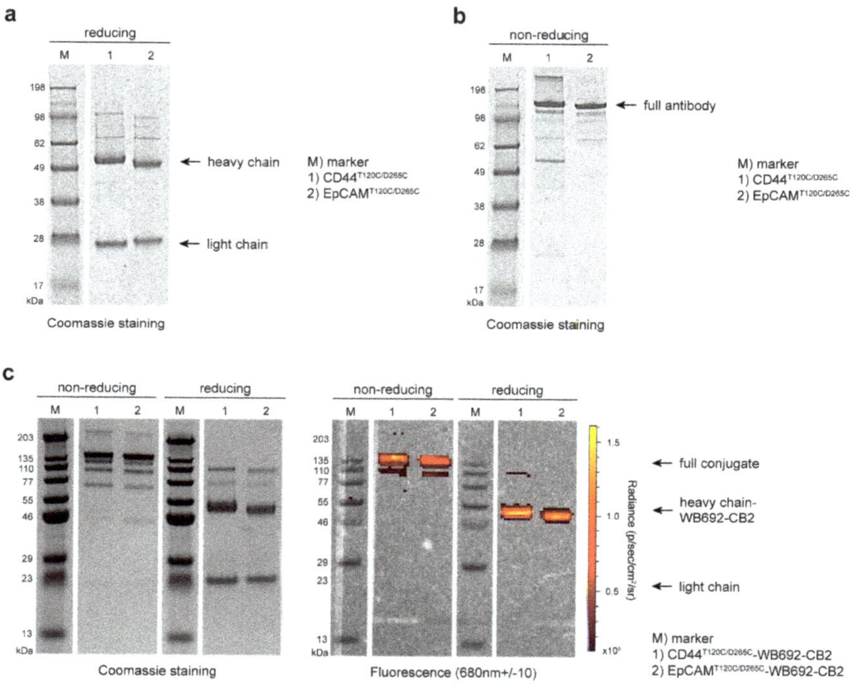

Figure 1. Generation of the anti-CD44 and anti-EpCAM antibody dye conjugates. (**a**) SDS gel of the purified antibodies CD44$^{T120C/D265C}$ and EpCAM$^{T120C/D265C}$ under reducing conditions; (**b**) SDS gel of the purified antibodies CD44$^{T120C/D265C}$ and EpCAM$^{T120C/D265C}$ under non-reducing conditions; (**c**) SDS gels of the purified antibody dye conjugates CD44$^{T120C/D265C}$-WB692-CB2 and EpCAM$^{T120C/D265C}$-WB692-CB2 under reducing and non-reducing conditions, and under white and red light.

Following dye conjugation, the antibody dye conjugates CD44$^{T120C/D265C}$-WB692-CB2 and EpCAM$^{T120C/D265C}$-WB692-CB2 were also analyzed by SDS-PAGE. Examination of the gels under red light (λ_{max} = 680 nm) confirmed the successful conjugation of the dye to the heavy chains of the antibodies (Figure 1c). Mean dye-to-antibody ratios were determined to be 2.82 ± 0.02 for the CD44$^{T120C/D265C}$-WB692-CB2 conjugate and 3.14 ± 0.17 for the EpCAM$^{T120C/D265C}$-WB692-CB2 conjugate.

In the subsequent step, Western Blot analyses confirmed the expression of the target proteins CD44 and EpCAM in the prostate cancer cell line PC3-PSMA and the prostate cancer stem cell-like (PCSC-like) cell line PC3-MM2, whereas the control cell line CHO showed no expression of either antigen (Figure 2a). The binding affinity of the conjugates $CD44^{T120C/D265C}$-WB692-CB2 and $EpCAM^{T120C/D265C}$-WB692-CB2 were then evaluated on both cell lines. The conjugates exhibited slightly lower binding affinities (higher Kd values) within the low nanomolar range compared to their uncoupled counterparts, $CD44^{T120C/D265C}$ and $EpCAM^{T120C/D265C}$ (Figure 2b,c,e,f). No binding was detected with both molecules on CD44 and EpCAM-negative CHO cells (Figure 2d,g). This proved that the dye conjugation did not affect or only minimally affected the avidity of the antibodies and did not result in unspecific binding of the conjugates.

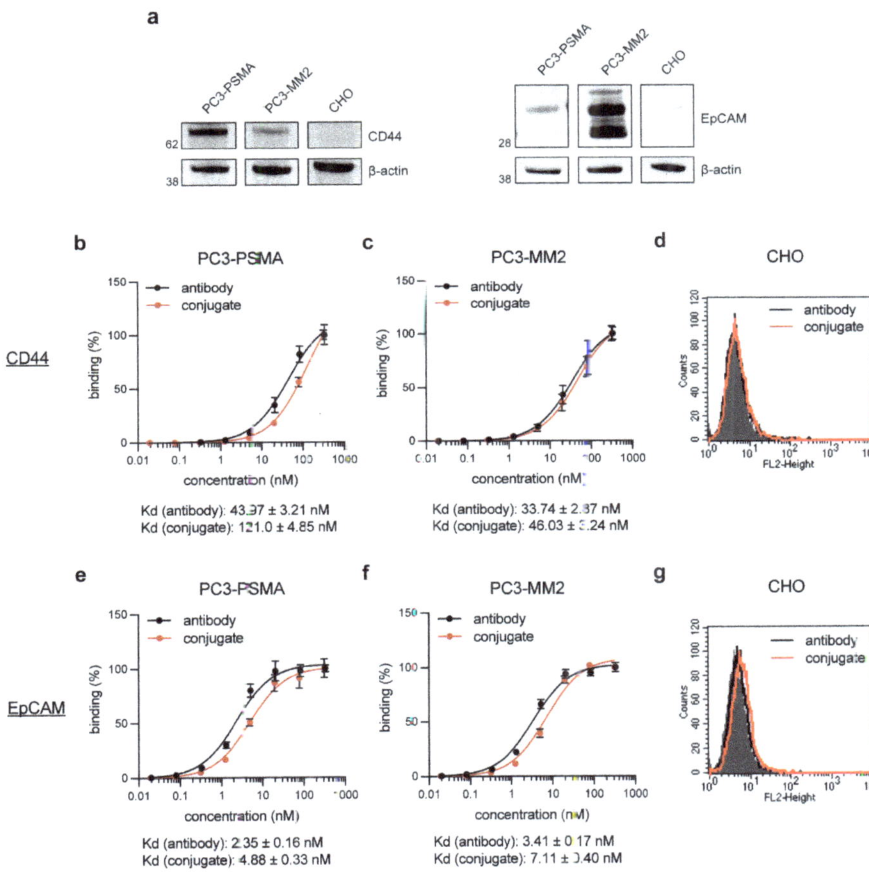

Figure 2. Cell binding of the antibodies and antibody dye conjugates. (**a**) CD44 and EpCAM expression of the target cells. Concentration-dependent binding of the antibody $CD44^{T120C/D265C}$ and the conjugate $CD44^{T120C/D265C}$-WB692-CB2 to (**b**) PC3-PSMA and (**c**) PC3-MM2 cells; (**d**) binding to CD44 negative CHO cells at a saturation concentration of 20.8 nM. Concentration-dependent binding of the antibody $EpCAM^{T120C/D265C}$ and the conjugate $EpCAM^{T120C/D265C}$-WB692-CB2 to (**e**) PC3-PSMA and (**f**) PC3-MM2 cells; (**g**) binding to EpCAM negative CHO cells at a saturation concentration of 20.8 nM. Mean values ± SD of three independent experiments. Abbreviation: Kd, dissociation constant.

The efficacy of PIT using the conjugates $CD44^{T120C/D265C}$-WB692-CB2 and $EpCAM^{T120C/D265C}$-WB692-CB2 was evaluated on the target cells 24 h after exposure with a light dose of 32 J/cm^2. Control cells remained untreated or were incubated with equimolar concentrations of antibody or free dye. The viability of PC3-PSMA and PC3-MM2 cells treated with the conjugate CD44$^{T120C/D265C}$-WB692-CB2 were significantly reduced to 29.9 ± 10.7% (p = 0.0027) and 27.1 ± 5.8% (p = 0.0102), respectively (Figure 3a,b). In contrast, no cytotoxic effects were observed in cells treated with the antibody or free dye. Cells from all treatment groups remained alive without irradiation. Moreover, CD44-negative CHO cells were completely unaffected by PIT (Figure 3c). Similar results were obtained with the conjugate EpCAM$^{T120C/D265C}$-WB692-CB2 in PC3-PSMA and PC3-MM2 cells after PIT. Cell viability was significantly reduced to 53.1 ± 15.4% (p = 0.0259; Figure 3d) and 7.1 ± 2.8% (p = 0.0015; Figure 3e), respectively, while EpCAM-negative CHO cells were not impaired by treatment (Figure 3f). These findings indicate that PIT using our dye WB692-CB2 is a highly selective and effective therapeutic strategy for targeting PC cells and PCSCs.

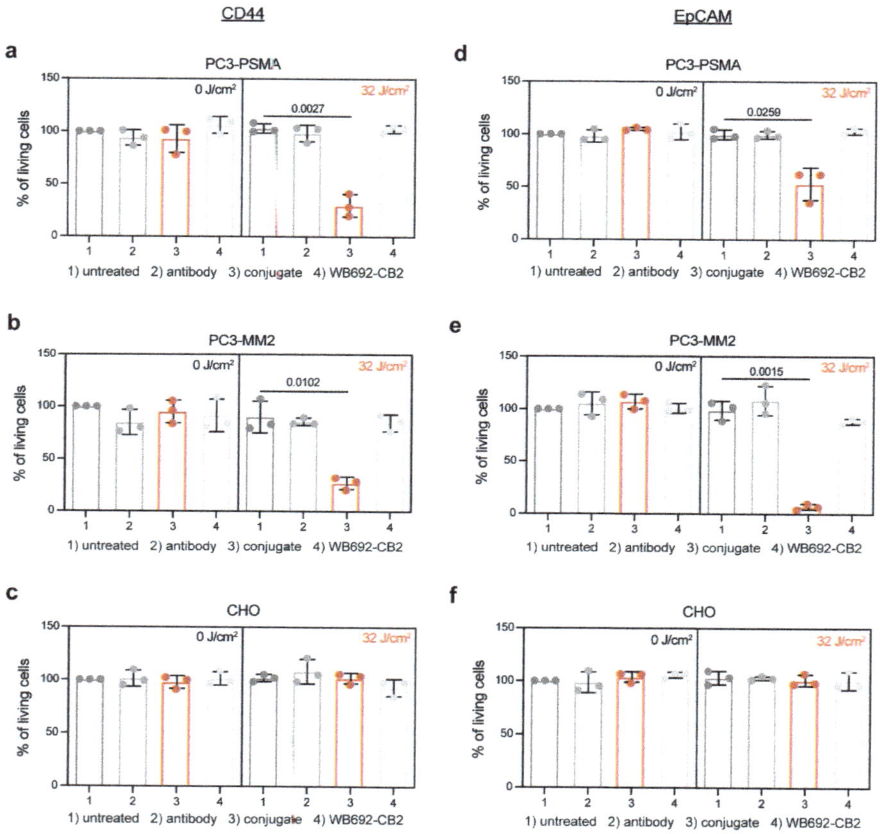

Figure 3. PIT of PC or PCSC-like cells with the conjugates CD44$^{T120C/D265C}$-WB692-CB2 and EpCAM$^{T120C/D265C}$-WB692-CB2. (**a**) Viability of PC3-PSMA, (**b**) PC3-MM2 or (**c**) CHO cells 24 h after PIT incubated with 10 µg/mL CD44$^{T120C/D265C}$, 10 µg/mL CD44$^{T120C/D265C}$-WB692-CB2, free dye or medium (untreated control), and irradiation with a light dose of 32 J/cm^2. Control cells were not irradiated (0 J/cm^2). Viability of (**d**) PC3-PSMA, (**e**) PC3-MM2 and (**f**) CHO cells 24 h after PIT incubated with 10 µg/mL EpCAM$^{T120C/D265C}$, 10 µg/mL EpCAM$^{T120C/D265C}$-WB692-CB2, free dye or medium (untreated control), and irradiated with a light dose of 32 J/cm^2. Control cells were not irradiated (0 J/cm^2). Mean values ± SD of three independent biological experiments. Statistical analyses by Student's t-test (unpaired, parametric with Welch's correction).

To develop a therapeutic approach against residual PC cells following RP, the efficacy of PIT using CD44$^{T120C/D265C}$-WB692-CB2 in combination with EpCAM$^{T120C/D265C}$-WB692-CB2 was evaluated using a reduced light irradiation dose of 16 J/cm^2. In the PC cell line PC3-PSMA, combinatorial PIT led to a significant reduction in cell viability compared to single treatments. Specifically, cell viability was significantly reduced from 52.5 ± 5.6% targeting CD44 and 88.1 ± 4.7% targeting EpCAM to only 28.9 ± 6.0% after PIT targeting CD44 in combination with EpCAM (p = 0.0001 and p < 0.0001, respectively; Figure 4a). Similar results were obtained in the PCSC-like PC3-MM2 cells, where the cell viability was significantly reduced from 64.0 ± 4.5% targeting CD44 and 58.4 ± 3.4% targeting EpCAM to 34.6 ± 6.5% after combinatorial PIT targeting CD44 and EpCAM (p = 0.0001 and p = 0.0006, respectively; Figure 4b).

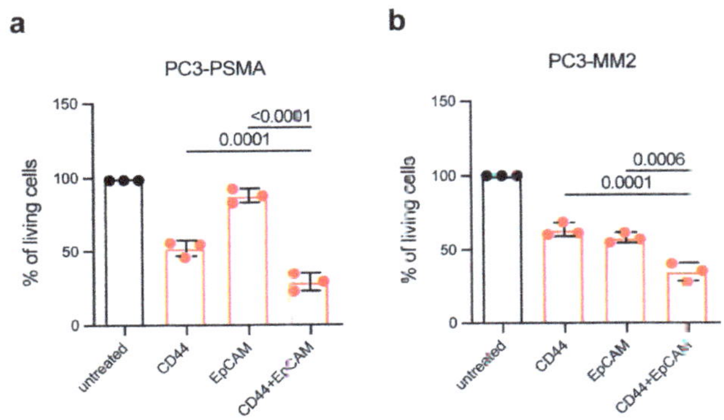

Figure 4. Combined PIT of PC cells and PCSC-like cells. (**a**) Viability of PC3-PSMA cells and (**b**) PC3-MM2 cells 24 h after PIT with 10 µg/mL CD44$^{T120C/D265C}$-WB692-CB2 or EpCAM$^{T120C/D265C}$-WB692-CB2 alone or in combination, and a light dose of 16 J/cm^2. Mean values ± SD of three independent biological experiments. Statistical analyses were performed using one-way ANOVA, followed by Tukey's multiple comparison test for post hoc analysis, with an adjusted significance level of p = 0.05.

Our results demonstrated the enhanced efficacy of combined PIT compared to single PIT, indicating that simultaneously targeting multiple tumor antigens may provide a more effective therapeutic strategy for PC.

4. Discussion

Tumor heterogeneity in PC remains a major challenge, contributing to therapeutic resistance and tumor recurrence. Consequently, there is an urgent need for the development of novel and innovative therapeutic approaches for PC. PIT has emerged as a new cornerstone in local cancer treatment with minimal off-target effects. This technique leverages photoactivatable photosensitizers conjugated to antibodies or antibody fragments to induce specific, immunogenic cancer cell death and systemic antitumor immunity [19]. We have developed the silicon phthalocyanine dye, WB692-CB2, which can be directly conjugated to free cysteines in antibodies via a maleimide linker [15]. Since all naturally occurring cysteines in an antibody molecule are used for the formation of intra- and interchain disulfide bridges, our dye can be specifically conjugated to engineered cysteines, which are not expected to affect the folding, and thus the avidity and specificity of the antibody. The cysteine-specific conjugation offers an advantage over the conventional coupling to lysines. A human IgG1 antibody contains approx. 80 to 90 lysines, leading to undefined and heterogeneous conjugate preparations, when coupling occurs at these sites. By specifically targeting engineered cysteines, we can achieve more homogenous antibody dye conjugate preparations. Our conjugation method allowed the attachment of approximately 2.8 to 3.1 dye molecules to the four engineered cysteines of one each of anti-CD44 and anti-EpCAM antibody molecules.

Our PIT experiments demonstrated that the anti-CD44 and anti-EpCAM conjugates, with their specific dye loads, effectively induced the cell death of both PC and PCSC-like cells within 24 h after irradiation using a light dose of 32 J/cm^2. Antigen-negative control cells were not affected by the treatment, which demonstrates the high selectivity of our approach. Despite differences in CD44 expression levels, the therapeutic effects of PIT targeting CD44 were similar in PC3-MM2 and PC3-PSMA cells. This discrepancy can be explained by the fact that the Western Blot data reflect total CD44 expression levels in the cells, but do not distinguish between intracellular and surface expression. Since PIT efficacy depends on the binding of the conjugates to surface antigens, it is possible that surface-accessible CD44 levels are more comparable between these cell lines than the total expression levels would suggest. Furthermore, intrinsic differences in susceptibility to PIT-induced cytotoxicity may also contribute to the equivalent therapeutic outcomes observed.

Combined PIT using both conjugates significantly enhanced cytotoxicity compared to single treatments in an additive manner. This could substantially enhance therapeutic outcomes compared to single treatments in the clinic. In addition to the combination of different conjugates against different target antigens, it is also possible to achieve better effects by increasing the light dose [15]. Moreover, engineering antibodies to incorporate additional cysteines could increase dye loading in future applications, which could also lead to an increased cytotoxicity. Further in vitro and in vivo studies are needed to validate the proof of principle for PIT using anti-CD44 and anti-EpCAM antibodies against PC, and to assess its potential as an effective option for future clinical use.

The un-conjugated dye exhibited no cytotoxicity, likely due to its inherent hydrophilicity, which enhances solubility and inhibits its diffusion across cellular membranes [15]. This characteristic is particularly advantageous, as it ensures that the dye exhibits cytotoxic effects only, when it is transported by the antibody into the target cells and is activated by light. This might reduce potential side effects in clinical practice and enhances the specificity of the therapeutic approach. Recently, we found that a main mechanism of PIT-induced cell death with our dye is pyroptosis [15]. Pyroptosis is a distinct type of pro-inflammatory programmed cell death, primarily driven by caspase-1 dependent cleavage of gasdermin D. This cleavage leads to the formation of non-selective pores in the plasma membrane, allowing the release of pro-inflammatory cytokines and influx of water. This is followed by cellular swelling, membrane rupture and release of damage-associated molecular patterns and tumor-associated neoantigens, which initiate an adaptive anti-cancer immune response [20]. Thus, locally applied PIT could potentially lead to systemic immunological antitumor effects. This means that, in clinical use, intraoperative PIT could also be effective against tumor cells that have already spread or metastasized. To further explore this, future studies can employ syngeneic tumor models in immunocompetent mice to investigate the immunogenic effects of PIT [21].

5. Conclusions

In the context of PC, WB692-CB2-based PIT can serve as a focal treatment approach, allowing for preservation of the prostate gland and minimizing side effects. It can be utilized during minimally invasive robotic-assisted RP to detect and treat residual tumor cells or lymph node metastases in areas where surgery cannot be continued in order to protect adjacent structures like the rectum or nerves responsible for urinary and sexual function. In clinical use, activation of the antibody dye conjugates during intraoperative PIT will be influenced by two main factors: firstly, by light accessibility of the tumor and, secondly, by tissue penetration of the conjugates. Light with a wavelength of around

700 nm, which can activate our dye, shows a maximum penetration depth of several millimeters in tissue, since it is only minimally absorbed by water or hemoglobin [22]. Light sources for illuminating the antibody dye conjugates could be easily implemented in robot surgical systems for RP in the future. The tissue penetration of the antibody dye conjugates depends on the antigen density, the tumor microenvironment, and the density and morphology of the tumor vessels [23]. Since the application of PIT in our case is limited to the outer margins after removal of the prostate, rather than deeper tumor areas, there is a chance that enough antibody dye conjugates have accumulated there to enable effective PIT. The use of multiple conjugates against antigens expressed on differentiated PC cells and PCSC cells, e.g., CD44 and EpCAM, could be an effective method to eradicate the heterogeneous population of residual cancer cells completely. This approach could lead to the prevention of local recurrence after RP and thus increase the therapeutic outcome of PC patients.

Author Contributions: Conceptualization, I.W. and P.W.; methodology, I.W., S.S.-S. and P.W.; formal analysis, I.W.; investigation, I.W., S.S.-S. and P.W.; resources, C.G. and P.W.; writing—original draft preparation, I.W. and P.W.; writing—review and editing, I.W., S.S.-S., C.G. and P.W.; visualization, I.W. and P.W.; supervision, P.W.; project administration, P.W.; funding acquisition, P.W. All authors have read and agreed to the published version of the manuscript.

Funding: This research was funded by the German Research Foundation, grant number WO 2178/3-1 to P.W., and supported by the Open Access Publication Fund of the University of Freiburg.

Institutional Review Board Statement: Not applicable.

Informed Consent Statement: Not applicable.

Data Availability Statement: The raw data supporting the conclusions of this article will be made available by the authors on request.

Acknowledgments: We thank P. Giangrande (University of Iowa, Iowa City, IA, USA) for the PC3-PSMA cells and the MD Anderson Cancer Center (Huston, TX, USA) for the PC3-MM2 cells. We acknowledge E.V. Wenzel and S. Dübel (University of Braunschweig) for providing the cloning vectors.

Conflicts of Interest: The authors declare no conflicts of interest.

References

1. Bray, F.; Laversanne, M.; Sung, H.; Ferlay, J.; Siegel, R.L.; Soerjomataram, I.; Jemal, A. Global cancer statistics 2022: Globocan estimates of incidence and mortality worldwide for 36 cancers in 185 countries. *CA Cancer J. Clin.* **2024**, *74*, 229–263. [CrossRef]
2. James, N.D.; Tannock, I.; N'Dow, J.; Feng, F.; Gillessen, S.; Ali, S.A.; Trujillo, B.; Al-Lazikani, B.; Attard, G.; Bray, F.; et al. The lancet commission on prostate cancer: Planning for the surge in cases. *Lancet* **2024**, *403*, 1683–1722. [CrossRef] [PubMed]
3. Rebello, R.J.; Oing, C.; Knudsen, K.E.; Loeb, S.; Johnson, D.C.; Reiter, R.E.; Gillessen, S.; Van der Kwast, T.; Bristow, R.G. Prostate cancer. *Nat. Rev. Dis. Primers* **2021**, *7*, 9. [CrossRef] [PubMed]
4. Ward, J.F.; Moul, J.W. Rising prostate-specific antigen after primary prostate cancer therapy. *Nat. Clin. Pract. Urol.* **2005**, *2*, 174–182. [CrossRef] [PubMed]
5. Karantanos, T.; Corn, P.G.; Thompson, T.C. Prostate cancer progression after androgen deprivation therapy: Mechanisms of castrate resistance and novel therapeutic approaches. *Oncogene* **2013**, *32*, 5501–5511. [CrossRef]
6. Ge, R.; Wang, Z.; Cheng, L. Tumor microenvironment heterogeneity an important mediator of prostate cancer progression and therapeutic resistance. *NPJ Precis. Oncol.* **2022**, *6*, 31. [CrossRef] [PubMed]
7. Haffner, M.C.; Zwart, W.; Roudier, M.P.; True, L.D.; Nelson, W.G.; Epstein, J.I.; De Marzo, A.M.; Nelson, P.S.; Yegnasubramanian, S. Genomic and phenotypic heterogeneity in prostate cancer. *Nat. Rev. Urol.* **2021**, *18*, 79–92. [CrossRef]
8. Marusyk, A.; Polyak, K. Tumor heterogeneity: Causes and consequences. *Biochim. Biophys. Acta* **2010**, *1805*, 105–117. [CrossRef]
9. Proietto, M.; Crippa, M.; Damiani, C.; Pasquale, V.; Sacco, E.; Vanoni, M.; Gilardi, M. Tumor heterogeneity: Preclinical models, emerging technologies, and future applications. *Front. Oncol.* **2023**, *13*, 1164535. [CrossRef]
10. Wolf, I.; Gratzke, C.; Wolf, P. Prostate cancer stem cells: Clinical aspects and targeted therapies. *Front. Oncol.* **2022**, *12*, 935715 [CrossRef]

11. Mohiuddin, T.M.; Zhang, C.; Sheng, W.; Al-Rawe, M.; Zeppernick, F.; Meinhold-Heerlein, I.; Hussain, A.F. Near infrared photoimmunotherapy: A review of recent progress and their target molecules for cancer therapy. *Int. J. Mol. Sci.* **2023**, *24*, 2655. [CrossRef]
12. Rowehl, R.A.; Crawford, H.; Dufour, A.; Ju, J.; Botchkina, G.I. Genomic analysis of prostate cancer stem cells isolated from a highly metastatic cell line. *Cancer Genom. Proteom.* **2008**, *5*, 301–310.
13. Lee, Y.C.; Jin, J.K.; Cheng, C.J.; Huang, C.F.; Song, J.H.; Huang, M.; Brown, W.S.; Zhang, S.; Yu-Lee, L.Y.; Yeh, E.T.; et al. Targeting constitutively activated β1 integrins inhibits prostate cancer metastasis. *Mol. Cancer Res. MCR* **2013**, *11*, 405–417. [CrossRef] [PubMed]
14. Botchkina, G.I.; Zuniga, E.S.; Rowehl, R.H.; Park, R.; Bhalla, R.; Bialkowska, A.B.; Johnson, F.; Golub, L.M.; Zhang, Y.; Ojima, I.; et al. Prostate cancer stem cell-targeted efficacy of a new-generation taxoid, sbt-1214 and novel polyenolic zinc-binding curcuminoid, cmc2.24. *PLoS ONE* **2013**, *8*, e69884. [CrossRef]
15. Wolf, I.; Storz, J.; Schultze-Seemann, S.; Esser, P.R.; Martin, S.F.; Lauw, S.; Fischer, P.; Peschers, M.; Melchinger, W.; Zeiser, R.; et al. A new silicon phthalocyanine dye induces pyroptosis in prostate cancer cells during photoimmunotherapy. *Bioact. Mater.* **2024**, *41*, 537–552. [CrossRef]
16. Birzele, F.; Cannarile, M.; Feuerhake, F.; Fischer, T.; Heil, F.; Honold, K.; Nopora, A.; Schmitt-Graeff, A.; Voss, E.; Weigand, S. Markers for Responsiveness to Anti-cd44 Antibodies. Patent WO2014198843A1, 12 June 2014.
17. Cizeau, J.; Macdonald, G.; Premsukh, A. Optimized Nucleic Acid Sequences for the Expression of vb4-845. Patent WO2009039630A1, 26 September 2008.
18. Steinwand, M.; Droste, P.; Frenzel, A.; Hust, M.; Dübel, S.; Schirrmann, T. The influence of antibody fragment format on phage display based affinity maturation of igg. *mAbs* **2014**, *6*, 204–218. [CrossRef] [PubMed]
19. Hsu, M.A.; Okamura, S.M.; De Magalhaes Filho, C.D.; Bergeron, D.M.; Rodriguez, A.; West, M.; Yadav, D.; Heim, R.; Fong, J.J.; Garcia-Guzman, M. Cancer-targeted photoimmunotherapy induces antitumor immunity and can be augmented by anti-pd-1 therapy for durable anticancer responses in an immunologically active murine tumor model. *Cancer Immunol. Immunother.* **2023**, *72*, 151–168. [CrossRef]
20. Hou, J.; Hsu, J.M.; Hung, M.C. Molecular mechanisms and functions of pyroptosis in inflammation and antitumor immunity. *Mol. Cell* **2021**, *81*, 4579–4590. [CrossRef] [PubMed]
21. Kato, T.; Okada, R.; Furusawa, A.; Inagaki, F.; Wakiyama, H.; Furumoto, H.; Okuyama, S.; Fukushima, H.; Choyke, P.L.; Kobayashi, H. Simultaneously combined cancer cell- and ctla4-targeted nir-pit causes a synergistic treatment effect in syngeneic mouse models. *Mol. Cancer Ther.* **2021**, *20*, 2262–2273. [CrossRef]
22. Ash, C.; Dubec, M.; Donne, K.; Bashford, T. Effect of wavelength and beam width on penetration in light-tissue interaction using computational methods. *Lasers Med. Sci.* **2017**, *32*, 1909–1918. [CrossRef] [PubMed]
23. Thurber, G.M.; Weissleder, R. Quantitating antibody uptake in vivo: Conditional dependence on antigen expression levels. *Mol. Imaging Biol.* **2011**, *13*, 623–632. [CrossRef] [PubMed]

antibodies

MDPI

Review

A Comprehensive Review About the Use of Monoclonal Antibodies in Cancer Therapy

Angel Justiz-Vaillant, Bijay Raj Pandit, Chandrashekhar Unakal, Sehlule Vuma and Patrick Eberechi Akpaka *

Department of Pathology/Microbiology & Pharmacology, The University of the West Indies, St. Augustine Campus, St. Augustine 330912, Trinidad and Tobago; angel.justiz-vaillant@uwi.edu (A.J.-V.); bijay.pandit@uwi.edu (B.R.P.); chandrashekhar.unakal@uwi.edu (C.U.); sehlule.vuma@uwi.edu (S.V.)
* Correspondence: eberechipatrick.akpaka@uwi.edu

Abstract: Monoclonal antibodies (mAbs) targeting various pathways in cancer therapy play crucial roles in enhancing the immune system's ability to recognise and eliminate tumour cells. These therapies are designed to either block inhibitory immune checkpoint pathways or to target specific tumour cell markers for direct destruction. Additionally, mAbs can modulate the tumour microenvironment, enhance antibody-dependent cellular cytotoxicity, and inhibit angiogenesis, further amplifying their therapeutic impact. Below is a summary of monoclonal antibodies targeting key pathways, along with their indications and mechanisms of action, which are reviewed based on therapeutic mechanisms.

Keywords: monoclonal antibody; immune cells; solid and haematological malignancies; personalised medicine

1. Introduction

Antibodies have become a cornerstone of cancer immunotherapy, offering a targeted approach to combat tumours [1]. Monoclonal antibodies (moAbs or mAbs) are produced to recognise antigens expressed on tumour cells, enabling precise intervention with minimal impact on healthy tissues [2]. These therapies work through various mechanisms, including immune system activation, inhibition of tumour growth, and delivery of cytotoxic agents [3].

A well-established application of antibodies is immune checkpoint blockade. Antibodies targeting "Programmed Cell Death Protein-1 (PD-1), Programmed Cell Death Ligand-1 (PD-L1), and Cytotoxic T-Lymphocyte-Associated Protein 4 (CTLA-4)" restore T-cell activity, allowing the immune system to identify and destroy tumours [4]. Another advanced strategy involves antibody–drug conjugates (ADCs), which integrate the precision of antibodies with the potency of cytotoxic agents. By directing these drugs specifically to cancer cells, ADCs reduce off-target effects [5].

Bispecific antibodies, designed to engage two different antigens simultaneously, are also gaining prominence. For example, bispecific T-cell engagers (BiTEs) like blinatumomab bring lymphocyte T into close proximity to tumour cells, facilitating immune-mediated destruction [6].

Additionally, antibodies can act as carriers for radioisotopes or other therapeutic agents in targeted delivery systems. This approach, known as radioimmunotherapy, has shown promise in certain haematological malignancies [7].

Despite their success, antibody therapies are not without challenges, including high production costs, potential resistance, and immune-related side effects [8]. Research is

 https://doi.org/10.3390/antib14020035

ongoing to improve antibody design, optimise dosing regimens, and identify biomarkers to predict patient response [9].

Antibody-based immunotherapy continues to revolutionise cancer treatment, offering hope for durable responses and improved survival in both haematological and solid tumours. Its future lies in further innovation and broader application across oncology [10].

2. Immunomodulatory Approaches (Checkpoint Inhibitors, Bispecific Antibodies, ADCs with Immune Modulation)

2.1. Immune Checkpoint Pathway

The immune checkpoint pathway plays a vital role in controlling immune responses, preventing overactivity, and maintaining self-tolerance [11]. However, cancer cells exploit these mechanisms to avoid destruction by the immune system [12]. Key checkpoints, such as CTLA-4 and PD-1, serve as major regulators in this process [13]. Their activation suppresses T-cell function, allowing tumour cells to proliferate without restraint [14]. Table 1 shows indications of immune checkpoint inhibitors, along with comments and references.

Table 1. Immune checkpoint inhibitors, targeting pathways like PD-1, PD-L1, and CTLA-4, have emerged as effective treatments for various solid tumours.

Indication	Comments	Reference
Melanoma	These therapies revolutionised treatment by significantly improving survival rates	[15]
Non-Small Cell Lung Cancer (NSCLC)	PD-1/PD-L1 inhibitors are standard for advanced or metastatic cases	[16]
Head and Neck Cancer	Immune checkpoint inhibitors have shown efficacy in recurrent or metastatic disease	[17]
Cervical Cancer	These treatments address advanced or recurrent cervical carcinoma, particularly in PD-L1-positive patients	[18]
Urothelial Carcinoma	Effective for metastatic urothelial cancers, especially in patients who are ineligible for chemotherapy	[19]

2.1.1. Mechanism of Action

Immune checkpoint inhibitors block the interaction between checkpoint proteins and their ligands (e.g., PD-1/PD-L1, CTLA-4/B7), restoring T-cell activity and enabling the immune system to recognise and attack tumour cells [20,21]. However, ongoing research focuses on overcoming resistance and expanding indications through combination therapies [22,23]. "CTLA-4 inhibitors (e.g., ipilimumab) block the interaction between CTLA-4 and its ligands B7-1/B7-2 on antigen-presenting cells", facilitating T-cell activation [24].

Immune checkpoint blockade enhances the activity of T cells that produce IFN-γ, which subsequently revitalises the antitumour capabilities of tumour-associated macrophages and dendritic cells as shown in Figure 1 [25].

2.1.2. The Future of Immune Checkpoint Inhibitors

Immune checkpoint inhibitors (ICIs) have transformed oncology by enabling the immune system to effectively combat cancer. The future of these therapies lies in overcoming current limitations, expanding indications, and optimising their efficacy. Addressing tumour resistance and the variability in patient responses remains a key challenge. Research is increasingly focused on identifying biomarkers, such as tumour mutational burden (TMB) and immune profiles, to personalise treatments and predict outcomes more accurately [26].

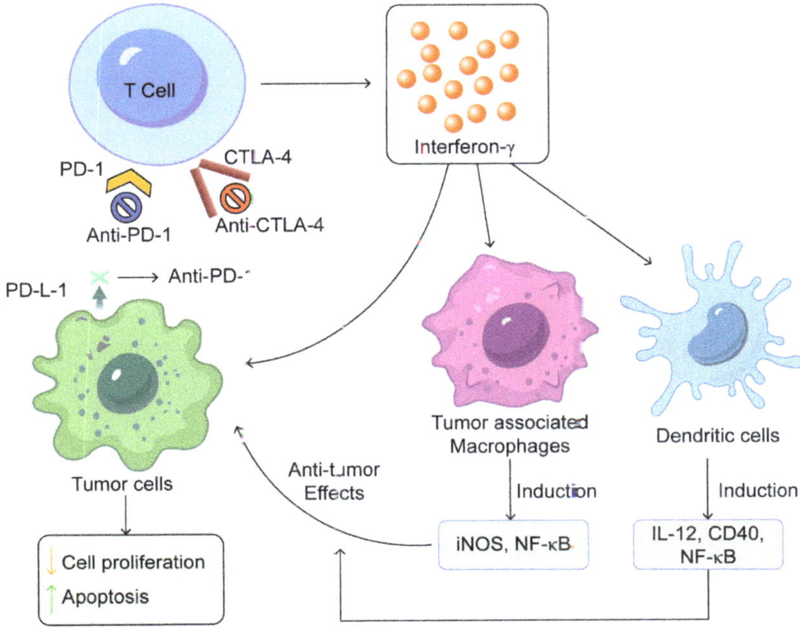

Figure 1. Summarises the mechanisms of action of immune checkpoint inhibitors. This was adapted from open access journal, Pandey, P. et al. *Pharmaceuticals* **2022**, *15* 335 [25]. Adaptation performed by MDPI author services, ID: English-91110.

Combination therapies hold significant potential for future cancer treatment [27]. Integrating immune checkpoint inhibitors (ICIs) with targeted therapies, chemotherapy, radiotherapy, or immunotherapies such as cancer vaccines and adoptive cell therapies aims to improve anti-tumour responses [28]. Furthermore, the advancement of next-generation ICIs targeting novel pathways, including "Lymphocyte Activation Gene-3 (LAG-3), T-cell Immunoglobulin and Mucin-Domain Containing-3 (TIM-3), and T-cell Immunoreceptor with Ig and ITIM domains (TIGIT)", provides new opportunities for patients with treatment-resistant cancers or those unresponsive to existing therapies [29,30].

Emerging applications for ICIs are also being explored in early-stage cancers, adjuvant settings, and even non-cancerous diseases, such as infectious diseases and autoimmune disorders [31]. Advances in technology, such as artificial intelligence, are aiding in drug discovery and patient stratification [32,33], while a growing focus on reducing immune-related adverse events seeks to improve tolerability. The future of ICIs lies in broader applications, greater precision, and more durable outcomes [34,35]. Table 2 shows examples of checkpoint inhibitors and their mechanism of action.

Table 2. Examples of checkpoint inhibitors and mechanism of action.

Antibody	Remarks	Reference
Nivolumab	Enhances immune activity against malignant cells by blocking PD-1 on T-cells.	[36,37]
Pembrolizumab	Another PD-1 inhibitor with broad indications, including melanoma and lung cancer.	[38]
Atezolizumab	PD-L1 inhibitor used in bladder cancer and non-small cell lung cancer.	[39,40]
Avelumab	PD-L1 inhibitor, approved for Merkel cell carcinoma and urothelial carcinoma.	[41,42]
Ipilimumab	First approved CTLA-4 inhibitor, used for melanoma and in combination with PD-1 inhibitors like nivolumab for other cancers including lung and breast cancer.	[43,44]

2.1.3. Use of Bispecific Antibodies: Mechanisms of Action

CD19 inhibition utilises bispecific antibodies, particularly bispecific T-cell engagers (BiTEs), to treat haematologic malignancies [45]. As a cell surface marker primarily found on B cells, CD19 serves as a key target in managing conditions such as B-cell acute lymphoblastic leukaemia (B-ALL) and non-Hodgkin lymphoma [46]. The selective targeting of CD19 has led to the development of novel therapeutic strategies in oncology.

Bispecific antibodies, for example, blinatumomab, serve a dual purpose by engaging CD19 and CD3 on malignant B-cells and T-cells, respectively. This unique mechanism facilitates the recruitment of T-cells directly to the vicinity of cancer cells, effectively bridging the gap between immune cells and tumours. Upon binding, T-cells become activated, which leads to the release of cytotoxic agents that induce apoptosis in the target tumour cells. This targeted engagement not only enhances the immune response against tumours but also minimises the impact on healthy cells, addressing a common concern associated with conventional therapies [47].

The clinical success of CD19 targeting has been evident, particularly with blinatumomab, which has achieved significant response rates in patients with relapsed or refractory B-cell malignancies. Research is also investigating the potential of CD19 inhibition in solid tumours, expanding its therapeutic applications. Blinatumomab is administered through a continuous intravenous infusion for 28 days, followed by a 14-day break. The dosage varies based on the treatment phase and patient weight. In adults, the initial dose is typically 9 mcg/day during the first week of cycle 1, increasing to 28 mcg/day in subsequent weeks and cycles [48].

2.1.4. The Action of T-Cell Engagers (TCEs)

The action of T-cell engagers (TCEs), a type of bispecific antibodies, is summarised in Figure 2. It involves their dual-binding capability, as they attach to tumour-associated antigens (TAAs) and CD3 receptors. This dual interaction facilitates the establishment of a cytolytic synapse. When TCEs engage with CD3, they initiate a signal that activates T-cell cytotoxicity, leading to the release of granzymes and perforins that cause tumour cell destruction. Moreover, this engagement promotes the expansion of the T-cell population [49].

2.2. *Antibody–Drug Conjugates (ADCs) with Immune Modulation*

Antibody–drug conjugates (ADCs) with immune modulation refer to ADCs that not only deliver cytotoxic payloads to tumour cells but also enhance anti-tumour immunity [50]. These ADCs can modulate the immune response through several mechanisms:

1. Inducing Immunogenic Cell Death (ICD): Some ADC payloads, such as microtubule inhibitors (e.g., MMAE) or DNA-damaging agents (e.g., PBD dimers), can trigger ICD, leading to the release of damage-associated molecular patterns (DAMPs) that activate dendritic cells and enhance antigen presentation;
2. Targeting the Tumour Microenvironment (TME): Some ADCs are designed to deplete regulatory T cells (Tregs) or myeloid-derived suppressor cells (MDSCs), by targeting specific markers like CD25 or CCR8;
3. Enhancing T-cell Infiltration: ADCs can modify the tumour microenvironment by reducing stromal barriers or disrupting tumour-associated vasculature, thereby facilitating better immune cell infiltration;
4. Activating Innate Immunity: ADCs can engage immune cells through Fcγ receptor interactions, promoting phagocytosis or antibody-dependent cellular cytotoxicity (ADCC);

5. Combining with Immune Checkpoint Inhibitors (ICIs): ADCs that induce immune responses may synergise with ICIs (e.g., anti-PD-1/PD-L1) to enhance anti-tumour immunity.

Examples of ADCs with immune modulation include the following:

- Enfortumab vedotin (EV, targeting Nectin-4): Induces ICD and enhances the efficacy of checkpoint inhibitors in urothelial carcinoma.

Having explored immunomodulatory strategies that reinvigorate the host immune response, we now examine therapies targeting oncogenic receptors such as HER2 and EGFR.

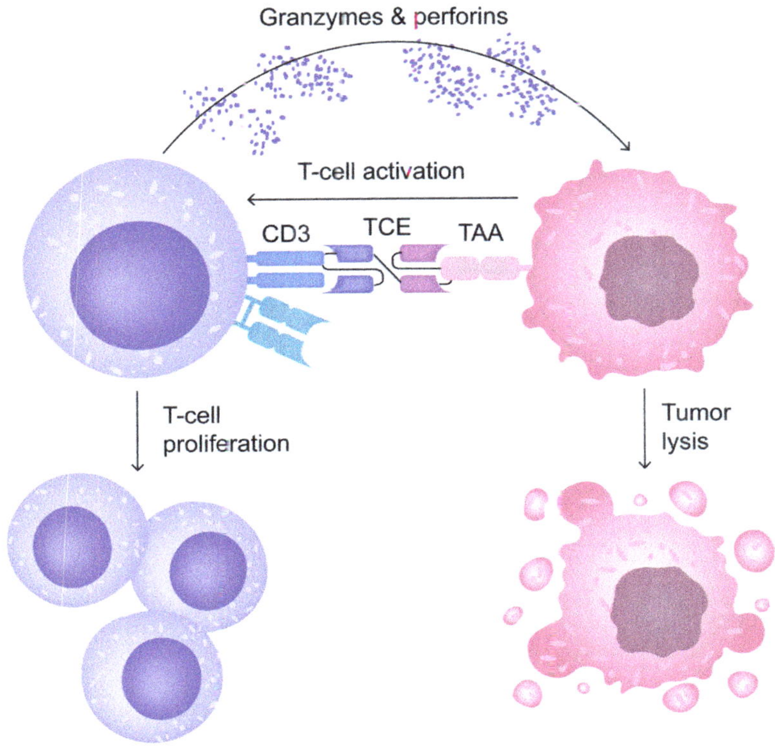

Figure 2. Mechanism of action of T-cell engagers (TCEs), a type of bispecific antibody. Adapted from open access journal referenced Cech, P. et al. *Cancers* **2024**, *16*, 1580 [49]. Adaptation performed by MDPI author services, ID: English-90959.

3. Receptor-Targeted Therapies (HER2, EGFR, IGF-1R Inhibitors)

3.1. HER2 Inhibition and Receptor Tyrosine Kinase Signalling

HER2 (human epidermal growth factor receptor 2) is a receptor tyrosine kinase involved in regulating cell growth, survival, and differentiation [51,52]. Overexpression or amplification of HER2 is commonly observed in aggressive cancers, including gastric cancers. HER2 inhibition aims to block this pathway, halting tumour growth and promoting cancer cell death [53].

The epidermal growth factor receptor (EGFR) family also comprises HER2, which activates critical intracellular signalling pathways. Upon dimerisation with other EGFR family members, HER2 triggers cascades such as the PI3K-AKT pathway (promoting cell survival and inhibiting apoptosis), the RAS-RAF-MEK-ERK pathway (stimulating cell proliferation), and the JAK-STAT pathway (regulating transcription linked to growth) [54]. Dysregulation of these pathways drives oncogenesis in HER2-positive cancers.

Therapies targeting HER2 include monoclonal antibodies like trastuzumab, which blocks receptor dimerisation, and pertuzumab, which inhibits heterodimerisation with HER3 [55]. Tyrosine kinase inhibitors (e.g., lapatinib) block HER2's intracellular signalling activity [56], while antibody–drug conjugates such as T-DM1 (trastuzumab emtansine) is an antibody–drug conjugate (ADC) used in cancer therapy, specifically for HER2-positive breast cancer, which deliver targeted chemotherapy [57]. However, resistance to HER2-targeted therapies remains a challenge, often driven by mutations, compensatory pathways, or phenotypic changes. Current research focuses on overcoming resistance through combination therapies, immune checkpoint inhibitors, and emerging approaches like HER2-specific chimeric antigen receptor-T (CAR-T) cell therapy [58,59].

3.2. Trastuzumab, Pertuzumab, and Margetuximab Are Monoclonal Antibodies Targeting HER2-Positive Cancers

Trastuzumab, developed by Genentech/Roche and approved in 1998, was the first HER2-targeting therapy. It binds to the HER2 extracellular domain, inhibiting dimerisation and activating antibody-dependent cellular cytotoxicity (ADCC). It significantly improves overall survival (OS) and progression-free survival (PFS) in HER2-positive breast and gastric cancers. Administered intravenously or subcutaneously, trastuzumab is combined with chemotherapy or pertuzumab. While revolutionary, its limitations include cardiotoxicity and resistance, addressed partly by biosimilars like Ogivri and Kanjinti, as biosimilars of trastuzumab, are used to treat HER2-positive breast and gastric cancers by inhibiting tumour growth [60]. Genentech, a biotechnology company based in the United States, is a subsidiary of Roche. It is a leader in tumour immunotherapy, advancing treatments like atezolizumab (anti-PD-L1) and combination therapies that counteract cancer immune evasion. Through biomarker-driven strategies, it enhances immuno-oncology efficacy for lung, breast, and other malignancies.

Pertuzumab, also from Genentech/Roche, was FDA-approved in 2012. It targets a different HER2 epitope, blocking HER2 dimerisation with other HER family receptors, such as HER3. Pertuzumab is most effective when combined with trastuzumab and docetaxel, as demonstrated in the CLEOPATRA trial. The CLEOPATRA trial was a phase III clinical study that investigated the effectiveness and safety of combining trastuzumab, pertuzumab, and docetaxel for treating HER2-positive metastatic breast cancer. Findings showed that this combination significantly prolonged progression-free and overall survival compared to trastuzumab and docetaxel alone, setting a new standard for first-line treatment in this cancer type. This combination further enhances OS and PFS, particularly in metastatic HER2-positive breast cancer. Pertuzumab's primary side effects include diarrhoea and neutropenia, and its patent is set to expire in 2027 [61].

Margetuximab, developed by MacroGenics and approved in 2020, introduces an engineered Fc domain that enhances binding to activating Fc receptors, boosting ADCC as shown in Figure 3. It is indicated for HER2-positive metastatic breast cancer after prior trastuzumab regimens. The SOPHIA trial highlighted its modest efficacy improvement over trastuzumab in resistant settings. Margetuximab's side effects include infusion-related reactions and fatigue, and it is used with chemotherapy. In addition, the SOPHIA trial was a phase III study that examined the effectiveness and safety of margetuximab in patients with HER2-positive metastatic breast cancer who had already undergone HER2-targeted treatments. It compared margetuximab plus chemotherapy with trastuzumab plus chemotherapy to assess improvements in survival outcomes. Results indicated potential benefits, particularly for individuals with the CD16A-158F allele, supporting its role as an alternative therapy [62].

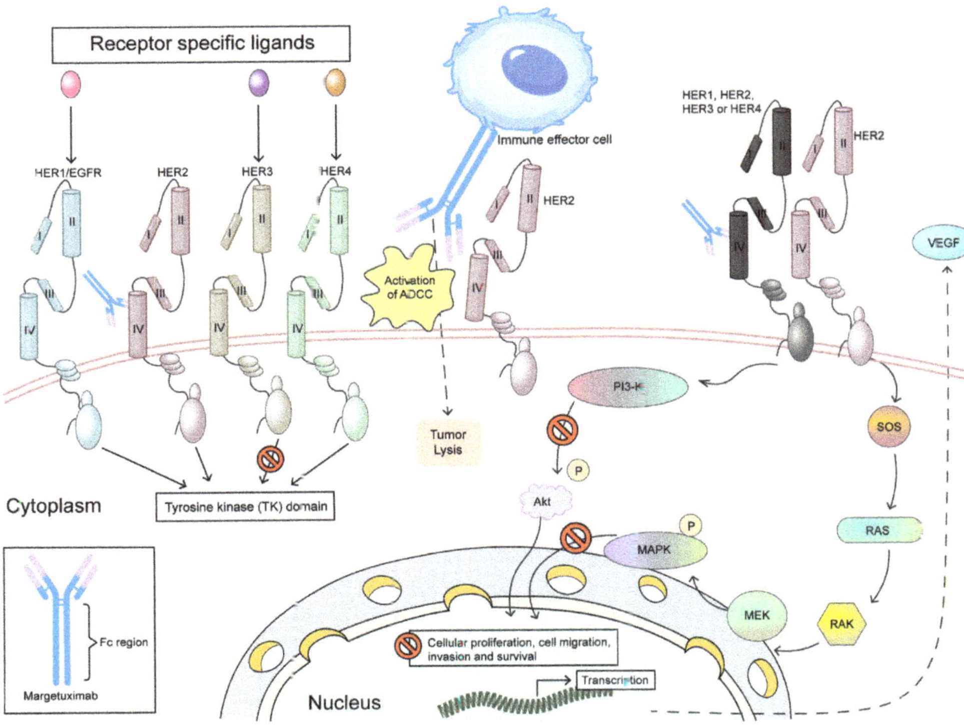

Figure 3. Margetuximab is a monoclonal antibody that has been modified at the Fc region to enhance its therapeutic effects. It specifically binds to the extracellular domain IV of the HER2 receptor, preventing the formation of HER2 homodimers and disrupting the ligand-independent interactions of HER2 with HER3, HER1, and HER4. This disruption strengthens antibody-dependent cellular cytotoxicity (ADCC), leading to the targeted elimination of tumour cells. Furthermore, by blocking the HER2-associated tyrosine kinase signalling pathway, margetuximab reduces cancer cell proliferation, migration, invasion, and survival. Adapted from open access journal referenced Alasmari, M.M. *Cancers* **2022**, *15*, 38 [62]. Adaptation performed by MDPI author services. ID: English-91110.

Together, these drugs revolutionise HER2-targeted therapies, with trastuzumab as a cornerstone, pertuzumab offering synergistic benefits, and margetuximab enhancing immune response for resistant cases. Despite their differences, their combined use exemplifies precision oncology, advancing outcomes for HER2-positive cancer patients. In our own words, precision oncology involves tailoring treatments based on a patient's genetic or molecular profile.

3.3. *Epidermal Growth Factor Receptor (EGFR) in Cancer Therapy*

The epidermal growth factor receptor (EGFR) plays a crucial role in inhibiting cancer cell growth and serves as a key target for treatment, particularly in malignancies characterised by EGFR overexpression or mutations. As a tyrosine kinase receptor, EGFR is integral to cellular signalling networks that regulate vital oncogenic processes, including cell growth and differentiation [63]. Genetic alterations in this pathway—such as gene amplification—are frequently associated with tumour initiation and progression. As a result, EGFR has become a significant immunotherapy target, particularly in colorectal cancer, non-small cell lung cancer (NSCLC), and head and neck squamous cell carcinoma [64].

3.3.1. Mechanism of Action

EGFR signalling is triggered when ligands such as epidermal growth factor (EGF) and transforming growth factor-alpha (TGF-α) bind to its extracellular domain [65]. This binding event induces receptor dimerisation, leading to autophosphorylation of intracellular tyrosine residues and activation of downstream pathways, including PI3K/AKT and RAS/RAF/MEK/ERK [66]. These pathways drive oncogenic processes such as uncontrolled proliferation, angiogenesis, tumour invasion, and resistance to apoptosis. Aberrant EGFR signalling is closely linked to increased cancer cell aggressiveness and poorer clinical outcomes [67].

EGFR inhibitors disrupt this signalling and are classified into monoclonal antibodies and tyrosine kinase inhibitors (TKIs) [68]. Monoclonal antibodies, such as panitumumab and cetuximab, target EGFR's extracellular domain, blocking ligand binding and preventing receptor activation [69]. By suppressing EGFR-dependent pathways, these antibodies slow tumour growth and enhance cancer cell susceptibility to treatments like chemotherapy and radiation therapy [70].

Conversely, TKIs, including gefitinib, erlotinib, and osimertinib, act on the intracellular tyrosine kinase domain of EGFR [71,72]. These small molecules competitively inhibit ATP binding, a crucial step for receptor phosphorylation and activation [73]. TKIs are especially effective in NSCLC patients with EGFR-activating mutations, such as exon 19 deletions or the L858R substitution in exon 21 [74]. These mutations increase EGFR kinase activity, making tumour cells heavily reliant on EGFR signalling for survival—a phenomenon known as "oncogene addiction" [75,76].

3.3.2. Anti-EGFR Antibody Characteristics

Anti-EGFR antibodies shown in Table 3 are monoclonal antibodies targeting the EGFR, inhibiting tumour growth and proliferation. These antibodies block ligand binding, preventing downstream signalling pathways like MAPK and PI3K/AKT. Primarily used in cancers such as colorectal and head and neck, they enhance therapeutic efficacy, especially in patients with wild-type RAS genes, and are critical in precision medicine [77,78].

Table 3. Comparison between anti-EGFR antibodies. Summary, indications, dosages, and side effects of each.

Antibodies	Remarks	Reference
Cetuximab	"Cetuximab is a chimeric monoclonal antibody that targets the epidermal growth factor receptor (EGFR), a transmembrane protein frequently overexpressed in several cancers". By binding to EGFR, it disrupts downstream signalling pathways involved in cell proliferation, survival, and angiogenesis. Cetuximab is primarily indicated for metastatic colorectal cancer (mCRC) and squamous cell carcinoma of the head and neck (SCCHN). Indications and dosage: 1. mCRC: Used either as monotherapy or in combination with chemotherapy regimens such as FOLFIRI or FOLFOX in patients with wild-type KRAS or NRAS tumours; 2. SCCHN: Administered with platinum-based chemotherapy for metastatic or recurrent cases or alongside radiation therapy for locally advanced disease; 3. Dosage: The initial loading dose is 400 mg/m^2, followed by weekly intravenous infusions of 250 mg/m^2. Side Effects: Common adverse reactions include an acneiform rash, hypomagnesaemia, infusion-related reactions, and diarrhoea. The presence of a skin rash is often associated with better treatment outcomes.	[79]

Table 3. *Cont.*

Antibodies	Remarks	Reference
Panitumumab	Panitumumab is a fully human monoclonal antibody that also targets EGFR, inhibiting ligand binding and subsequent activation of cancer-promoting pathways. It is primarily employed in the treatment of *NRAS* mCRC or wild-type *KRAS*. Being fully human, panitumumab has a reduced risk of immunogenicity compared to chimeric antibodies like cetuximab. Indications and Dosages 1. mCRC: Panitumumab is approved as monotherapy or in combination with chemotherapy regimens for patients with wild-type *RAS* tumours. The standard dosage is 6 mg/kg administered every two weeks via intravenous infusion. Side Effects Panitumumab's adverse effects resemble those of cetuximab, including acneiform dermatitis, hypomagnesaemia, and infusion-related reactions. Rare but serious side effects include pulmonary fibrosis and severe diarrhoea.	[80,81]
Amivantamab	Amivantamab is a bispecific monoclonal antibody that simultaneously targets EGFR and MET, two proteins involved in oncogenic signalling. This dual action is particularly beneficial for tumours with EGFR exon 20 insertions, frequently observed in non-small cell lung cancer (NSCLC). In addition to inhibiting these pathways, amivantamab also triggers antibody-dependent cellular cytotoxicity (ADCC). Indications and Dosage 1. NSCLC: Used for advanced or metastatic NSCLC with EGFR exon 20 insertion mutations, particularly after disease progression on platinum-based chemotherapy; 2. Dosage: The initial loading dose is weight-based—1050 mg for patients under 80 kg and 1400 mg for those 80 kg or above. It is administered weekly for the first four weeks, followed by biweekly intravenous infusions. Side Effects: Common side effects include infusion-related reactions, rash, paronychia, and gastrointestinal symptoms such as nausea and diarrhoea. Infusion reactions tend to occur most frequently with initial doses, necessitating premedication with antihistamines and corticosteroids.	[82]

3.3.3. FOLFIRI and FOLFOX

They are chemotherapy regimens commonly used to treat colorectal cancer. They are named based on the combination of drugs included in each regimen:

1. FOLFIRI:

FOL—Leucovorin (folinic acid);

F—5-Fluorouracil (5-FU);

IRI—Irinotecan

FOLFIRI is often used as a first- or second-line treatment for metastatic colorectal cancer.

2. FOLFOX:

FOL—Leucovorin (folinic acid);

F—5-Fluorouracil (5-FU);

OX—Oxaliplatin.

FOLFOX is frequently used as an adjuvant therapy after surgery or as a first-line treatment for advanced colorectal cancer. Both regimens aim to inhibit cancer cell growth by targeting DNA replication and repair. The choice between FOLFIRI and FOLFOX depends on factors such as the patient's overall health, tumour characteristics, and prior treatments [83].

4. VEGF/VEGFR Inhibition Cancer Therapy

4.1. Targeting Angiogenesis in Cancer: VEGF/VEGFR Pathway

Angiogenesis plays a fundamental role in cancer growth and progression. Tumours exploit this process to establish a blood supply, ensuring the delivery of oxygen and nutrients essential for their expansion and metastasis. The vascular endothelial growth factor (VEGF) signalling pathway, which involves VEGF and its receptor VEGFR, is a key regulator of angiogenesis [84]. The development of monoclonal antibodies (mAbs) and other inhibitors that target this pathway has led to significant advancements in cancer treatment.

4.2. Angiogenesis and the VEGF/VEGFR Pathway

The VEGF family comprises several ligands, including VEGF-A, VEGF-B, VEGF-C, and VEGF-D, which interact with VEGFRs—specifically VEGFR-1, VEGFR-2, and VEGFR-3—on endothelial cells. Among these, VEGF-A is the primary driver of tumour-induced angiogenesis. When VEGF-A binds to VEGFR-2, it triggers downstream signalling cascades such as PI3K/AKT and MAPK, promoting endothelial cell proliferation, migration, and the formation of new blood vessels [85]. Many tumours overexpress VEGF to accelerate angiogenesis, fostering their growth and metastatic potential [86].

Blocking VEGF/VEGFR signalling interferes with tumour vascularisation, leading to the regression of blood vessels and a decrease in the supply of oxygen and nutrients to cancerous tissues [87]. This therapeutic approach, often referred to as a "starvation" strategy, has demonstrated effectiveness in several types of cancer.

4.3. Indications for VEGF/VEGFR Inhibitors

A review explores the clinical effects of combining ramucirumab, a VEGFR-2 inhibitor, with docetaxel in patients with non-small cell lung cancer (NSCLC) who have previously received immunotherapy. The study evaluates the impact of this combination on tumour progression and overall survival. The findings highlight the potential of anti-angiogenic therapy alongside chemotherapy as a second-line treatment for advanced NSCLC, offering valuable insights into its efficacy and safety profile [88]. VEGF/VEGFR inhibitors are used to treat various cancers by targeting angiogenesis. Bevacizumab is approved for colorectal, lung, renal, ovarian, and glioblastoma. Tyrosine kinase inhibitors (TKIs) like sunitinib and axitinib treat renal cell carcinoma (RCC) and hepatocellular carcinoma. Ramucirumab is used for gastric, lung, and hepatocellular cancers. These inhibitors reduce tumour vascularisation, limiting oxygen and nutrient supply, thereby slowing tumour progression. Beyond oncology, they are explored for retinal diseases like age-related macular degeneration. Their efficacy is often enhanced in combination with chemotherapy or immunotherapy [89].

Under normal oxygen levels, HIF-α is hydroxylated by PHD, bound by VHL, and degraded via ubiquitination. In hypoxia, HIF-α stabilises, translocates to the nucleus, and dimerizes with HIF-β, activating genes with hypoxia-responsive elements (HREs) that regulate angiogenesis, survival, metabolism, and proliferation to adapt to low oxygen conditions.

Having described the VEGF/VEGFR inhibition cancer therapy, we now look at emerging strategies including tumour cell surface marker: Trop-2, next generation antibodies, and IgE antibodies against tumours.

5. Emerging Strategies (Trop-2 ADCs, Combinatorial Approaches, Next-Generation Abs)

5.1. Tumour Cell Surface Marker: Trop-2

Trop-2 is a protein found in high amounts on the surface of certain cancer cells, including those in breast, bladder, lung, and stomach cancers. It supports tumour growth by helping cells multiply, spread, and avoid death. This makes it a useful target for treatments that deliver toxic drugs directly to cancer cells. Blocking Trop-2 may slow disease progression, as its presence is linked to more aggressive cancers and poorer outcomes. Research continues to explore its role in therapy development.

5.2. Mechanism of Action: Trop-2 ADCC

Trop-2 is a surface protein highly expressed in various cancers, including breast, lung, and gastrointestinal tumours. It promotes cancer progression by enhancing cell growth, migration, and survival. Due to its role in tumour aggressiveness and poor prognosis, Trop-2 has become a promising target for antibody–drug conjugates and other therapies. Research highlights its potential for improving cancer treatment by selectively targeting malignant cells while sparing healthy tissues, making it a valuable focus for ongoing clinical development. The ADC strategy involves using monoclonal antibodies (mAbs) conjugated to cytotoxic agents. One of the prominent ADCs targeting Trop-2 is sacituzumab govitecan, approved for TNBC and urothelial carcinoma. The ADC binds specifically to Trop-2 on tumour cell surfaces, internalises into the cell, and releases the cytotoxic payload. This mechanism spares normal tissues while maximising tumour cell death, reducing off-target toxicity. SN-38, the active metabolite of irinotecan, serves as the cytotoxic payload in sacituzumab govitecan, where it interferes with DNA replication, leading to tumour cell apoptosis [90–93].

Sacituzumab govitecan (Trodelvy) is an FDA-approved antibody–drug conjugate (ADC) designed to target Trop-2, a protein that is frequently overexpressed in multiple cancers. Regulatory approval has been granted for its use in treating [94]:

- Metastatic triple-negative breast cancer (mTNBC): Approved in 2020 for individuals who have previously undergone at least two treatment regimens for metastatic disease;
- HR-positive/HER2-negative metastatic breast cancer: Approved in 2023 for patients who have received prior endocrine therapy along with at least two systemic treatments;
- Locally advanced or metastatic urothelial carcinoma: Approved in 2021 for patients who have been previously treated with platinum-based chemotherapy and a PD-1/PD-L1 inhibitor.

5.3. Other Trop-2 ADCs in Development

Datopotamab deruxtecan (Dato-DXd) is another Trop-2 ADC that utilises a topoisomerase inhibitor as its cytotoxic component, delivered through a tumour-selective, stable tetra-peptide-based cleavable linker. Datopotamab deruxtecan (brand name Datroway) was granted FDA approval on 17 January 2025, for adult patients with unresectable or metastatic hormone receptor (HR)-positive, HER2-negative breast cancer who have previously undergone endocrine therapy and chemotherapy for advanced disease [95,96].

These agents demonstrate promise across various Trop-2 overexpressing malignancies in early-phase trials. Despite advancements in Trop-2-directed ADC treatments, significant hurdles remain. These include fluctuations in Trop-2 expression, resistance development, ADC stability concerns, and the absence of reliable biomarkers to predict therapeutic success. Future studies should prioritise refining biomarker assays to accurately identify patients who will respond best to Trop-2-targeted ADCs. This could involve integrating methods like immunohistochemistry, flow cytometry, and molecular diagnostics. Establishing a clear link between Trop-2 expression levels and treatment response may aid in selecting suitable patients and adjusting dosages where necessary. Improving antibody specificity is crucial to targeting Trop-2 on cancer cells while reducing activity in healthy tissues, thereby minimising unintended toxicity. Predicting toxicity risks from combination therapies can be enhanced through gene editing techniques and 3D tumour culture models.

Additionally, advanced data analytics, including machine learning, can help anticipate potential side effects based on patient-specific information. However, as with most cytotoxic treatments, the long-term effectiveness of monotherapy remains limited due to resistance development [97].

5.4. Next Generation Antibodies

With advancements in next-generation antibody technologies, the development of highly targeted therapies for tumours has significantly evolved. Innovations such as single-chain variable fragments, nanobodies, bispecific antibodies, Fc-engineered variants, antibody mimetics, and antibody–drug conjugates have enhanced precision in targeting tumour-specific antigens while minimising off-target effects [98].

These strategies improve therapeutic efficacy, reduce resistance, and enable combination approaches with immune checkpoint inhibitors and other modalities. The emergence of antibody biosimilars has also expanded accessibility, ensuring broader clinical applications in oncology and improving patient outcomes in various malignancies [98,99]. Monoclonal antibody development techniques include hybridoma technology, B cell immortalisation, bacterial display, yeast display, phage display, ribosome display, mammalian cell display, DNA-encoded antibodies, RNA-encoded antibodies, transgenic animal technology, single B cell technology, and artificial intelligence/machine learning [98].

Next-generation antibody technologies are transforming tumour defence, yet several emerging advancements remain underdiscussed. One promising approach is computationally designed de novo antibodies, which use AI-driven algorithms to create highly specific antibodies against tumour-associated antigens, even those with minimal immune visibility. This enables the rapid development of personalised therapies targeting unique cancer profiles. AI-driven cancer vaccines enhance precision by optimising epitope design, mRNA/DNA instructions, and patient-specific responses. Despite challenges like tumour heterogeneity and ethical concerns, AI integration promises personalised immunotherapies, advancing cancer treatment through interdisciplinary collaboration and innovation [100].

Synthetic antigen receptors (SARs) are an alternative to chimeric antigen receptors (CARs) and offer greater precision in tumour recognition, reducing off-target toxicity. Additionally, self-assembling antibody scaffolds allow for multi-target engagement, improving efficacy against tumours with heterogeneous antigen expression. This is particularly relevant in solid tumours, where immune evasion is a major challenge [101,102]. Glycoengineered antibodies are also gaining attention, as modified glycosylation patterns enhance immune activation, leading to stronger antibody-dependent cellular cytotoxicity against tumour cells [103].

5.5. IgE Antibodies Against Tumours

IgE immunotherapy is an emerging approach in cancer treatment that utilises FcεRI-expressing immune cells, such as mast cells, basophils, and eosinophils, to enhance immune surveillance and antibody-dependent cellular cytotoxicity (ADCC). Unlike traditional IgG-based therapies, IgE can bypass immune resistance and activate potent anti-tumour responses. A key example is MOv18 IgE, targeting folate receptor-α in ovarian cancer, which has entered clinical trials. Preclinical studies further suggest that IgE-based therapies promote long-term anti-tumour immunity. Given its tumour specificity and distinct mechanisms of action, IgE immunotherapy presents a novel and complementary strategy in oncology, warranting further research into its clinical applications. Several IgE-based therapies (e.g., MOv18 IgE targeting folate receptor-α in ovarian cancer) are currently under clinical investigation [104].

FRα is a promising target for cancer therapy due to its high tumour specificity and stability post-chemotherapy. MOv18 IgE and IMGN853 have shown early potential, though robust clinical data are lacking. Farletuzumab's Phase III trials had mixed results, while FRα-specific CAR-T cells offer a novel approach, pending resolution of efficacy and toxicity challenges [105].

The study examines the potential of IgE antibodies, specifically MOv18 IgE, as a more effective cancer therapy compared to traditional IgG antibodies. IgE antibodies exhibit a significantly higher affinity for Fc receptors on immune cells such as natural killer (NK) cells and macrophages, which enhances their ability to mediate immune responses against cancer cells. Preclinical studies have shown that IgE can induce superior anti-tumour activity compared to IgG in various animal models, suggesting that IgE's unique binding properties may translate to improved clinical efficacy. To investigate the safety and efficacy of MOv18 IgE, a chimeric monoclonal IgE targeting the human folate receptor-alpha (FRα) was subjected to a Phase I clinical trial. Patients with solid tumours expressing FRα were administered escalating doses of MOv18 IgE, following skin prick tests to mitigate potential allergic reactions. The trial involved 24 patients, primarily with ovarian or endometrial cancers, who received up to six weekly infusions. The trial reported dose-limiting toxicities, including instances of urticaria and anaphylaxis at specific doses. The safety profile was deemed tolerable, with cutaneous reactions being the most common adverse events. Despite these challenges, there was evidence of anti-tumour activity among patients, with some exhibiting stable disease or responses. In summary, the trial highlights the promise of IgE therapy in cancer treatment, particularly emphasising the need for careful monitoring and patient selection to ensure safety while exploring the potential benefits of this novel treatment approach. Further studies are required to ascertain the full clinical efficacy and identify the maximum tolerated dose of MOv18 IgE [106]. IgE promotes anti-tumour immunity by activating Fc receptors on immune cells, inducing ADCC, ADCP, and degranulation, which release toxic mediators. It enhances antigen presentation and may directly inhibit cancer cell growth by blocking receptor signalling, akin to IgG antibodies (Figure 4) [107].

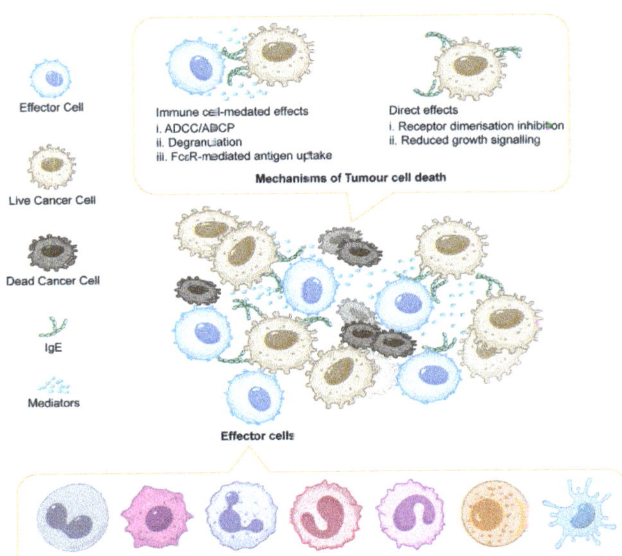

Figure 4. IgE mediated mechanisms of tumour cell death. Adapted from open access journal referenced Sutton, B.J. et al. *Antibodies* **2019**, *8*, 19 [107]. Adaptation made by MDPI author services. ID: English-91110.

We explore chimeric antibodies in CAR-T cells, which enhance tumour targeting by combining the specificity of monoclonal antibodies with T-cell activation.

6. Integration of Monoclonal Antibodies with Chimeric Antigen Receptor (CAR) T-Cell Therapy

The development of monoclonal antibodies (mAbs) has significantly transformed cancer therapy by providing precise targeting mechanisms for various malignancies. When combined with chimeric antigen receptor (CAR) T-cell therapy, mAbs enhance immunotherapeutic effectiveness, offering tailored approaches to both haematological and solid tumour treatment [108].

Monoclonal antibodies function by specifically recognising and binding to antigens present on cancer cell surfaces. This targeted interaction either inhibits tumour growth directly or facilitates immune-mediated destruction. In CAR T-cell therapy, T cells are genetically engineered to express receptors that identify particular antigens, thereby increasing their capacity to recognise and eliminate malignant cells. This synergistic approach ultimately leads to better clinical outcomes [108,109].

For haematological malignancies such as acute lymphoblastic leukaemia (ALL) and certain lymphomas, CAR T-cell therapy has yielded exceptional results. CD19, a frequently targeted antigen, has been pivotal in these advancements [110,111]. CAR T-cell therapies like Kymriah and Yescarta, which specifically target CD19-positive cells, have demonstrated high remission rates in patients with refractory diseases, establishing them as standard treatments for select haematological cancers [112,113].

Additionally, the use of mAbs can further strengthen CAR T-cell therapy. For example, combining CAR T-cells with mAbs that inhibit co-inhibitory molecules, such as PD-1, can enhance T-cell activity and improve antitumour responses [114]. This strategy is particularly valuable in overcoming the immunosuppressive effects of the tumour microenvironment, which is a major challenge in treating solid tumours.

Solid tumour treatment with CAR T-cell therapy is complicated due to tumour heterogeneity and an immunosuppressive microenvironment that hinders T-cell efficacy [115]. While conventional mAbs have faced limitations in these settings due to antigen variability and tumour aggressiveness [116], integrating CAR technology with mAbs presents a novel strategy to improve treatment outcomes.

For solid tumours, targetable antigens include overexpressed proteins unique to specific cancer types. CAR T-cells can be designed to recognise markers such as HER2 or tumour-associated antigens (TAAs) like carcinoembryonic antigen (CEA) [117]. By leveraging mAbs in conjunction with CAR T-cells, enhanced T-cell activation can be achieved, offering a dual-targeting approach against tumours [118].

Another promising area of research involves bispecific antibodies, which can bind simultaneously to both tumour cells and T cells, promoting T-cell activation and recruitment to tumour sites [119]. These bispecific constructs, which target both the T-cell receptor and tumour-specific antigens, hold the potential for improving CAR T-cell efficacy, particularly in solid tumours.

However, challenges persist in incorporating mAbs into CAR T-cell therapies. Identifying tumour-specific antigens that ensure selective targeting without affecting normal tissues remains critical [120]. Additionally, the immunosuppressive tumour microenvironment necessitates further strategies to sustain T-cell function and improve therapeutic efficacy.

Ongoing research continues to refine these approaches. Investigators are exploring combination therapies that incorporate mAbs with checkpoint inhibitors, cytokines, and other immunotherapeutic agents. Efforts to enhance CAR construct designs and prolong T-cell persistence through genetic modifications are also being pursued [121].

In conclusion, integrating monoclonal antibodies with CAR T-cell therapy represents a promising advancement in cancer immunotherapy, offering solutions for both haematological and solid tumours. As research progresses, these therapies may become even more effective, leading to improved patient survival and treatment outcomes. Future clinical trials and ongoing studies will be crucial in determining the optimal approaches for harnessing the full potential of these immunotherapeutic strategies [122], as illustrated in Figure 5.

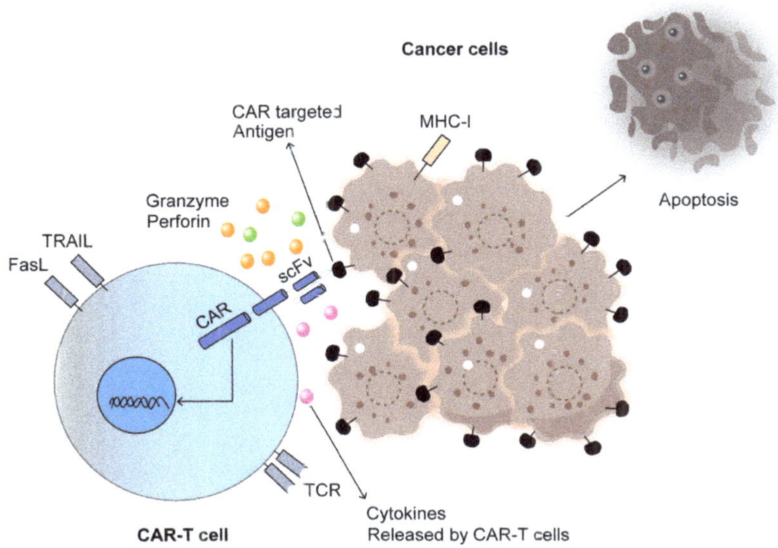

Figure 5. CAR-T cell immunotherapy for cancer. Adapted from Open access journal referenced by Majumder, A. *Cancers* **2023**, *15*, 39 [122]. Adaptation performed by MDPI author service. ID: English-90959.

We summarise below some issues in cancer antibody immunotherapy including recent FDA approval of next-generation antibodies for cancer and we finalise with a subtopic that has brought attention in oncology: IgA and cancer.

7. Other Issues in Cancer Immunotherapy

7.1. Recent FDA Approval of Next-Generation Antibodies for Cancer

Next-generation antibodies for cancer, including bispecific and antibody–drug conjugates, have received FDA approvals, enhancing targeted therapies. These advancements improve efficacy, reduce toxicity, and offer promising treatment options for various malignancies. Recent FDA approvals of next-generation antibodies, such as immune checkpoint inhibitors and bispecific antibodies, revolutionise cancer treatment. These therapies enhance immune response, target tumour cells precisely, and improve patient overall survival with reduced side effects. These innovations not only target cancer cells with higher precision but also help overcome resistance mechanisms. FDA-approved therapies continue to evolve, providing new hope for patients with previously hard-to-treat cancers.

"Advanced, unresectable, or metastatic non–small cell lung cancer (NSCLC) harbouring a neuregulin 1 (NRG1) gene fusion with disease progression on or after prior systemic therapy. Advanced, unresectable, or metastatic pancreatic adenocarcinoma harbouring an NRG1 gene fusion with disease progression on or after prior systemic therapy”. Zenocutuzumab-zbco (Bizengri) [123] was approved by the FDA in 2024 against these tumours.

“Relapsed or refractory multiple myeloma (RRMM) in adult patients who have received at least four prior lines of therapy, including a proteasome inhibitor, an immunomodulatory agent, and an anti-CD38 monoclonal antibody”. Regarding elranatamab-bcmm (Elrexfio) [124], FDA approved in 2023 against this tumour. “Relapsed or refractory multiple myeloma (RRMM) in adult patients who have received at least four prior lines of therapy”. Talquetamab-tgvs (Talvey) [125] was approved by the FDA in 2023 against this tumour.

“Relapsed or refractory diffuse large B-cell lymphoma (DLBCL) not otherwise specified or large B-cell lymphoma (LBCL) arising from follicular lymphoma, after two or more lines of systemic therapy”. Glofitamab-gxbm (Columvi) [126] was approved by the FDA in 2023 for the treatment of this tumour.

“Relapsed or refractory diffuse large B-cell lymphoma (DLBCL) after two or more lines of systemic treatment, including DLBCL caused by indolent lymphoma and advanced B-cell lymphoma after two or more systemic treatments”. Epcoritamab-bysp (Epkinly) [127] was approved by the FDA in 2023 against this tumour.Yes

“Folate receptor alpha (FRα)-positive, platinum-resistant epithelial ovarian, fallopian tube, or primary peritoneal cancer who have received one to three prior treatments”. Mirvetuximab soravtansine (Elahere) [128] was approved by the FDA in 2022 against this tumour.

7.2. Exploring IgA Antibodies in Cancer Therapy: A Focus on Mucosal Surface Tumour

IgA antibodies are essential for immune defence at mucosal surfaces, including the respiratory, gastrointestinal, and urogenital tracts. They are the first line of defence against pathogens, preventing infections at these sites. Recently, IgA antibodies have gained attention in cancer immunotherapy, particularly for targeting mucosal cancers, such as those of the colon, oesophagus, and head and neck. These cancers often arise in areas rich in mucosal tissues, making IgA antibodies a promising tool for localised therapy. The potential of IgA antibodies in cancer therapy lies in their ability to recognise tumour-associated antigens present on mucosal surfaces. When these antibodies bind to such antigens, they can trigger immune responses at the tumour site, activating mechanisms such as complement activation and antibody-dependent cellular cytotoxicity (ADCC). These processes lead to the destruction of tumour cells while sparing normal, healthy tissue. This targeted approach reduces the likelihood of systemic toxicity, a common issue with many conventional cancer treatments. Researchers are currently focusing on engineering IgA antibodies to enhance their therapeutic efficacy. Advances include increasing their affinity for cancer cell antigens, improving their stability in circulation, and optimising their ability to induce immune responses specifically at mucosal sites. This targeted strategy could make IgA-based therapies particularly effective for cancers that are difficult to treat with systemic therapies alone. In addition, IgA antibodies have the advantage of being able to provide both local and systemic protection against cancer, which could result in better patient outcomes. As research in this area continues, IgA antibodies hold significant promise for improving cancer treatment, particularly in patients with mucosal cancers, offering a more specific and less toxic alternative to current therapies. The connection to immunoglobulin A (IgA) in this context could refer to the role of IgA in the mucosal immune system, particularly in the lungs. IgA is involved in the first line of defence against respiratory pathogens, including bacteria like *Chlamydia pneumoniae*. The infection may influence the immune response in lung cancer patients, potentially affecting the tumour microenvironment, immune evasion, or the body's ability to fight cancer. Investigating this connection might offer insights into how infections like *Chlamydia pneumoniae* modulate cancer progression through immune pathways, particularly involving IgA. Another connection between immunoglobulin A (IgA) and cancer in this context might relate to IgA's role in the mucosal immune system, which is particularly important in cancers like cervical cancer, where the tumour is often exposed to external pathogens. IgA is a key antibody in defending mucosal surfaces (such as the cervix) from infections, and its levels may change in response to immune checkpoint blockade therapy. Since ICIs can alter immune responses, this study could provide valuable insights into how IgA and other immune markers in peripheral blood might influence treatment responses, potentially serving as biomarkers for prognosis or treatment efficacy in cervical cancer patients. IgA plays a crucial role in the immunity of the respiratory and gastrointestinal tracts, where EBV is often initially encountered. The study's focus on anti-EBV antibody responses could explore how sex differences in IgA levels or activity contribute to the immune system's ability to control or eliminate EBV, a well-known virus directly involved in the immunopathogenesis of several cancers, including those of the head and neck. A stronger or weaker immune response could influence susceptibility to EBV-related cancers or affect the progression of existing malignancies. It could also inform how immune checkpoint blockade therapies or other immunotherapies might vary by sex in terms of efficacy, particularly in cancers associated with EBV [129–132].

7.3. Engineered IgA for Cancer Immunotherapy: Targeting Tumours in Mucosal Tissues

Secretory immunoglobulin A (IgA), the predominant antibody in mucosal secretions, can be engineered for cancer immunotherapy to target tumours in mucous membranes. Unlike IgG-based therapies, IgA naturally interacts with the polymeric immunoglobulin receptor (pIgR) and the Fc alpha receptor (FcαRI), facilitating efficient immune activation in mucosal tissues. Engineered IgA antibodies can enhance tumour cell destruction by promoting neutrophil-mediated antibody-dependent cellular cytotoxicity (ADCC) and phagocytosis. Additionally, IgA can be designed to improve stability, reduce degradation in mucosal environments, and enhance tumour specificity. Combining IgA with other immune modulators, such as checkpoint inhibitors, may further strengthen anti-tumour responses. Research suggests that IgA-based therapies could be particularly effective against mucosal malignancies, including colorectal, lung, and cervical cancers. Figure 6 shows the use of secretory IgA against tumour cells. While challenges such as manufacturing and immune regulation remain, IgA engineering offers a promising avenue for mucosal cancer immunotherapy, harnessing the body's natural defence mechanisms in these specialised environments. IgA antibodies hold promise as anti-cancer therapeutics, but their clinical translation faces challenges due to limitations in half-life and production. Engineering strategies have improved IgA developability, achieving production levels comparable to IgG and enhancing glycan homogeneity. However, extending IgA's serum half-life remains a challenge, with FcRn-binding sites and IgG fusion strategies showing potential. Neutrophil-engaging bispecific molecules offer promise but require further refinement. While a long half-life is beneficial, IgA's six-day duration provides a potent, short-lived impact on tumours, positioning it between prolonged IgG exposure and small molecule therapies, expanding its potential for targeted cancer treatment [133].

The alternative pathway of the complement system is continuously active at a low level (tick-over mechanism) and can be amplified if C3b is deposited on tumour cell surfaces. However, tumour cells often express complement regulatory proteins (e.g., CD55, CD46, CD59) that prevent complement-mediated attack, reducing the effectiveness of this pathway. If these regulatory mechanisms are overcome (e.g., via therapeutic interventions or immune dysregulation), the alternative pathway can promote complement-dependent cytotoxicity (CDC) against tumour cells (Figure 6).

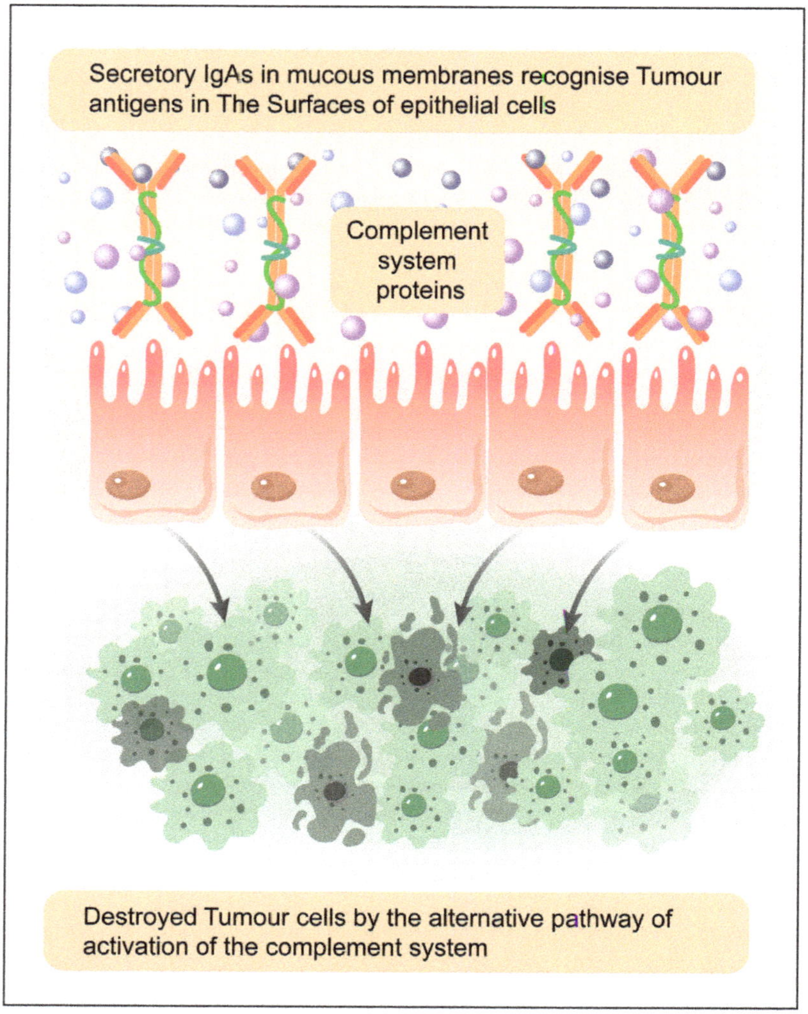

Figure 6. Complement activation via the alternative and lectin pathways can occur against tumour cells in mucosal secretions, but its effectiveness depends on several factors, including tumour cell surface properties, immune evasion mechanisms, the presence of secretory IgA (sIgA), and epithelial interactions. Elaborated by authors with the help of MDPI author services. MDPI author service ID: English-90959.

Tumour cells often exhibit altered glycosylation patterns, which can be recognised by mannose-binding lectin (MBL) or ficolins, leading to complement activation. The MBL-associated serine proteases (MASPs) then cleave C4 and C2, forming the C3 convertase, which can initiate downstream complement activation and contribute to tumour cell lysis or immune modulation. sIgA is not a strong activator of complement, but it can interact with epithelial cells, tumour-associated antigens, and innate immune receptors, indirectly influencing tumour cell recognition. Epithelial cells in mucosal tissues may contribute to tumour immunity by modulating complement activity and immune responses in the tumour microenvironment [134–136].

The potential outcomes of complement activation against tumours:

1. Direct Tumour Cell Lysis: If complement activation overcomes regulatory mechanisms, membrane attack complex (MAC) formation can lead to tumour cell death;
2. Opsonization and Phagocytosis: C3b deposition enhances recognition by immune cells, facilitating phagocytosis and antigen presentation;
3. Pro-Inflammatory Responses: Complement components (e.g., C5a, C3a) can recruit immune cells to the tumour microenvironment, promoting an anti-tumour response;
4. Immune Evasion: Many tumour cells resist complement attack by overexpressing complement inhibitors, limiting complement-mediated destruction.

8. Conclusions

Monoclonal antibodies (mAbs) have significantly transformed cancer therapy by providing highly specific, targeted treatment. These laboratory-engineered molecules identify and bind to antigens on cancer cells, leading to tumour inhibition through direct cytotoxicity, immune activation, and disruption of cancer-promoting pathways.

A key immune-mediated mechanism of mAbs is antibody-dependent cellular cytotoxicity, where natural killer cells bind to the Fc region of antibody-coated cancer cells, triggering apoptosis. This approach is central to therapies like rituximab for B-cell malignancies and trastuzumab for HER2-positive breast cancer. Additionally, checkpoint inhibitors such as pembrolizumab and nivolumab restore T-cell function by blocking immune checkpoint proteins, enhancing the immune response against tumours. Another category of antibody–drug conjugates like ado-trastuzumab emtansine delivers cytotoxic agents directly to cancer cells while minimising damage to healthy tissue.

Despite their success, monoclonal antibodies face challenges, including resistance due to antigen loss, tumour microenvironment modifications, or activation of alternative survival pathways. Tumour evasion strategies, such as immune checkpoint modulation and recruitment of immunosuppressive cells, also limit efficacy. Furthermore, high costs, complex production, and the need for specialised administration reduce accessibility, particularly in resource-limited settings. Immune-related adverse effects, especially with checkpoint inhibitors, can also pose risks.

To overcome these limitations, research is advancing novel strategies. Bispecific antibodies, targeting two antigens simultaneously, enhance tumour targeting and immune activation by directly linking immune cells to cancer cells. Next-generation antibody–drug conjugates with improved linker stability and potent cytotoxic payloads ensure precise tumour targeting while reducing off-target effects. Engineering Fc regions to enhance antibody-dependent cellular cytotoxicity is another approach to improve immune-mediated tumour clearance.

Combination therapies, integrating monoclonal antibodies with checkpoint inhibitors, CAR T-cell therapy, or small-molecule inhibitors, are showing promising results in improving efficacy and reducing resistance. Advances in personalised medicine enable better patient selection and tailored treatments, maximising therapeutic benefits.

Additionally, multispecific antibodies, targeting multiple pathways, are emerging as an effective approach, particularly in cancers with complex immune evasion mechanisms. Enhancing complement activity by blocking tumour-associated complement inhibitors may further strengthen immune responses, particularly in mucosal cancers. As research progresses, the integration of these approaches is expected to drive the next phase of advanced cancer therapies, offering more personalised, durable, and effective treatment options.

Author Contributions: The manuscript was conceptualised by A.J.-V. and P.E.A. Methodology and software, B.R.P., S.V. and C.U. Validation, A.J.-V. Investigation, All authors; writing—original draft preparation, A.J.-V. and P.E.A.; writing—review and editing, P.E.A. All authors have read and agreed to the published version of the manuscript.

Funding: This study did not receive any external funding.

Institutional Review Board Statement: Not applicable.

Informed Consent Statement: Not applicable.

Data Availability Statement: The dataset supporting the findings of this study is included within the manuscript and its referenced sources, ensuring comprehensive access to the relevant data for further examination and analysis.

Conflicts of Interest: The authors declare no conflicts of interest.

References

1. Ghemrawi, R.; Abuamer, L.; Kremesh, S.; Hussien, G.; Ahmed, R.; Mousa, W.; Khair, M. Revolutionizing Cancer Treatment: Recent Advances in Immunotherapy. *Biomedicines* **2024**, *12*, 2158. [CrossRef] [PubMed]
2. Raghani, N.R.; Chorawala, M.R.; Mahadik, M.; Patel, R.B.; Prajapati, B.G.; Parekh, P.S. Revolutionizing Cancer Treatment: Comprehensive Insights into Immunotherapeutic Strategies. *Med. Oncol.* **2024**, *41*, 51. [CrossRef] [PubMed]
3. Chen, Z.; Kankala, R.K.; Yang, Z.; Li, W.; Xie, S.; Li, H.; Chen, A.Z.; Zou, L. Antibody-based drug delivery systems for cancer therapy: Mechanisms, challenges, and prospects. *Theranostics* **2022**, *12*, 3719. [CrossRef] [PubMed]
4. Su, X.; Li, J.; Xu, X.; Ye, Y.; Wang, C.; Pang, G.; Liu, W.; Liu, A.; Zhao, C.; Hao, X. Strategies to enhance the therapeutic efficacy of anti-PD-1 antibody, anti-PD-L1 antibody and anti-CTLA-4 antibody in cancer therapy. *J. Transl. Med.* **2024**, *22*, 751. [CrossRef]
5. Jia, G.; Jiang, Y.; Li, X. Targeted Drug Conjugates in Cancer Therapy: Challenges and Opportunities. *Pharm. Sci. Adv.* **2024**, *2*, 100048. [CrossRef]
6. Ponziani, S.; Di Vittorio, G.; Pitari, G.; Cimini, A.M.; Ardini, M.; Gentile, R.; Giansanti, F. Antibody-Drug Conjugates: The New Frontier of Chemotherapy. *Int. J. Mol. Sci.* **2020**, *21*, 5510. [CrossRef]
7. Lan, H.R.; Chen, M.; Yao, S.Y.; Chen, J.X.; Jin, K.T. Bispecific Antibodies Revolutionizing Breast Cancer Treatment: A Comprehensive Overview. *Front. Immunol.* **2023**, *14*, 1266450. [CrossRef]
8. Caers, J.; Duray, E.; Vrancken, L.; Marcion, G.; Bocuzzi, V.; De Veirman, K.; D'Huyvetter, M. Radiotheranostic Agents in Hematological Malignancies. *Front. Immunol.* **2022**, *13*, 911080. [CrossRef]
9. Fourie Zirkelbach, J.; Shah, M.; Vallejo, J.; Cheng, J.; Ayyoub, A.; Liu, J.; Hudson, R.; Sridhara, R.; Ison, G.; Amiri-Kordestani, L.; et al. Improving dose-optimization processes used in oncology drug development to minimize toxicity and maximize benefit to patients. *J. Clin. Oncol.* **2022**, *40*, 3489–3500. [CrossRef]
10. Mohite, P.; Yadav, V.; Pandhare, R.; Maitra, S.; Saleh, F.M.; Saleem, R.M.; Al-Malky, H.S.; Kumarasamy, V.; Subramaniyan, V.; Abdel-Daim, M.M.; et al. Revolutionizing Cancer Treatment: Unleashing the Power of Viral Vaccines, Monoclonal Antibodies, and Proteolysis-Targeting Chimeras in the New Era of Immunotherapy. *ACS Omega* **2024**, *9*, 7277–7295. [CrossRef]
11. Carbone, F.; Russo, C.; Colamatteo, A.; La Rocca, C.; Fusco, C.; Matarese, A.; Procaccini, C.; Matarese, G. Cellular and molecular signaling towards T cell immunological self-tolerance. *J. Biol. Chem.* **2024**, *300*, 107134. [CrossRef] [PubMed]
12. Jhunjhunwala, S.; Hammer, C.; Delamarre, L. Antigen presentation in cancer: Insights into tumour immunogenicity and immune evasion. *Nat. Rev. Cancer* **2021**, *21*, 298–312. [CrossRef] [PubMed]
13. Zhang, H.; Dai, Z.; Wu, W.; Wang, Z.; Zhang, N.; Zhang, L.; Zeng, W.J.; Liu, Z.; Cheng, Q. Regulatory mechanisms of immune checkpoints PD-L1 and CTLA-4 in cancer. *J. Exp. Clin. Cancer Res.* **2021**, *40*, 184. [CrossRef] [PubMed]
14. Verma, N.K.; Wong, B.H.; Poh, Z.S.; Udayakumar, A.; Verma, R.; Goh, R.K.; Duggan, S.P.; Shelat, V.G.; Chandy, K.G.; Grigoropoulos, N.F. Obstacles for T-lymphocytes in the tumour microenvironment: Therapeutic challenges, advances and opportunities beyond immune checkpoint. *eBioMedicine* **2022**, *83*, 104216. [CrossRef]
15. Chang, C.Y.; Park, H.; Malone, D.C.; Wang, C.Y.; Wilson, D.L.; Yeh, Y.M.; Lo-Ciganic, W.H. Immune Checkpoint Inhibitors and Immune-Related Adverse Events in Patients with Advanced Melanoma: A Systematic Review and Network Meta-Analysis. *JAMA Netw. Open* **2020**, *3*, e201611. [CrossRef]
16. Wang, D.D.; Shaver, L.G.; Shi, F.Y.; Wei, J.J.; Qin, T.Z.; Wang, S.Z.; Kong, Y.J. Comparative efficacy and safety of PD-1/PD-L1 immunotherapies for non-small cell lung cancer: A network meta-analysis. *Eur. Rev. Med. Pharmacol. Sci.* **2021**, *25*, 2866–2884.
17. Chen, L.; Mo, D.C.; Hu, M.; Zhao, S.J.; Yang, Q.W.; Huang, Z.L. PD-1/PD-L1 inhibitor monotherapy in recurrent or metastatic squamous cell carcinoma of the head and neck: A meta-analysis. *Am. J. Otolaryngol.* **2022**, *43*, 103324. [CrossRef]
18. Galicia-Carmona, T.; Arango-Bravo, E.A.; Coronel-Martínez, J.A.; Cetina-Perez, L.; Vanoye-Carlo, E.G.; Villalobos-Valencia, R.; García-Pacheco, J.A.; Cortés-Esteban, P. Advanced, recurrent, and persistent cervical cancer management: In the era of immunotherapy. *Front. Oncol.* **2024**, *14*, 1392639. [CrossRef]

19. Martini, A.; Raggi, D.; Fallara, G.; Nocera, L.; Schultz, J.G.; Belladelli, F.; Marandino, L.; Salonia, A.; Briganti, A.; Montorsi, F.; et al. Immunotherapy versus chemotherapy as first-line treatment for advanced urothelial cancer: A systematic review and meta-analysis. *Cancer Treat. Rev.* **2022**, *104*, 102360. [CrossRef]
20. Lin, Q.; Wang, X.; Hu, Y. The opportunities and challenges in immunotherapy: Insights from the regulation of PD-L1 in cancer cells. *Cancer Lett.* **2023**, *569*, 216318. [CrossRef]
21. Skafi, N.; Fayyad-Kazan, M.; Badran, B. Immunomodulatory Role for MicroRNAs: Regulation of PD-1/PD-L1 and CTLA-4 Immune Checkpoints Expression. *Gene* **2020**, *754*, 144888. [CrossRef]
22. Garg, P.; Malhotra, J.; Kulkarni, P.; Horne, D.; Salgia, R.; Singhal, S.S. Emerging Therapeutic Strategies to Overcome Drug Resistance in Cancer Cells. *Cancers* **2024**, *16*, 2478. [CrossRef] [PubMed]
23. Nishimura, C.D.; Pulanco, M.C.; Cui, W.; Lu, L.; Zang, X. PD-L1 and B7-1 cis-interaction: New mechanisms in immune checkpoints and immunotherapies. *Trends Mol. Med.* **2021**, *27*, 207–219. [CrossRef]
24. Korman, A.J.; Garrett-Thomson, S.C.; Lonberg, N. The Foundations of Immune Checkpoint Blockade and the Ipilimumab Approval Decennial. *Nat. Rev. Drug Discov.* **2022**, *21*, 509–528. [CrossRef] [PubMed]
25. Pandey, P.; Khan, F.; Qari, H.A.; Upadhyay, T.K.; Alkhateeb, A.F.; Oves, M. Revolutionization in Cancer Therapeutics via Targeting Major Immune Checkpoints PD-1, PD-L1 and CTLA-4. *Pharmaceuticals* **2022**, *15*, 335. [CrossRef] [PubMed]
26. Budczies, J.; Kazdal, D.; Menzel, M.; Beck, S.; Kluck, K.; Altbürger, C.; Schwab, C.; Allgäuer, M.; Ahadova, A.; Kloor, M.; et al. Tumour mutational burden: Clinical utility, challenges and emerging improvements. *Nat. Rev. Clin. Oncol.* **2024**, *21*, 725–742. [CrossRef]
27. Sacchi de Camargo Correia, G.; Zhao, Y.; Manochakian, R.; Lou, Y. The Role of Immunotherapy Sensitizers and Novel Immunotherapy Modalities in the Treatment of Cancer. *Front. Oncol.* **2024**, *14*, 1336546. [CrossRef]
28. He, P.; Ma, L.; Xu, B.; Wang, Y.; Li, X.; Chen, H.; Li, Y. Research Progress and Future Directions of Immune Checkpoint Inhibitor Combination Therapy in Advanced Gastric Cancer. *Ther. Adv. Med. Oncol.* **2024**, *16*, 17588359241266156. [CrossRef]
29. Cai, L.; Li, Y.; Tan, J.; Xu, L.; Li, Y. Targeting LAG-3, TIM-3, and TIGIT for cancer immunotherapy. *J. Hematol. Oncol.* **2023**, *16*, 101. [CrossRef]
30. Kong, X.; Zhang, J.; Chen, S.; Wang, X.; Xi, Q.; Shen, H.; Zhang, R. Immune checkpoint inhibitors: Breakthroughs in cancer treatment. *Cancer Biol. Med.* **2024**, *21*, 451–472. [CrossRef]
31. Shukla, N.; Hanna, N. Neoadjuvant and Adjuvant Immunotherapy in Early-Stage Non-Small Cell Lung Cancer. *Lung Cancer Targets Ther.* **2021**, *12*, 51–60. [CrossRef] [PubMed]
32. Jin, W.; Luo, Q. When artificial intelligence meets PD-1/PD-L1 inhibitors: Population screening, response prediction and efficacy evaluation. *Comput. Biol. Med.* **2022**, *145*, 105499. [CrossRef]
33. Gupta, R.; Srivastava, D.; Sahu, M.; Tiwari, S.; Ambasta, R.K.; Kumar, P. Artificial Intelligence to Deep Learning: Machine Intelligence Approach for Drug Discovery. *Mol. Divers.* **2021**, *25*, 1315–1360. [CrossRef]
34. Wang, Y.; Tong, Z.; Zhang, W.; Zhang, W.; Buzdin, A.; Mu, X.; Li, X. FDA-Approved and Emerging Next-Generation Predictive Biomarkers for Immune Checkpoint Inhibitors in Cancer Patients. *Front. Oncol.* **2021**, *11*, 683419. [CrossRef]
35. Marei, H.E.; Hasan, A.; Pozzoli, G.; Cenciarelli, C. Cancer Immunotherapy with Immune Checkpoint Inhibitors (ICIs): Potential, Mechanisms of Resistance, and Strategies for Reinvigorating T Cell Responsiveness When Resistance Is Acquired. *Cancer Cell Int.* **2023**, *23*, 64. [CrossRef] [PubMed]
36. Kjeldsen, J.W.; Lorentzen, C.L.; Martinenaite, E.; Ellebaek, E.; Donia, M.; Holmstroem, R.B.; Klausen, T.W.; Madsen, C.O.; Ahmed, S.M.; Weis-Banke, S.E.; et al. A phase 1/2 trial of an immune-modulatory vaccine against IDO/PD-L1 in combination with nivolumab in metastatic melanoma. *Nat. Med.* **2021**, *27*, 2212–2223. [CrossRef]
37. Yi, M.; Zheng, X.; Niu, M.; Zhu, S.; Ge, H.; Wu, K. Combination strategies with PD-1/PD-L1 blockade: Current advances and future directions. *Mol. Cancer* **2022**, *21*, 28. [CrossRef] [PubMed]
38. Marcus, L.; Fashoyin-Aje, L.A.; Donoghue, M.; Yuan, M.; Rodriguez, L.; Gallagher, P.S.; Philip, R.; Ghosh, S.; Theoret, M.R.; Beaver, J.A.; et al. FDA approval summary: Pembrolizumab for the treatment of tumour mutational burden–high solid tumours. *Clin. Cancer Res.* **2021**, *27*, 4685–4689. [CrossRef]
39. Liu, S.V.; Reck, M.; Mansfield, A.S.; Mok, T.; Scherpereel, A.; Reinmuth, N.; Garassino, M.C.; De Castro Carpeno, J.; Califano, R.; Nishio, M.; et al. Updated overall survival and PD-L1 subgroup analysis of patients with extensive-stage small-cell lung cancer treated with atezolizumab, carboplatin, and etoposide (IMpower133). *J. Clin. Oncol.* **2021**, *39*, 619–630. [CrossRef]
40. Catalano, M.; Shabani, S.; Venturini, J.; Ottanelli, C.; Voltolini, L.; Roviello, G. Lung cancer immunotherapy: Beyond common immune checkpoints inhibitors. *Cancers* **2022**, *14*, 6145. [CrossRef]
41. Hu, P.; Dai, H.I.; Bourdage, J.; Zhou, D.; Trang, K.; Kowalski, K.; Bello, C.; Hibma, J.; Khandelwal, A.; Cowan, K.; et al. Immunogenicity of avelumab in patients with metastatic Merkel cell carcinoma or advanced urothelial carcinoma. *Clin. Transl. Sci.* **2024**, *17*, e13730. [CrossRef] [PubMed]
42. Gui, T.; Ma, L. The Evaluation of Effects, Mechanisms, and Clinical Data of Avelumab. *Highlights Sci. Eng. Technol.* **2023**, *36*, 23–28. [CrossRef]

43. Adams, S.; Othus, M.; Patel, S.P.; Miller, K.D.; Chugh, R.; Schuetze, S.M.; Chamberlin, M.D.; Haley, B.J.; Storniolo, A.M.; Reddy, M.P.; et al. A multicenter phase II trial of ipilimumab and nivolumab in unresectable or metastatic metaplastic breast cancer: Cohort 36 of dual anti–CTLA-4 and anti–PD-1 blockade in rare tumours (DART, SWOG S1609). *Clin. Cancer Res.* **2022**, *28*, 271–278. [CrossRef]
44. Hellmann, M.D.; Paz-Ares, L.; Bernabe Caro, R.; Zurawski, B.; Kim, S.W.; Carcereny Costa, E.; Park, K.; Alexandru, A.; Lupinacci, L.; de la Mora Jimenez, E.; et al. Nivolumab plus ipilimumab in advanced non–small-cell lung cancer. *N. Engl. J. Med.* **2019**, *381*, 2020–2031. [CrossRef] [PubMed]
45. Tian, Z.; Liu, M.; Zhang, Y.; Wang, X. Bispecific T cell engagers: An emerging therapy for management of hematologic malignancies. *J. Hematol. Oncol.* **2021**, *14*, 75. [CrossRef]
46. Aamir, S.; Anwar, M.Y.; Khalid, F.; Khan, S.I.; Ali, M.A.; Khattak, Z.E. Systematic Review and Meta-analysis of CD19-Specific CAR-T Cell Therapy in Relapsed/Refractory Acute Lymphoblastic Leukemia in the Pediatric and Young Adult Population: Safety and Efficacy Outcomes. *Clin. Lymphoma Myeloma Leuk* **2021**, *21*, e334–e347. [CrossRef]
47. Subklewe, M.; Magno, G.; Gebhardt, C.; Bücklein, V.; Szelinski, F.; Arévalo, H.J.; Hänel, G.; Dörner, T.; Zugmaier, G.; von Bergwelt-Baildon, M.; et al. Application of blinatumomab, a bispecific anti-CD3/CD19 T-cell engager, in treating severe systemic sclerosis: A case study. *Eur. J. Cancer* **2024**, *204*, 114071. [CrossRef]
48. Lantz, J.; Pham, N.; Jones, C.; Reed, D.; El Chaer, F.; Keng, M. Blinatumomab in Practice. *Curr. Hematol. Malig. Rep.* **2024**, *19*, 1–8. [CrossRef]
49. Cech, P.; Skórka, K.; Dziki, L.; Giannopoulos, K. T-Cell Engagers-The Structure and Functional Principle and Application in Hematological Malignancies. *Cancers* **2024**, *16*, 1580. [CrossRef]
50. Gholap, A.D.; Gupta, J.S.; Kamandar, P.A.; Banchhod, G.V.; Hatvate, N.T. Antibody-drug conjugates for cancer therapy: An up-to-date review on the chemistry and pharmacology. In *Comprehensive Analytical Chemistry*; Elsevier: Amsterdam, The Netherlands, 2023; Volume 103, pp. 105–190.
51. Elshazly, A.M.; Gewirtz, D.A. An overview of resistance to Human epidermal growth factor receptor 2 (Her2) targeted therapies in breast cancer. *Cancer Drug Resist.* **2022**, *5*, 472. [CrossRef]
52. Talukdar, S.; Emdad, L.; Das, S.K.; Fisher, P.B. EGFR: An essential receptor tyrosine kinase-regulator of cancer stem cells. *Adv. Cancer Res.* **2020**, *147*, 161–188. [PubMed]
53. Cheng, X. A comprehensive review of HER2 in cancer biology and therapeutics. *Genes* **2024**, *15*, 903. [CrossRef]
54. Pan, L.; Li, J.; Xu, Q.; Gao, Z.; Yang, M.; Wu, X.; Li, X. HER2/PI3K/AKT pathway in HER2-positive breast cancer: A review. *Medicine* **2024**, *103*, e38508. [CrossRef]
55. Oh, D.Y.; Bang, Y.J. HER2-targeted therapies—A role beyond breast cancer. *Nat. Rev. Clin. Oncol.* **2020**, *17*, 33–48. [CrossRef]
56. Singh, D.D.; Lee, H.J.; Yadav, D.K. Clinical updates on tyrosine kinase inhibitors in HER2-positive breast cancer. *Front. Pharmacol.* **2022**, *13*, 1089066. [CrossRef] [PubMed]
57. Tarantino, P.; Tolaney, S.M. The dawn of the antibody–drug conjugates era: How T-DM1 reinvented the future of chemotherapy for solid tumours. *Cancer Res.* **2022**, *82*, 3659–3661. [CrossRef] [PubMed]
58. Kronig, M.N.; Wehrli, M.; Salas-Benito, D.; Maus, M.V. Hurdles race for CAR T-cell therapy in digestive tract cancer. *Immunol. Rev.* **2023**, *320*, 100–119. [CrossRef]
59. Gámez-Chiachio, M.; Sarrió, D.; Moreno-Bueno, G. Novel therapies and strategies to overcome resistance to anti-HER2-targeted drugs. *Cancers* **2022**, *14*, 4543. [CrossRef]
60. Swain, S.M.; Shastry, M.; Hamilton, E. Targeting HER2-positive breast cancer: Advances and future directions. *Nat. Rev. Drug Discov.* **2023**, *22*, 101–126. [CrossRef]
61. Swain, S.M.; Miles, D.; Kim, S.B.; Im, Y.H.; Im, S.A.; Semiglazov, V.; Ciruelos, E.; Schneeweiss, A.; Loi, S.; Monturus, E.; et al. CLEOPATRA study group Pertuzumab, trastuzumab, and docetaxel for HER2-positive metastatic breast cancer (CLEOPATRA): End-of-study results from a double-blind, randomised, placebo-controlled, phase 3 study. *Lancet Oncol.* **2020**, *21*, 519–530. [CrossRef]
62. Alasmari, M.M. A Review of Margetuximab-Based Therapies in Patients with HER2-Positive Metastatic Breast Cancer. *Cancers* **2022**, *15*, 38. [CrossRef]
63. Zhou, G.; Fu, S.; Zhang, Y.; Li, S.; Guo, Z.; Ouyang, D.; Ying, T.; Lu, Y.; Zhao, Q. Antibody recognition of human epidermal growth factor receptor-2 (HER2) juxtamembrane domain enhances anti-tumour response of chimeric antigen receptor (CAR)-T cells. *Antibodies* **2024**, *13*, 45. [CrossRef] [PubMed]
64. Orofiamma, L.A.; Vural, D.; Antonescu, C.N. Control of cell metabolism by the epidermal growth factor receptor. *Biochim. Biophys. Acta Mol. Cell Res.* **2022**, *1869*, 119359. [CrossRef] [PubMed]
65. Liu, H.; Zhang, B.; Sun, Z. Spectrum of EGFR aberrations and potential clinical implications: Insights from integrative pan-cancer analysis. *Cancer Commun.* **2020**, *40*, 43–59. [CrossRef]

66. Murphrey, M.B.; Quaim, L.; Rahimi, N.; Varacallo, M.A. Biochemistry, Epidermal Growth Factor Receptor. In *StatPearls [Internet]*; StatPearls Publishing: Treasure Island, FL, USA, 2025. Available online: https://www.ncbi.nlm.nih.gov/books/NBK482459/ (accessed on 9 March 2025).
67. Li, Q.; Li, Z.; Luo, T.; Shi, H. Targeting the PI3K/AKT/mTOR and RAF/MEK/ERK pathways for cancer therapy. *Mol. Biomed.* **2022**, *3*, 47. [CrossRef] [PubMed]
68. Cheng, W.L.; Feng, P.H.; Lee, K.Y.; Chen, K.Y.; Sun, W.L.; Van Hiep, N.; Luo, C.S.; Wu, S.M. The role of EREG/EGFR pathway in tumour progression. *Int. J. Mol. Sci.* **2021**, *22*, 12828. [CrossRef]
69. Esteban-Villarrubia, J.; Soto-Castillo, J.J.; Pozas, J.; San Román-Gil, M.; Orejana-Martín, I.; Torres-Jiménez, J.; Carrato, A.; Alonso-Gordoa, T.; Molina-Cerrillo, J. Tyrosine kinase receptors in oncology. *Int. J. Mol. Sci.* **2020**, *21*, 8529. [CrossRef]
70. Cai, W.Q.; Zeng, L.S.; Wang, L.F.; Wang, Y.Y; Cheng, J.T.; Zhang, Y.; Han, Z.W.; Zhou, Y.; Huang, S.L.; Wang, X.W.; et al. The latest battles between EGFR monoclonal antibodies and resistant tumour cells. *Front. Oncol.* **2020**, *10*, 1249. [CrossRef]
71. Uribe, M.L.; Marrocco, I.; Yarden, Y. EGFR in cancer: Signaling mechanisms, drugs, and acquired resistance. *Cancers* **2021**, *13*, 2748. [CrossRef]
72. Liu, C.Y.; Lin, H.F.; Lai, W.Y.; Lin, Y.Y.; Lin, T.W.; Yang, Y.P.; Tsai, F.T.; Wang, C.L.; Luo, Y.H.; Chen, Y.M.; et al. Molecular target therapeutics of EGF-TKI and downstream signaling pathways in non-small cell lung cancers. *J. Chin. Med. Assoc.* **2022**, *85*, 409–413. [CrossRef]
73. Abourehab, M.A.; Alqahtani, A.M.; Youssif, B.G.; Gouda, A.M. Globally approved EGFR inhibitors: Insights into their syntheses, target kinases, biological activities, receptor interactions, and metabolism. *Molecules* **2021**, *26*, 6677. [CrossRef] [PubMed]
74. Yang, Y.; Li, S.; Wang, Y.; Zhao, Y.; Li, Q. Protein tyrosine kinase inhibitor resistance in malignant tumours: Molecular mechanisms and future perspective. *Signal Transduct. Target. Ther.* **2022**, *7*, 329. [CrossRef] [PubMed]
75. Batra, U.; Biswas, B.; Prabhash, K.; Krishna, M.V. Differential clinicopathological features, treatments and outcomes in patients with Exon 19 deletion and Exon 21 L858R EGFR mutation-positive adenocarcinoma non-small-cell lung cancer. *BMJ Open Respir. Res.* **2023**, *10*, e001492. [CrossRef] [PubMed]
76. Passaro, A.; Mok, T.; Peters, S.; Popat, S.; Ahn, M.J.; De Marinis, F. Recent advances on the role of EGFR tyrosine kinase inhibitors in the management of NSCLC with uncommon, non exon 20 insertions, EGFR mutations. *J. Thorac. Oncol.* **2021**, *16*, 764–773. [CrossRef]
77. Ferrara, M.G.; Di Noia, V.; D'Argento, E.; Vita, E.; Damiano, P.; Cannella, A.; Ribelli, M.; Pilotto, S.; Milella, M.; Tortora, G.; et al. Oncogene-addicted non-small-cell lung cancer: Treatment opportunities and future perspectives. *Cancers* **2020**, *12*, 1196. [CrossRef]
78. Graham, J.; Muhsin, M.; Kirkpatrick, P. Cetuximab. *Nat. Rev. Drug Discov.* **2004**, *3*, 549–550.
79. Kast, J.; Dutta, S.; Upreti, V.V. Panitumumab: A review of clinical pharmacokinetic and pharmacology properties after over a decade of experience in patients with solid tumours. *Adv. Ther.* **2021**, *38*, 3712–3723. [CrossRef]
80. Liu, T.; Jiang, S.; Teng, X.; Zhong, L.; Liu, M.; Jin, Y.; Dong, M. A comparison of panitumumab and cetuximab in the treatment of KRAS wild-type metastatic colorectal cancer: A systematic review and meta-analysis. *Immunopharmacol. Immunotoxicol.* **2023**, *45*, 1–9. [CrossRef]
81. John, A.; Noronha, V.; Singh, A.; Menon, N.; Prabhash, K. Amivantamab: A narrative drug review. *Cancer Res. Stat. Treat.* **2023**, *6*, 261–271. [CrossRef]
82. Papassotiriou, I.; Kapogiannatos, A.; Makatsoris, C.; Bakogeorgou, S.; Mantogiannakou, I.; Roussou, E.; Souras, G.; Liakas, D.; Sergentanis, T.N.; Gavriatopoulou, M.; et al. Efficacy and safety of amivantamab in advanced or metastatic EGFR-mutant non-small cell lung cancer: A systematic review. *J. Clin. Med.* **2024**, *13*, 5489. [CrossRef]
83. Mohelnikova-Duchonova, B.; Melichar, B.; Soucek, P. FOLFOX/FOLFIRI pharmacogenetics: The call for a personalized approach in colorectal cancer therapy. *World J. Gastroenterol. WJG* **2014**, *20*, 10316. [CrossRef] [PubMed]
84. Wang, X.; Bove, A.M.; Simone, G.; Ma, B. Molecular bases of VEGFR-2-mediated physiological function and pathological role. *Front. Cell Dev. Biol.* **2020**, *8*, 599281. [CrossRef]
85. Yang, Y.; Cao, Y. The impact of VEGF on cancer metastasis and systemic disease. In *Seminars in Cancer Biology*; Academic Press: Cambridge, MA, USA, 2022; Volume 86, pp. 251–261.
86. Al-Ostoot, F.H.; Salah, S.; Khamees, H.A.; Khanum, S.A. Tumour angiogenesis: Current challenges and therapeutic opportunities. *Cancer Treat. Res. Commun.* **2021**, *28*, 100422. [CrossRef]
87. Liu, Y.; Li, H.M.; Wang, R. Effectiveness and safety of adding bevacizumab to platinum-based chemotherapy as first-line treatment for advanced non-small-cell lung cancer: A meta-analysis. *Front. Med.* **2021**, *8*, 616380. [CrossRef] [PubMed]
88. Garon, E.B.; Visseren-Grul, C.; Rizzo, M.T.; Puri, T.; Chenji, S.; Reck, M. Clinical outcomes of ramucirumab plus docetaxel in the treatment of patients with non-small cell lung cancer after immunotherapy: A systematic literature review. *Front. Oncol.* **2023**, *13*, 1247879. [CrossRef]
89. Patel, S.A.; Nilsson, M.B.; Le, X.; Cascone, T.; Jain, R.K.; Heymach, J.V. Molecular mechanisms and future implications of VEGF/VEGFR in cancer therapy. *Clin. Cancer Res.* **2023**, *29*, 30–39. [CrossRef] [PubMed]

90. Guillaume, Z.; Auvray, M.; Vano, Y.; Oudard, S.; Helley, D.; Mauge, L. Renal Carcinoma and Angiogenesis: Therapeutic Target and Biomarkers of Response in Current Therapies. *Cancers* **2022**, *14*, 6167. [CrossRef]
91. Garcia, J.; Hurwitz, H.I.; Sandler, A.B.; Miles, D.; Coleman, R.L.; Deurloo, R.; Chinot, O.L. Bevacizumab (Avastin®) in cancer treatment: A review of 15 years of clinical experience and future outlook. *Cancer Treat. Rev.* **2020**, *86*, 102017. [CrossRef]
92. Bennouna, J.; Falchero, L.; Schott, R.; Bonnetain, F.; Coudert, M.; Ben Hadj Yahia, B.; Chouaid, C. Bevacizumab in combination with platinum-based chemotherapy in patients with advanced non-squamous non-small cell lung cancer with or without brain metastases: A French cohort study (EOLE). *Oncology* **2018**, *94*, 55–64. [CrossRef]
93. Tong, Y.; Fan, X.; Liu, H.; Liang, T. Advances in Trop-2 targeted antibody-drug conjugates for breast cancer: Mechanisms, clinical applications, and future directions. *Front. Immunol.* **2024**, *15*, 1495675. [CrossRef]
94. Pramanik, D. Development of Antibody-Drug Conjugates: Future Perspective Towards Solid Tumour Treatment. *Anti-Cancer Agents Med. Chem.* **2023**, *23*, 642–657. [CrossRef] [PubMed]
95. Okajima, D.; Yasuda, S.; Suzuki, T.; Kitamura, M.; Yamaguchi, J.; Maejima, T.; Karibe, T.; Toki, T.; Phillips, P.; Agatsuma, T. Datopotamab deruxtecan (Dato-DXd) enhances antitumour response to PD-1/PD-L1 inhibitors in TROP2-expressing tumours in mice. *Cancer Res.* **2023**, *83* (Suppl. 7), 2932. [CrossRef]
96. Mode, D. Dato-DXd Approved in US for HR+ Breast Cancer. 2025. Available online: https://www.fda.gov/drugs/resources-information-approved-drugs/fda-approves-datopotamab-deruxtecan-dlnk-unresectable-or-metastatic-hr-positive-her2-negative-breast (accessed on 9 March 2025).
97. Jiang, Y.; Zhou, H.; Liu, J.; Ha, W.; Xia, X.; Li, J.; Chao, T.; Xiong, H. Progress and Innovative Combination Therapies in Trop-2-Targeted ADCs. *Pharmaceuticals* **2024**, *17*, 652. [CrossRef] [PubMed]
98. Singh, R.; Chandley, P.; Rohatgi, S. Recent advances in the development of monoclonal antibodies and next-generation antibodies. *Immunohorizons* **2023**, *7*, 886–897. [CrossRef]
99. Lou, H.; Cao, X. Antibody variable region engineering for improving cancer immunotherapy. *Cancer Commun.* **2022**, *42*, 804–827. [CrossRef] [PubMed]
100. Kumar, A.; Dixit, S.; Srinivasan, K.; Vincent, P.D.R. Personalized cancer vaccine design using AI-powered technologies. *Front. Immunol.* **2024**, *15*, 1357217. [CrossRef]
101. Raja, A.; Kasana, A.; Verma, V. Next-generation therapeutic antibodies for cancer treatment: Advancements, applications, and challenges. *Mol. Biotechnol.* **2024**, 1–21. [CrossRef]
102. Zhang, C.; Zhuang, Q.; Liu, J.; Liu, X. Synthetic biology in chimeric antigen receptor T (CAR T) cell engineering. *ACS Synth. Biol.* **2022**, *11*, 1–15. [CrossRef]
103. Dent, M.; Mayer, K.L.; Verjan Garcia, N.; Guo, H.; Kajiura, H.; Fujiyama, K.; Matoba, N. Impact of glycoengineering and antidrug antibodies on the anticancer activity of a plant-made lectin-Fc fusion protein. *Plant Biotechnol. J.* **2022**, *20*, 2217–2230. [CrossRef]
104. Karagiannis, S.N.; Josephs, D.H.; Bax, H.J.; Spicer, J.F. Therapeutic IgE antibodies: Harnessing a macrophage-mediated immune surveillance mechanism against cancer. *Cancer Res.* **2017**, *77*, 2779–2783. [CrossRef]
105. Bergamini, A.; Ferrero, S.; Leone Roberti Maggiore, U.; Scala, C.; Pella, F.; Vellone, V.G.; Petrone, M.; Rabaiotti, E.; Cioffi, R.; Candiani, M.; et al. Folate receptor alpha antagonists in preclinical and early stage clinical development for the treatment of epithelial ovarian cancer. *Expert Opin. Investig. Drugs* **2016**, *25*, 1405–1412. [CrossRef] [PubMed]
106. Spicer, J.; Basu, B.; Montes, A.; Banerji, U.; Kristeleit, R.; Miller, R.; Veal, G.J.; Corrigan, C.J.; Till, S.J.; Figini, M.; et al. Safety and anti-tumour activity of the IgE antibody MOv18 in patients with advanced solid tumours expressing folate receptor-alpha: A phase I trial. *Nat. Commun.* **2023**, *14*, 4180. [CrossRef]
107. Sutton, B.J.; Davies, A.M.; Bax, H.J.; Karagiannis, S.N. IgE Antibodies: From Structure to Function and Clinical Translation. *Antibodies* **2019**, *8*, 19. [CrossRef]
108. Zahavi, D.; Weiner, L. Monoclonal antibodies in cancer therapy. *Antibodies* **2020**, *9*, 34. [CrossRef] [PubMed]
109. Damián-Blanco, P.; Ahuexoteco-Sánchez, S.; Carbajal-Gallardo, A.A.; Coctecon-Chávelas, F.C.; Rodríguez-Nava, C.; Vences-Velázquez, A.; Medina-Flores, Y.; Mata-Ruíz, O.; Lloret-Sánchez, L.; Cortés-Sarabia, K. Use of monoclonal antibodies in cancer immunotherapy: Types and mechanisms of action. *Boletín Médico Hosp. Infant. México* **2023**, *80*, 153–164. [CrossRef]
110. Sheykhhasan, M.; Manoochehri, H.; Dama, P. Use of CAR T-cell for acute lymphoblastic leukemia (ALL) treatment: A review study. *Cancer Gene Ther.* **2022**, *29*, 1080–1096. [CrossRef] [PubMed]
111. Haslauer, T.; Greil, R.; Zaborsky, N.; Geisberger, R. CAR T-cell therapy in hematological malignancies. *Int. J. Mol. Sci.* **2021**, *22*, 8996. [CrossRef] [PubMed]
112. Yáñez-Muñoz, R.J.; Grupp, S.A. CAR-T in the clinic: Drive with care. *Gene Ther.* **2018**, *25*, 157–161. [CrossRef]
113. Subklewe, M.; von Bergwelt-Baildon, M.; Humpe, A. Chimeric antigen receptor T cells: A race to revolutionize cancer therapy. *Transfus. Med. Hemotherapy* **2019**, *46*, 15–24. [CrossRef]
114. Sterner, R.C.; Sterner, R.M. CAR-T cell therapy: Current limitations and potential strategies. *Blood Cancer J.* **2021**, *11*, 69. [CrossRef]
115. Khosravi, G.R.; Mostafavi, S.; Bastan, S.; Ebrahimi, N.; Gharibvand, R.S.; Eskandari, N. Immunologic tumour microenvironment modulators for turning cold tumours hot. *Cancer Commun.* **2024**, *44*, 521–553. [CrossRef] [PubMed]

116. Shah, A.; Rauth, S.; Aithal, A.; Kaur, S.; Ganguly, K.; Orzechowski, C.; Varshney, G.C. Jain, M.; Batra, S.K. The current landscape of antibody-based therapies in solid malignancies. *Theranostics* **2021**, *11*, 1493. [CrossRef] [PubMed]
117. Flugel, C.L.; Majzner, R.G.; Krenciute, G.; Dotti, G.; Riddell, S.R.; Wagner, D.L.; Abou-el-Enein, M. Overcoming on-target, off-tumour toxicity of CAR T cell therapy for solid tumours. *Nat. Rev. Clin. Oncol.* **2023**, *20*, 49–62. [CrossRef]
118. Huang, Y.; Qin, Y.; He, Y.; Qiu, D.; Zheng, Y.; Wei, J.; Zhang, L.; Yang, D.H.; Li, Y. Advances in molecular targeted drugs in combination with CAR-T cell therapy for hematologic malignancies. *Drug Resist Updat* **2024**, *74*, 101082. [CrossRef]
119. Sharma, R.; Suravarjhula, L.; Banerjee, M.; Kumar, G.; Kumar, N. Chimeric antigen receptor T-cell therapy in cancer. A critical review. *Curr. Drug Res. Rev.* **2023**, *15*, 241–261. [CrossRef]
120. Ebrahimi, N.; Akbari, M.; Ghanaatian, M.; Roozbahani Moghaddam, P.; Adelian, S.; Borjian Boroujeni, M.; Yazdani, E.; Ahmadi, A.; Hamblin, M.R. Development of neoantigens: From identification in cancer cells to application in cancer vaccines. *Expert Rev. Vaccines* **2022**, *21*, 941–955. [CrossRef]
121. Liu, Z.; Lei, W.; Wang, H.; Liu, X.; Fu, R. Challenges and strategies associated with CAR-T cell therapy in blood malignancies. *Exp. Hematol. Oncol.* **2024**, *13*, 22. [CrossRef] [PubMed]
122. Majumder, A. Evolving CAR-T-cell therapy for cancer treatment: From scientific discovery to cures. *Cancers* **2023**, *16*, 39. [CrossRef]
123. Blair, H.A. Zenocutuzumab: First Approval. *Drugs* **2025**, *85*, 591–597. [CrossRef]
124. Rais, T.; Khan, A.; Riaz, R. Elrexfio™(elranatamab-bcmm): The game-changer in treatment of multiple myeloma. *Rare Tumours* **2023**, *15*, 20363613231207483. [CrossRef]
125. Shaver, J.; Horton, D.; Halford, Z. Targeting GPRC5D With Talquetamab: A New Frontier in Bispecific Antibody Therapy for Relapsed/Refractory Multiple Myeloma. *Ann. Pharmacother.* **2024**, *59*, 10600280241271192. [CrossRef] [PubMed]
126. Pelosci, A. European Commission Approves Glofitamab in Relapsed/Refractory DLBCL. *Cancer Netw.* **2023**.
127. Riaz, R.; Khan, A.; Siddiqui, T. Epcoritamab-bysp (Epkinly)–A phenomenal breakthrough in the treatment of diffuse large B-cell lymphoma. *Rare Tumours* **2023**, *15*, 20363613231193566. [CrossRef]
128. Parvez, S.; Prakash, K.; Soni, D. A Review on Chemistry, Mechanism, Clinical Trials Studies of ELAHERE (Mirvetuximab Soravtansine-Gynx): A Novel FDA Approved Drug to Treat Ovarian Cancer. *World J. Pharm. Pharm. Sci.* **2023**, *12*, 891–898.
129. Zhou, X.; Wu, Y.; Zhu, Z.; Lu, C.; Zhang, C.; Zeng, L.; Xie, F.; Zhang, L.; Zhou, F. Mucosal immune response in biology, disease prevention and treatment. *Signal Transduct. Target. Ther.* **2025**, *10*, 7. [CrossRef] [PubMed]
130. Liu, M.; Ye, H.; Lin, T.; Chen, Z.; Xu, X.; Zhuang, J.; Yang, Y.; Chen, X.; Chen, C.; Lin, M.; et al. The association between Chlamydia pneumoniae infection and prognosis in lung cancer patients: A prospective study. *BMC Infect. Dis.* **2025**, *25*, 119. [CrossRef] [PubMed]
131. Gong, W.; Wang, Z.; Wei, Y.; Wang, M.; Li, K.; Chen, X.; Huang, X.; Zhou, L.; Gan, Q.; Xu, X.; et al. Dynamic changes in peripheral blood immunophenotyping and its prognostic value in cervical cancer patients undergoing immune checkpoint blockade therapy. *Discov. Oncol.* **2025**, *16*, 167. [CrossRef]
132. Jackson, S.S.; Francis, J.; Pfeiffer, R.M.; Proietti, C.; Coghill, A.E.; Yu, K.J.; Sarathkumara, Y.D.; Hsu, W.-L.; Argirion, I.; Wang, C.-P.; et al. Sex differences in anti-EBV antibody responses. *J. Infect. Dis.* **2025**, jiaf057. [CrossRef]
133. van Tetering, G.; Evers, M.; Chan, C.; Stip, M.; Leusen, J. Fc Engineering Strategies to Advance IgA Antibodies as Therapeutic Agents. *Antibodies* **2020**, *9*, 70. [CrossRef]
134. Meyer, S.; Leusen, J.H.; Boross, P. Regulation of complement and modulation of its activity in monoclonal antibody therapy of cancer. In *MAbs*; Taylor & Francis: Abingdon, UK, 2014; Volume 6, pp. 1133–1144.
135. Dobó, J.; Kocsis, A.; Farkas, B.; Demeter, F.; Cervenak, L.; Gál, P. The Lectin Pathway of the Complement System-Activation, Regulation, Disease Connections and Interplay with Other (Proteolytic) Systems. *Int. J. Mol. Sci.* **2024**, *25*, 1566. [CrossRef]
136. Cedzyński, M.; Świerzko, A.S. Components of the Lectin Pathway of Complement in Solid Tumour Cancers. *Cancers* **2022**, *14*, 1543. [CrossRef] [PubMed]

antibodies

MDPI

Article

A Targeted Integration-Based CHO Cell Platform for Simultaneous Antibody Display and Secretion

Jessica P. Z. Ng [1,†], Mariati Mariati [1,†], Jiawu Bi [2], Matthew Wook Chang [3,4] and Yuansheng Yang [1,*]

1 Bioprocessing Technology Institute (BTI), Agency for Science, Technology and Research (A*STAR), 20 Biopolis Way, #03-01 Centros, Singapore 138668, Singapore; jessica_ng@bti.a-star.edu.sg (J.P.Z.N.); mariati@bti.a-star.edu.sg (M.M.)

2 Institute of Molecular Cell Biology (IMCB), Agency for Science, Technology and Research (A*STAR), 61 Biopolis Dr, #07-01 Proteos, Singapore 138673, Singapore; bi_jiawu@imcb.a-star.edu.sg

3 Synthetic Biology Translation Research Programme, Yong Loo Lin School of Medicine, National University of Singapore, Singapore 117465, Singapore; bchcmw@nus.edu.sg

4 NUS Synthetic Biology for Clinical and Technological Innovation (SynCTI), National University of Singapore, 28 Medical Dr, #02-07 Centre for Life Sciences, Singapore 117456, Singapore

* Correspondence: yang_yuansheng@bti.a-star.edu.sg

† These authors contributed equally to this work.

Abstract: Objective: We developed a targeted integration-based CHO cell platform for simultaneous antibody display and secretion, enabling a streamlined transition from antibody library screening to production without requiring the re-cloning of antibody genes. Methods: The platform consists of a CHO master cell line with a single-copy landing pad, a helper vector expressing FLPe recombinase, and bi-functional targeting vectors. Recombinase-mediated cassette exchange was utilized to integrate targeting vectors into the landing pad. Bi-functional vectors were designed by incorporating a minimal furin cleavage sequence (mFCS), RRKR, and various 2A peptides between the heavy chain (HC) and a membrane anchor. Results: Incomplete cleavage at the mFCS and 2A sites facilitated the expression of both membrane-bound and secreted antibodies, while mutations in the 2A peptide produced a range of display-to-secretion ratios. However, a fraction of secreted antibodies retained 2A residues attached to the HC polypeptides. Further analysis demonstrated that modifying the first five amino acids of the 2A peptide significantly influenced furin cleavage efficiency, resulting in different display-to-secretion ratios for targeting vectors containing mFCS-2A variant combinations. To overcome this, we designed nine-amino-acid FCS variants that, when placed between the HC and membrane anchor, provided a range of display-to-secretion ratios and eliminated the issue of attached 2A residues in the secreted antibodies. Vectors with lower display levels proved more effective at distinguishing cells expressing high-affinity antibodies with closely matched binding affinities. The platform also demonstrated high sensitivity in isolating high-affinity antibody-expressing cells and supported robust antibody production. Conclusion: This targeted integration-based CHO platform enables efficient, in-format screening and production of antibodies with tunable display-to-secretion profiles. It provides a powerful and scalable tool for accelerating the development of functional, manufacturable therapeutic antibodies.

Keywords: CHO cells; antibodies; recombinase-mediated cassette exchange; manufacturability; functionality; simultaneous display and secretion

1. Introduction

Therapeutic immunoglobulin G (IgG) monoclonal antibodies have become the best-selling biopharmaceutical products on the market [1]. However, their development remains a technically challenging, time-consuming, and expensive process. The general antibody development workflow begins with the discovery phase, followed by production in mammalian cells and subsequent preclinical and clinical studies before market approval. Traditionally, antibody discovery has relied on either in vivo immunization of animals or in vitro display-based technologies. In vivo immunization triggers an immune response in animals, followed by the use of hybridoma or single B cell technologies to discover high-affinity, highly specific monoclonal antibodies [1,2]. On the other hand, in vitro display technologies, such as phage-, bacteria-, and yeast-display, identify high-affinity antibodies through high-throughput screening of immunized, naïve, and synthetic or semi-synthetic libraries [3]. Following this discovery, the identified antibodies are then cloned and expressed through transient transfections in mammalian cells to generate small quantities for affinity, specificity, and in vitro cell-based testing. Subsequently, lead antibodies are produced in large quantities using stably transfected Chinese hamster ovary (CHO) cells for further developability assessment, animal studies, and clinical trials. CHO cells have become the preferred choice for the mass production of therapeutic antibodies due to their high yield and ability to properly fold and assemble complex proteins with human-like glycosylations.

In the current process of antibody development, the transition from antibody discovery to production involves shifts in production strategies and host cells, such as transitioning from transient to stable production and changing from mouse cells and HEK293 cells to CHO cells. In addition, antibodies identified through microbial system-based display are usually in the form of single-chain variable fragment (scFv) or fragment antigen binding (Fab) fragments and lack natural post-translational modifications, necessitating conversion into full-length antibodies, the final format for production in mammalian cells. These modifications often result in changes to antibody function, expression level, and biophysical properties, which can increase both costs and timelines, along with the risk of failure [4,5]. Therefore, the development of novel technologies is needed to bridge the gap between antibody discovery and production.

The mammalian cell-based display technologies, which allow the screening of antibodies in their full-length format with authentic post-translational modifications, have become a more attractive alternative to microbial-based display systems [3,6–10]. Typically, these technologies display antibodies on the cell surface by genetically fusing the C-terminus of the heavy chain constant region to a membrane anchor domain, such as the human platelet-derived growth factor receptor (PDGFR) [11–13] or the glycophospholipid (GPI) transmembrane domains [14–16]. By double-staining the cell surface with antigens and antibodies against the constant regions of displayed antibodies, mammalian cell display enables flow cytometry-based screening of both binding affinity and expression levels of full-length antibodies [8]. Additionally, cell surface staining signals provide insights into biophysical properties such as aggregation propensity, polyreactivity, and thermostability [17–19]. Mammalian cell display captures all the expressed antibodies on the cell surface. However, screening for antibodies against difficult-to-express membrane proteins [20] and additional functionalities, such as cell activation and internalization, requires mammalian cell libraries that secrete antibodies [7,21–25]. Various microfluidic technologies—based on compartmentalization in droplets [24,26–28], nanostructures like nanopens [29], or microcapillaries [30]—have been developed to enable screening for these functionalities in combination with secretion-based mammalian cell libraries. In addition, large-scale antibody production for animal testing and developability assessments (e.g., stability, viscosity, and solubility) also requires secretion-mode expression.

To streamline the transition between display and secretion modes without re-cloning antibody genes, various technologies have been developed to enable the integration of antibody display and secretion within the same mammalian cell. These technologies can be grouped into two categories: (1) switchable display-to-secretion systems, which allow antibody expression to toggle between display-only and secretion-only modes, and (2) simultaneous display and secretion systems, which enable antibodies to be expressed concurrently both on the cell surface and in a secreted form. Switchable systems include approaches such as amber suppression [31], inducible fusion of protein A dual Z-domain with PDGFR transmembrane domain (ZZ-PDGFR) [7], chemically induced inhibition of furin cleavage [32], antibiotic-promoted translational readthrough of stop codons [33], the enzyme-cleavable surface-tethered all-purpose screening system (ECSTASY) [34], and "antibody-membrane switching" via recombinase-mediated DNA recombination [35–37]. The first four methods require the addition and removal of specific components—such as non-natural amino acids, gene expression inducers, furin enzyme inhibitors, or antibiotics—to toggle between display and secretion modes. This complexity can introduce workflow challenges and lead to cell-to-cell variability in display due to inconsistent induction. The latter two methods face limitations in achieving complete switching from display to secretion, due to low DNA recombination or enzyme cleavage efficiency. Additionally, switchable display-to-secretion systems lack flexibility in controlling display levels of antibodies, which may need to be adjusted depending on the application [38].

Simultaneous display and secretion systems include methods such as alternative splicing [22,23,39–41], leaky stop codons [42], minimal furin cleavage sequence (mFCS) [43], and 2A peptides [44]. Each approach comes with its own challenges. For instance, leaky stop codons may lead to low membrane-bound antibody display and a lack of control of display-to-secretion ratios. Similarly, adjusting the display-to-secretion ratio is challenging with the mFCS. Furin, a naturally occurring protease in the Golgi apparatus, sporadically cleaves the mFCS. Cleavage results in the secretion of antibodies, while the absence of cleavage leads to cell surface-displayed antibodies. The mFCS motif, composed of the conserved sequence R-X-K/R-R, offers limited engineering flexibility for tuning cleavage efficiency and thus the ratio of display-to-secretion [45,46]. Different strengths of splicing signals and 2A peptides have been developed to adjust the ratio of display to secretion [39,41,44]. However, cryptic splicing sites within the heavy chain (HC) gene may lead to the expression of incorrect products. Meanwhile, 2A peptides, which are 18–22 amino acids long, enable ribosomal skipping during translation to produce secreted antibodies. When skipping does not occur, antibodies remain surface-bound. Despite their utility, 2A peptides cleave at the glycine-proline bond, leaving residual fragments (16–20 amino acids) on secreted antibodies, potentially compromising the antibody's biophysical and functional properties [47].

One more key consideration for developing technologies that streamline the transition from antibody discovery to production is the strategy for introducing antibody genes into mammalian cells. An ideal approach should meet several criteria: (1) high delivery efficiency to maximize library size, (2) expression of a single antibody gene per cell to allow efficient enrichment of clones with desired properties, (3) sustained expression to allow multiple rounds of enrichment and transition from antibody screening to production, and (4) high-level expression for generating enough material for testing complexed developability and functional properties. Transient expression, which has been historically used for the generation of mammalian cell libraries [48–50], fails to meet these requirements, as it provides short-term and low expression levels, and limited control over single-gene introduction per cell. Episomal plasmid vectors, which replicate without integrating into the chromosome, extend expression duration but still result in low expression levels and lack control over single-gene insertion per cell [15,39]. In contrast, stable integration offers both long-term and high-level expression. Methods for stable integration that have been used in the construction of mammalian cell libraries include random plasmid integration [15,35,38,43,51], viral-based transduction [12,20,36,52], transposons [22,23,53], and targeted integration. Random integration has low efficiency and poorly controls single-gene insertion per cell. Viral transduction and transposon-based methods enhance integration efficiency, yet they struggle with multiple-gene integration. In addition, these two approaches lack precise control over insertion sites, leading to transcriptional variation across different cells that complicates flow cytometry sorting. Targeted integration, achievable through recombinase- [13,18,19,37,54,55] or nuclease-mediated methods [56–58], has become the most favorable approach for constructing mammalian cell libraries. This strategy offers precise integration at desired loci, enabling transcriptional uniformity, effective single-gene insertion, and high expression levels, ultimately supporting efficient antibody screening and production.

Previous studies have utilized episomal plasmid vectors [39], transposon expression systems [22,23], viral transduction [44], and random integration [41,43] for achieving simultaneous display and secretion of monoclonal antibodies. However, the reported secreted titers from these technologies, ranging from 0.1 mg/L to 80 mg/L, are relatively low and insufficient to meet the requirements for testing antibody properties that demand larger quantities of antibodies. To address this limitation, we have previously developed a targeted integration-based CHO cell expression system that supports single-copy integration of targeting plasmid vectors into a predefined active genomic site through FLPe-aided recombinase-mediated cassette exchange (RMCE) [59]. This system enables rapid antibody production in stably transfected cell pools, reaching titers of several hundred mg/L in fed-batch cultures. In this study, we enhanced this system with a set of bi-functional vectors that enable simultaneous antibody display and secretion at different ratios. With the display function, this expression system facilitates high-throughput screening to efficiently identify high-affinity antibodies from a library. The secretion function allows CHO cells expressing high-binding antibodies to be further screened for functionality using microfluidic technologies. Subsequently, identified CHO cells expressing desirable antibodies can be directly used as production cell lines, generating sufficient material for further developability assessments and functional studies. By integrating both capabilities within the same CHO cells, this platform offers a powerful, streamlined solution for antibody discovery to production.

2. Materials and Methods

2.1. Construction of Targeting Vectors Expressing Single Antibodies and Antibody Libraries

The secretion vector, containing the sequence FRT3-LC-IRES-HC-IRES-FRT, was synthesized by GenScript and served as the base vector for constructing other targeting vectors (Figure 1A). Additional targeting vectors in Figure 1A, Figure 3A and Figure 4A were created by inserting synthesized elements between the NsiI site at the 5′ end of the HC and the EcoRI site upstream of the second EMCV IRES in the base vector. Sequences for FLPe recombinase recognition sites (FRT3 and FRT), EMCV IRES, trastuzumab IgG1 LC and HC, P2A, F2A, E2A, T2A, and GPI have been previously described [14–16,47,59,60].

For the high- and low-display targeting vectors expressing antibodies of varying binding affinities, the high-affinity antibodies utilized the trastuzumab sequence (Figure 5A). Medium- and low-affinity humanized anti-HER2 antibodies were developed in-house. Their binding affinities were quantified using bio-layer interferometry assays on the Octet system (LakePharma). Targeting vectors expressing medium- and low-affinity antibodies were constructed by replacing the variable light chain (VL) and variable heavy chain (VH) regions in the trastuzumab-expressing high and low display vectors, using the SacII and BsiWI sites for VL and the NgoMIV and SalI sites for VH.

A humanized anti-HER2 antibody library was generated by shuffling the CDR of the murine anti-HER2 4D5 antibody with human germline framework sequences, as previously described [61]. The humanized VL and VH libraries were cloned into the low-display targeting vector containing the RRKRATNFP variant, using the SacII and BsiWI sites for the VL library and the NgoMIV and SalI sites for the VH library.

2.2. Generation and Characterization of Stably Transfected Cell Pools

The CHO K1 master cell line (MCL) was previously established by integrating a landing pad vector into CHO K1 cells (ATCC, Manassas, VA, USA) [59]. Protocols for culturing the MCL and generating stable cell pools via RMCE have been detailed in earlier work [59]. Stable cell pools were characterized for growth and productivity using 5-day batch or 12-day fed-batch cultures. Cultures were initiated at a seeding density of 3×10^5 cells/mL in 30 mL protein-free medium consisting of HyQ-PF (GE Healthcare Life Sciences, Waltham, MA, USA) and CD-CHO (Thermo Fisher Scientific, Waltham, MA, USA) at 1:1 ratio, supplemented with 0.05% Pluronic F-68 (Thermo Fisher Scientific, Waltham, MA, USA), 1 g/L sodium bicarbonate (Sigma-Aldrich, St. Louis, MO, USA), and 6 mM L-glutamine (Sigma-Aldrich, St. Louis, MO, USA). For fed-batch cultures, 10% of Ex-cell Advanced CHO Feed 1 media (Sigma-Aldrich, St. Louis, MO, USA) was added every other day starting on day 3. Cell viability and density were assessed using a Vi-Cell XR viability analyzer (Beckman Coulter, Brea, CA, USA), and antibody secretion titers were measured with an IMMAGE 800 immunochemistry system (Beckman Coulter, Brea, CA, USA). Integrated viable cell density (IVCD) was calculated using the trapezoidal method. Specific antibody productivity (qMab) was determined by dividing the antibody concentration by the IVCD.

2.3. Cell Staining, Flow Cytometry Analysis, and FACS Enrichment

To evaluate targeting vectors with varying antibody display levels (Figure 1A), 1×10^7 cells were collected, washed with 1×PBS, and incubated with 500 μL of a 1:100 dilution of Anti-Human IgG (γ-chain specific)–FITC (Sigma-Aldrich, St. Louis, MO, USA) on ice in the dark for 30 min. After staining, cells were washed twice with 1X PBS and resuspended at 5×10^6 cells/mL for flow cytometric analysis using a FACSCalibur™ Flow Cytometer (BD Biosciences, Franklin Lakes, NJ, USA).

Both single-staining and double-staining methods were used to differentiate cells expressing antibodies with varying binding affinities. For single staining, 1×10^7 cells were washed with 1× PBS and incubated with 50 μL of 10 μg/mL FITC-labeled Human Her2/ErbB2 Protein, His Tag (HER2-FITC) (ACROBiosystems, Newark, DE, USA) at room temperature in the dark for 30 min. For double staining, the same protocol was followed. Besides HER2-FITC, 50 μL of 100 μg/mL R-phycoerythrin (R-PE) AffiniPure F(ab')$_2$ Fragment Goat Anti-Human IgG, Fc Gamma-specific (HC-PE) (Jackson ImmunoResearch, West Grove, PA, USA) was also added during incubation. After staining, cells were either washed immediately before flow cytometry analysis (protocol with less dissociation) or resuspended in 1 mL of pre-warmed 1× PBS and incubated at 37 °C for 15 min to enhance antibody dissociation (protocol with enhanced dissociation). Cell washing was performed using 1 mL of 37 °C 1× PBS, followed by resuspension to a final concentration of 5×10^6 cells/mL. Stained cells were analyzed using a FACSCalibur™ Flow Cytometer or sorted with a BD FACSAria™ III Cell Sorter (BD Biosciences). The sorting collection media comprised regular culture media supplemented with 1× Antibiotic-Antimycotic (Gibco, Waltham, MA, USA). Flow cytometry data were processed using FlowJo™ 10.7.2 (Tree Star Inc., Ashland, OR, USA).

2.4. Protein Purification, SDS-PAGE, and Peptide Mapping Analysis

Culture supernatants were purified using protein A on a GE AKTA Purifier 100 FPLC System (GE Healthcare, Chicago, IL, USA). The purified antibodies were analyzed under reducing conditions using SDS-PAGE, followed by peptide mapping of their amino acid sequences via NanoLC-MS/MS analysis. Protocols for these processes have been detailed in previous studies [47].

2.5. TOPO Cloning for HC and LC Sequences

RNA was extracted using the RNeasy Mini Kit (Qiagen, Hilden, Germany) following the manufacturer's protocol. cDNA synthesis was performed with 2.5 μg of RNA using the EvoScript Universal cDNA Master (Roche, Basel, Switzerland). The LC and HC cDNA were independently amplified using the 2X Platinum SuperFi II Green PCR Master Mix (Thermo Fisher Scientific, Waltham, MA, USA) with primers listed in Supplementary Table S1. The purified PCR products were used for TA cloning with the TOPO™ TA Cloning™ Kit (Thermo Fisher Scientific, Waltham, MA, USA) for sequencing.

3. Results

3.1. Evaluation of mFCS, 2A Peptides, and Their Combinations for Simultaneous Display and Secretion of Full-Length IgG Antibody

Previous studies have utilized either mFCS or 2A peptides to achieve the simultaneous display and secretion of full-length IgG antibodies in mammalian cells [43,44]. However, engineering mFCS to adjust the display-to-secretion ratio presents significant challenges, while antibodies secreted using the 2A peptide approach retain residual 2A sequences, which may impact their properties. To achieve simultaneous antibody display and secretion while avoiding the secretion of products containing 2A residues, we designed a set of targeting vectors in which the LC gene is positioned as the first cistron and linked to the HC gene through an encephalomyocarditis virus (EMCV) internal ribosome entry site (IRES) element. The HC is further connected to a GPI membrane anchor via an mFCS (R-R-K-R) combined with one of four 2A peptides: foot-and-mouth disease virus 2A peptide (F2A), equine rhinitis A virus 2A peptide (E2A), porcine teschovirus-1 2A peptide (P2A), or Thosea asigna virus 2A peptide (T2A). The HC stop codon is removed, ensuring the HC, R-R-K-R, 2A, and GPI anchor are in a single reading frame. An additional EMCV IRES sequence is included downstream of the GPI sequence. The entire expression cassette is flanked by a pair of FLPe recombination targeting sites: wild-type FRT and mutated FRT3. These four vectors were named RRKR-P2A, RRKR-F2A, RRKR-T2A, and RRKR-E2A, respectively. For comparison, seven additional targeting vectors were designed with the same basic structure as those vectors described above. These variants differ in how the HC is linked to the GPI anchor: using only a 2A peptide, through R-R-K-R alone, directly fused to the GPI, or without the GPI anchor entirely. These seven vectors were named P2A, F2A, T2A, E2A, RRKR, display, and secretion, respectively (Figure 1A).

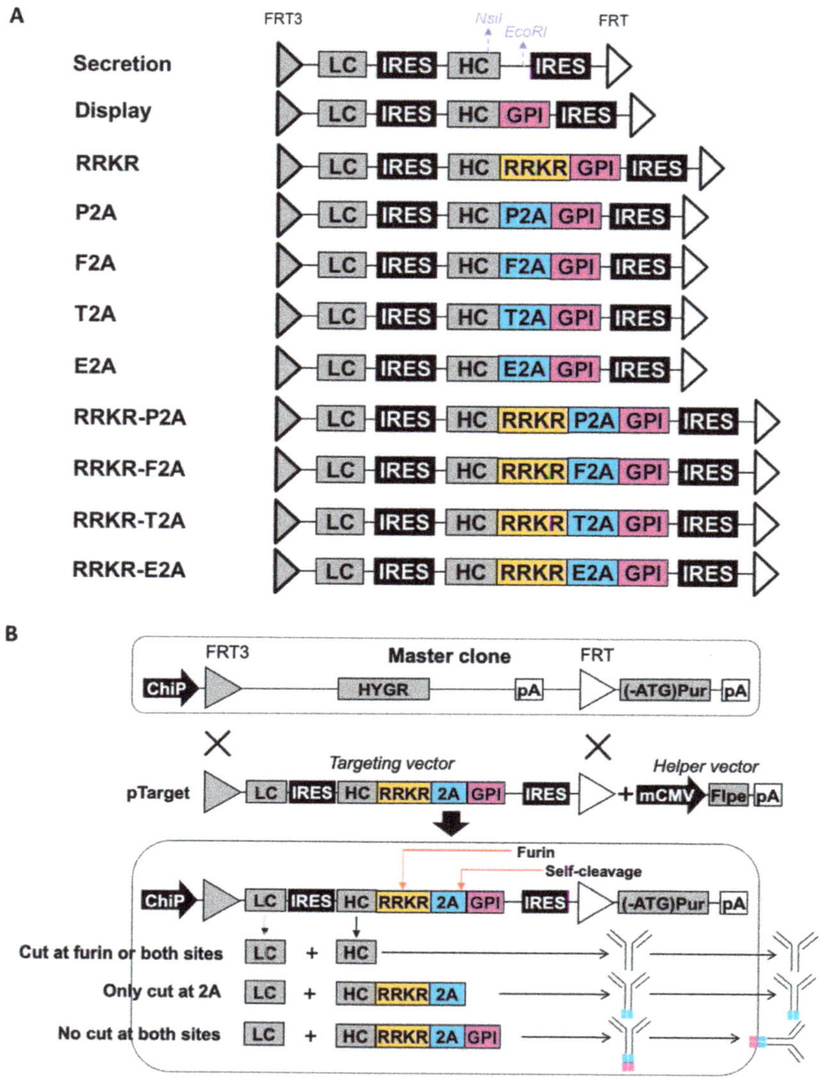

Figure 1. Targeting vectors, helper vectors, and recombinase-mediated cassette exchange (RMCE) for simultaneous antibody display and secretion. (**A**) Schematic representation of targeting vectors designed for antibody secretion only, display only, and simultaneous display and secretion. (**B**) Overview of RMCE-mediated integration of targeting vectors to enable simultaneous antibody display and secretion. Arrows indicate the initial translation, assembly, and transport of antibodies, either for secretion into the extracellular space or for display on the cell surface. The blue region represents the 2A peptide, while the pink region denotes the GPI membrane anchoring domain. ChiP, chimeric promoter comprising mCMV (cytomegalovirus) enhancer, human CMV core promoter and intron; FRT3 and FRT, mutated and wildtype flippase recognition targets; FLPe, enhanced recombinase flippase; HYGR, hygromycin resistance gene; pA, simian virus 40 polyadenylation signal; (-ATG) Pur, puromycin resistance gene lacking start codon ATG; IRES, wild-type encephalomyocarditis virus internal ribosomal entry site; mCMV, murine CMV enhancer and promoter; LC, light chain cDNA; HC, heavy chain cDNA; F2A, foot-and-mouth disease virus 2A peptide; E2A, equine rhinitis A virus 2A peptide; P2A, porcine teschovirus-1 2A peptide; T2A, Thosea asigna virus 2A peptide; GPI, glycosidylphosphatidylinositol membrane anchor from human decay-accelerating factor.

All targeting vectors were evaluated in stably transfected pools generated via RMCE using our previously developed CHO master clone. To establish stable pools, each targeting vector was co-transfected with a helper vector expressing FLPe into the CHO master cells, followed by selection in a medium containing puromycin. The CHO master clone contains a single-copy landing pad that expresses the hygromycin resistance (HYGR) gene flanked by FRT3 and FRT recombination sites matching those in the targeting vectors. Downstream of the FRT site, a start codon-deficient puromycin resistance gene ((ATG-)Pur) is located. The targeting vectors were designed without promoters and polyadenylation signals (pA). Accurate RMCE-mediated integration of the targeting vector into the landing pad activates the expression of the LC, HC (and its associated components), and the puromycin resistance gene, allowing the cells to survive selection. CHO master cells with incorrect integration fail to express these genes and are eliminated during selection. Simultaneous display and secretion functionality could be achieved in stable cell pools transfected with RRKR-2A vectors. These vectors enable antibody processing through furin cleavage at the RRKR site and/or ribosomal skipping at the 2A peptide site. This design results in the generation of three possible antibody forms: (1) secreted antibodies without attachment of 2A residues when cleavage occurs at the furin cleavage site or at both the furin cleavage site and the 2A peptide, (2) surface-displayed antibodies when neither furin cleavage nor ribosomal skipping of the 2A peptide occurs, and (3) secreted undesired antibodies when only ribosomal skipping occurs at the 2A peptide, resulting in antibodies with a redundant trailing amino acid sequence (Figure 1B).

Once stable pools were established, they were analyzed for surface antibody display using flow cytometry after staining with fluorescein isothiocyanate (FITC)-conjugated IgG specific to the heavy chain (FITC-HC) (Figure 2A). Antibody secretion levels were also evaluated for each stable pool in 5-day batch cultures. Two control pools were generated: one using the secretion vector (which lacks GPI and secretes all antibodies into the medium) and another using the display vector (which directly links the HC to GPI, resulting in all antibodies being displayed on the cell surface). The normalized display level was quantified as the mean fluorescence intensity (MFI) of pools generated using a specific targeting vector divided by the MFI from the display control. Similarly, the normalized secretion level was calculated as the specific productivity (qMab) of the pool generated using a specific targeting vector divided by the qMab from the secretion control (Figure 2B).

All targeting vectors exhibited homogeneous expression levels across cells, as indicated by sharp flow cytometry histograms. As expected, compared to non-transfected cells (blank), the secretion vector displayed negligible antibody levels on the cell surface, while the display vector demonstrated high levels of surface antibody display with no secretion. Interestingly, the RRKR vector achieved a display level slightly higher than the display control while exhibiting no secretion. This observation contrasts with a previous study that used the RIRR sequence to link the HC to a PDGFR membrane anchor, enabling simultaneous display and secretion of antibodies [43]. A potential explanation for this discrepancy lies in the differences in the membrane anchor and mFCS sequences used. Among the 2A vectors, P2A, F2A, T2A, and E2A displayed varying levels of surface antibody display, achieving 95%, 80%, 20%, and 10% of the display control, respectively. When combined with RRKR, all targeting vectors showed reduced display levels except for RRKR-E2A. The display levels observed from the RRKR-2A combinations did not directly correlate with those of their respective 2A vectors without RRKR. Specifically, RRKR-P2A exhibited the lowest display level, achieving only 2% of the display control. RRKR-T2A, RRKR-F2A, and RRKR-E2A exhibited higher display levels, achieving 4%, 18%, and 30% of the display control, respectively. A general inverse relationship between secretion and display levels was observed across the 2A and RRKR-2A-containing targeting vectors, with secretion levels ranging from 44% to 88% of the secretion control.

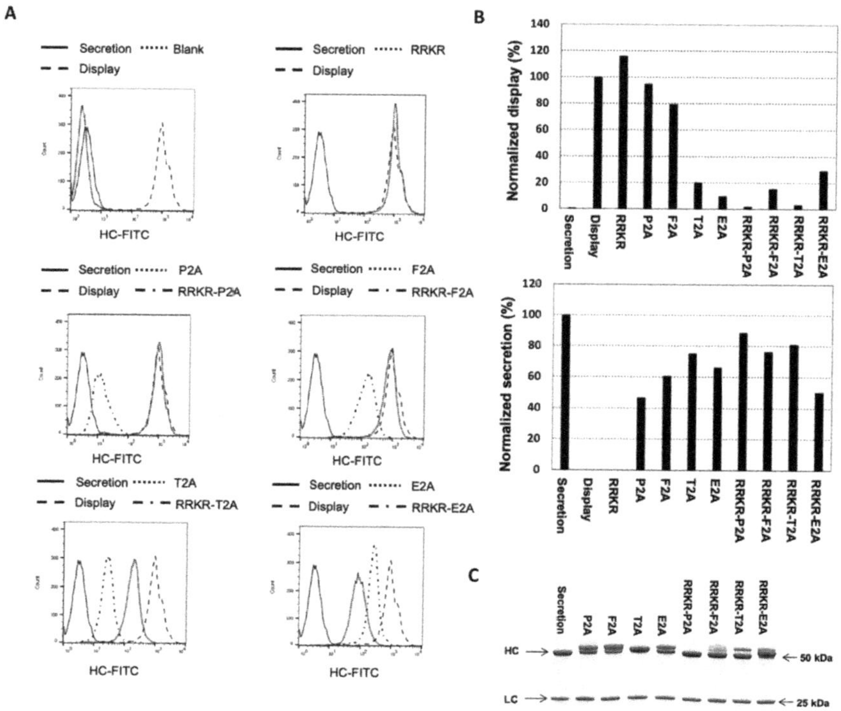

Figure 2. Evaluation of targeting vectors containing minimal furin cleavage sequence (RRKR), 2A peptides, and their combinations for antibody display and secretion. Stable cell pools were generated by co-transfecting CHO master cells with a specified targeting vector and a helper vector expressing FLPe recombinase, followed by selection in a puromycin-containing medium. Transfected cells were subsequently selected for correct integration of the targeting vectors into the landing pad. Non-transfected cells (blank), stable pools generated using the secretion vector, and those using the display vector served as controls. (**A**) Flow cytometry analysis of antibody levels displayed on the cell surface. Cells were stained with FITC-conjugated IgG specific to the heavy chain. (**B**) Normalized display and secretion levels relative to the control pools. Normalized display levels were calculated as the mean fluorescence intensity (MFI) of FITC-HC divided by that of the display control. Normalized secretion levels were determined as the qMab value from a 5-day batch culture relative to the secretion control. (**C**) SDS-PAGE analysis of purified secreted antibodies under reducing conditions.

The secreted antibodies from different targeting vectors were purified using protein A and analyzed under reducing conditions on a denaturing SDS-PAGE gel (Figure 2C). As expected, the HC and LC polypeptides expressed from the secretion control vector were observed at approximately 50 kDa and 25 kDa, respectively. The LC polypeptides secreted from all other targeting vectors were also observed at approximately 25 kDa. However, for all four 2A peptide-containing targeting vectors, one or two bands with sizes larger than the HC polypeptide of the secretion control were observed. Peptide mapping analysis revealed that these HC species were attached with residual 2A or 2A-GPI peptide fragments (Supplementary Figure S1). The amount of HC-2A-GPI polypeptide present in the medium was small, which could have resulted from cell death. When RRKR was included upstream of F2A, T2A, and E2A, two distinct HC bands were still observed. However, the sizes of the lower bands were smaller than those from the corresponding 2A-only targeting vectors. Peptide mapping indicated that the upper bands corresponded to HC polypeptides with residual 2A fragments, while the lower bands represented correctly processed HC polypeptides. Notably, only the RRKR-P2A vector produced an HC polypeptide with the correct size, completely free of residual 2A fragments, as confirmed by peptide mapping (Supplementary Figure S1). However, the display level from this vector was very low, potentially insufficient for antibody library screening.

3.2. Point Mutation of P2A in RRKR-P2A for Obtaining Different Display-to-Secretion Ratios

To achieve different display-to-secretion ratios using the RRKR-P2A vector, we introduced point mutations in P2A, substituting each amino acid with amino acid G, P, or A (Figure 3). G and P influence secondary structure differently—G increases flexibility, while P imposes conformational constraints. A, on the other hand, has a minimal impact on secondary structure. The mutated targeting vectors were assessed for surface display and antibody secretion as described previously. Nine-point mutations—A1G, T2G, A1P, T2P, N3P, F4P, S5P, N3A, and F4A—enhanced display levels compared to the wild-type RRKR-P2A vector. The extent of the increase varies depending on the mutation. Notably, A1P and T2G mutations raised display levels to 80% and 50% of the display control vector, respectively. Correspondingly, secretion levels from these vectors dropped to 20% and 60% of the secretion control vector, respectively. The HC polypeptides secreted from the A1P vector were larger than those from the secretion control vector. The T2G vector produced two HC bands. Peptide mapping revealed that most HC species from A1P and T2G represented HC polypeptides attached to P2A residues, while a minor proportion retained P2A-GPI residues (Supplementary Figure S1). For the other seven mutations, display levels ranged from 9% to 20% of the display vector, and secretion levels were slightly reduced compared to the secretion vector. Secreted HC polypeptides from these vectors exhibited the correct size on SDS PAGE. However, peptide mapping indicated a small proportion of species retained residual 2A or 2A-GPI fragments (Supplementary Figure S1).

Interestingly, only mutations within the first five amino acids of P2A in the RRKR-P2A vectors increased display levels. Previous studies suggest that mutations at other positions also dramatically reduce 2A peptide cleavage efficiency [44,62]. To explore whether mutations outside the first five amino acids affect P2A cleavage efficiency, one set of targeting vectors containing mutated P2A sequences without mFCS was tested. Analysis of these P2A-only vectors revealed that mutations both within and outside the first five amino acids decreased secretion levels compared to the wild-type P2A vector, likely due to reduced P2A cleavage efficiency. Similarly to the wild-type P2A vector, HC polypeptides expressed from the mutated P2A exhibited two bands. However, these mutations did not significantly further increase display levels compared to the display control vector, suggesting that the number of antibodies displayed on the cell surface had reached saturation. One possible explanation for the lack of increased display levels in mutations outside the first five amino acids for the RRKR-P2A vectors is that these mutations affected only P2A cleavage efficiency. Antibody secretion still occurred via cleavage at RRKR, resulting in the removal of both P2A and GPI from HC. In contrast, the nine mutations within the first five amino acids that increased display levels likely influenced cleavage efficiencies at both RRKR and P2A sites.

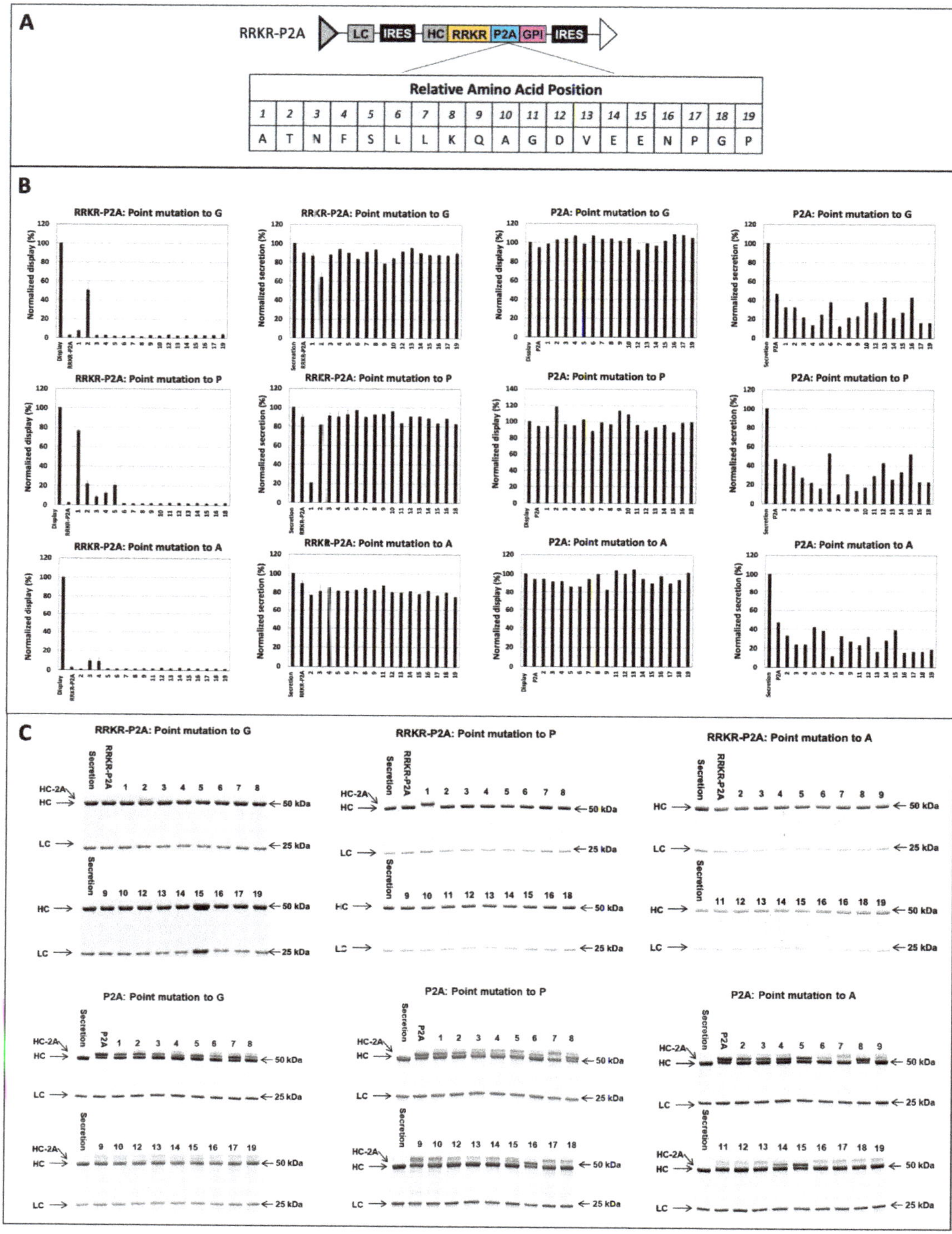

Figure 3. Evaluation of targeting vectors containing P2A variants for antibody display and secretion levels. (**A**) Schematic representation of the targeting vector containing the RRKR-P2A combination and

the amino acid sequence of the P2A peptide. Positional scanning mutagenesis was performed within the P2A using substitutions with P, G, and A. Positions 1 to 19 indicate the relative locations of amino acids within the P2A sequence. (**B**) Normalized antibody display and secretion levels in stable pools. These pools were generated by co-transfecting CHO master cells with each targeting vector containing a P2A variant and a helper vector expressing FLPe recombinase, followed by selection in a puromycin-containing medium. Control cell pools were established using the display vector and secretion vector for display and secretion benchmarks, respectively. Stable cell pools were stained with FITC-conjugated IgG specific to the heavy chain of surface-displayed antibodies. Normalized display levels (%) were calculated as the FITC-HC mean fluorescence intensity relative to the display control pool. A 5-day batch culture was performed, and normalized secretion levels (%) were determined as qMab relative to the secretion control pool. (**C**) SDS-PAGE analysis of secreted antibodies under reducing conditions.

3.3. Engineering Furin Cleavage Sequence (FCS) for Obtaining Different Display-to-Secretion Ratios

Previous studies indicate that the five amino acids downstream of the mFCS play a critical role in furin cleavage efficiency [45,46]. To examine the impact of the first five amino acids from the wild-type or mutated P2A on furin cleavage, we designed one set of nine amino-acid FCS variants and linked them to the HC and GPI to enable simultaneous antibody display and secretion. These variants incorporated the first five amino acids from the N-terminus of the wild-type and mutated P2A sequences downstream of RRKR. Corresponding RRKR-P2A vectors with the same mutations were included for comparison. The RRKR*ATNFS* vector, incorporating the first five amino acids (italicized) of the wild-type P2A, exhibited low display and high secretion levels, closely resembling the performance of the wild-type RRKR-P2A vector (Figure 4). All targeting vectors containing the nine FCS variants with the first five amino acids of the mutated P2A showed increased display and decreased secretion compared to the wild-type RRKR-P2A vector. The changes in display and secretion levels closely aligned with those observed in the corresponding mutated RRKR-P2A vectors. Notably, two exceptions, RRKRGTNFS and RRKRAPNFS, displayed significantly higher display levels—approximately five-fold and three-fold higher, respectively—than their RRKR-(GTNFS)P2A and RRKR-(APNFS)P2A counterparts. Correspondingly, secretion levels in the RRKRGTNFS and RRKRAPNFS vectors decreased compared to their P2A-containing counterparts, although the reduction was less pronounced than the increase in display levels. These findings suggest that point mutations in P2A predominantly affect cleavage efficiency at RRKR rather than P2A. However, the discrepancies in display and secretion levels for RRKRGTNFS and RRKRAPNFS vectors relative to their P2A-containing counterparts indicate that mutations A1G and T2P may influence cleavage efficiencies at both RRKR and P2A.

Fed-batch cultures of the pools generated using the ten FCS variants were carried out until day 12. Four variants—RRKRPTNFS, RRKRAGNFS, RRKRAPNFS, and RRKRATNFP—produced titers ranging from 50 to 275 mg/L, with titers inversely correlated to their display levels. Interestingly, the remaining variants did not further enhance antibody titers despite reduced display levels. The harvested antibodies were purified using protein A and analyzed via SDS-PAGE under reducing conditions. HC polypeptides expressed from RRKRAGNFS and RRKRPTNFS vectors, which had the highest display levels, exhibited two distinct bands, with the upper bands having higher molecular weights than those of the secretion control HC. Peptide mapping analysis of the RRKRAGNFS sample revealed that a small proportion of HC polypeptides from the top band remained attached to GPI (Supplementary Figure S1). These aberrant species in the medium may have resulted from cell death. In contrast, HC polypeptides from the other seven FCS variants exhibited the correct molecular weight. Peptide mapping analysis of secreted samples from three targeting vectors—RRKRATNFP, RRKRGTNFS, and RRKRAPNFS—with relative display levels of 9.4%, 23.9%, and 39.4%, respectively, confirmed that the HC polypeptides were correctly processed, with both FCS and GPI successfully removed. The remaining targeting vectors, which exhibited relative display levels below 5%, may not be suitable for antibody screening due to their insufficient display levels.

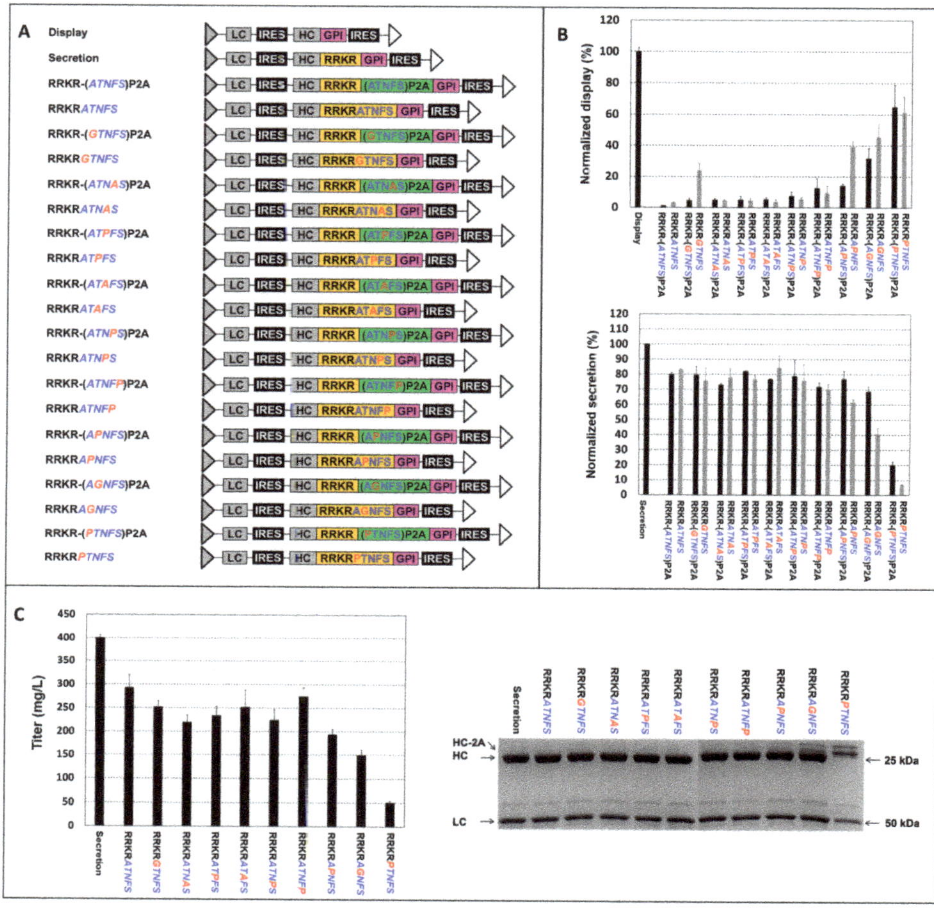

Figure 4. Evaluation of targeting vectors with furin cleavage sequence (FCS) variants or RRKR-P2A combinations for antibody display and secretion levels. (**A**) Schematic representation of targeting vectors incorporating either nine-amino-acid FCS variants or RRKR-P2A combinations. Amino acids highlighted in blue represent the original ATNFS sequence, while those in red indicate the mutations. (**B**) Normalized antibody display and secretion levels relative to control cell pools. Stable pools were generated by co-transfecting CHO master cells with a specified targeting vector and a helper vector expressing FLPe recombinase, followed by selection in a puromycin-containing medium. Display and secretion control pools were established using the display vector and secretion vector, respectively. Flow cytometry analysis was performed on cells stained with FITC-conjugated IgG specific to the heavy chain (FITC-HC) of surface-displayed antibodies. Normalized display levels (%) were calculated as the FITC-HC mean fluorescence intensity relative to the display control pool. Normalized secretion levels (%) were determined based on the qMab values of fed-batch cultures, relative to the secretion control pool. (**C**) Titers of secreted antibodies on day 12 of fed-batch cultures and SDS-PAGE analysis of secreted antibodies under reducing conditions. Each point in panels (**B**,**C**) represents the average and standard deviation of measurements from two independent stable pools.

3.4. Comparison of Low- and High-Display-Level Vectors for Discriminating Antibodies with Similar Binding Affinities

To evaluate the impact of display levels on the ability to discriminate cells expressing antibodies with close binding affinities, we generated two sets of stable pools using two targeting vectors –RRKRATNFP vector exhibiting display levels of 9.4% (low-display) and RRKRAPNFS vector exhibiting display levels of 39.4% (high-display) relative to the display control, respectively. Each set of stable pools expressed three antibodies against human epidermal growth factor receptor 2 (HER2) antigen with high (KD = 3.5×10^{-10} M), medium (KD = 2.0×10^{-9} M), and low (KD = 6.7×10^{-9} M) binding affinities (Figure 5A). The binding affinities of these antibodies were quantified by Octet® Bio-Layer Interferometry assay. Each stable pool was either single-stained with HER2 antigen labeled with FITC (HER2-FITC) or double-stained with HER2-FITC and R-phycoerythrin (R-PE)-conjugated IgG specific to the heavy chain (HC-PE) staining reagents. Optimal concentrations for each staining reagent, yielding a low signal-to-noise ratio and improved discrimination between cells expressing different antibodies, were identified through titration curves.

The non-transfected cells with double staining served as the blank control, while a stable pool generated using the low-display vector expressing the high-affinity antibody, singly stained with either HER2-FITC or HC-PE, served as the positive control (Figure 5B). Single-staining with HER2-FITC failed to differentiate between the three high-display vector-generated pools expressing high, medium, and low-affinity antibodies (Figure 5B, Less Dissociation). For the three low-display-vector-generated pools, those expressing low and medium binding affinities exhibited distinct levels of fluorescence intensity, but high- and medium-binding affinity pools showed significant overlap. Under double-staining conditions, high-display vector-generated pools continued to exhibit poor discrimination between populations, whereas low-display vector-generated pools demonstrated clear separation between low- and medium-affinity pools and marginal discrimination between medium- and high-affinity pools. Both high-display and low-display vector-generated pools showed overestimated relative affinities compared to measurements obtained using Octet®. A better correlation was observed between the Octet® measurements and flow cytometry results for low-display vector-generated pools than for high-display vector-generated pools (Figure 5C, Less Dissociation).

The flow cytometry approach measures binding affinity by incubating cells with staining reagents for 30 min, followed by a brief washing step and immediate analysis. The flow cytometry approach measures binding affinity by incubating cells with staining reagents for 30 min, followed by a brief washing step and immediate analysis. Antibodies with high binding affinity typically exhibit rapid association and slow dissociation. However, this slow dissociation may limit the sensitivity of flow cytometry in distinguishing antibodies with similar high affinities. To improve the resolution of the assay, we modified the staining protocol to enhance dissociation, thereby increasing the sensitivity of flow cytometry in differentiating closely related high-affinity antibodies. This was achieved by incubating cells in the washing buffer at 37 °C for 15 min following antigen staining. Various parameters were tested, including multiple rounds of washing, different incubation times, and larger incubation volumes, to optimize dissociation conditions. Incubating cells at 37 °C for 15 min after the first wash step was found to be most effective, consistent with a previous study that reported enhanced antibody dissociation kinetics at this temperature [63].

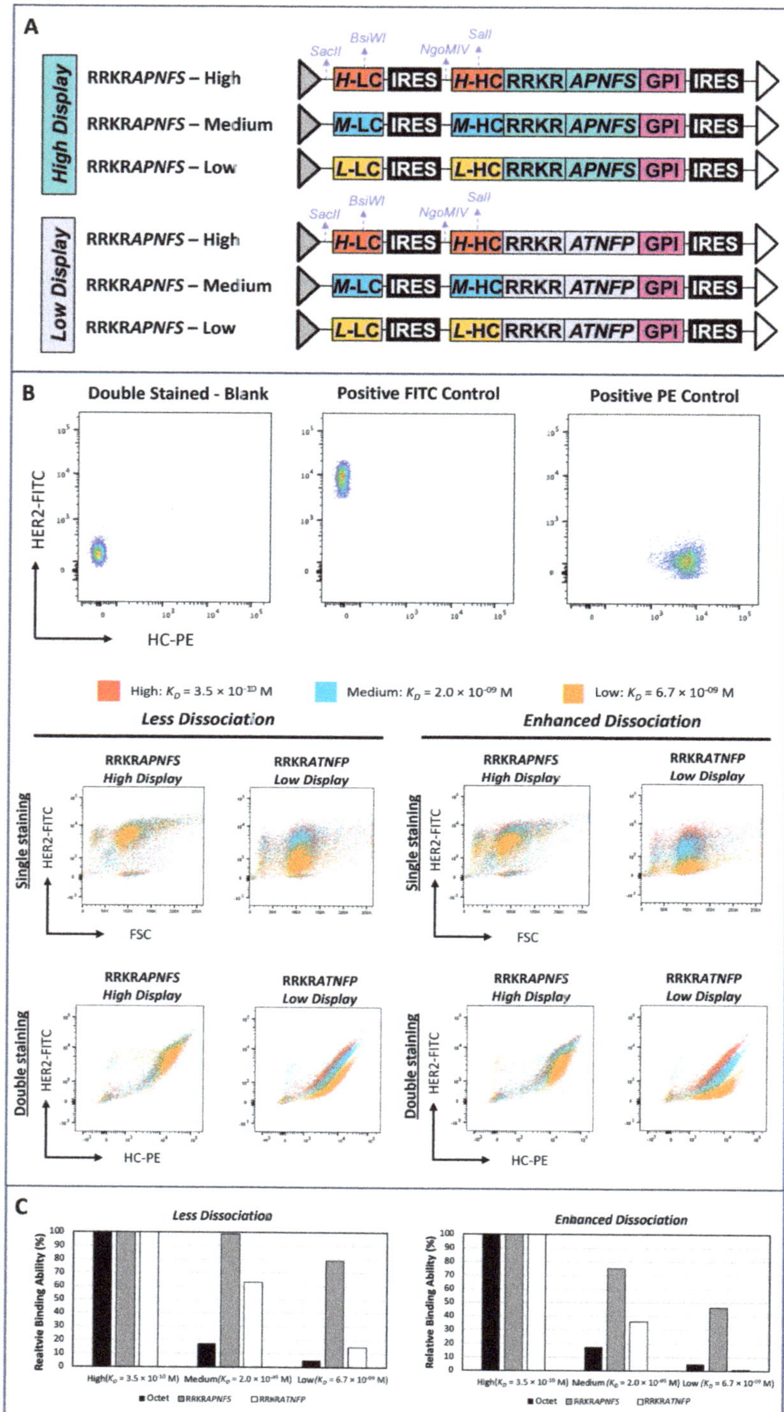

Figure 5. Comparison of low- and high-display level vectors for the discrimination of antibodies with varying binding affinities. (**A**) Schematic representation of high- and low-display level targeting

vectors expressing antibodies with high, medium, and low binding affinities. (**B**) Flow cytometry analysis of stable pools expressing different levels of binding affinity antibodies. For the controls, each dot—colored according to the heatmap—represents an individual cell (or 'event'), with yellow/red indicating high density and blue/green indicating lower density. Different stable pools were generated by co-transfecting CHO master cells with a specified targeting vector and a helper vector expressing FLPe recombinase, followed by selection in puromycin-containing medium. Each stable pool was stained with both FITC-labeled Human Her2/ErbB2 protein (HER2-FITC), R-PE AffiniPure F(ab')$_2$ Fragment Goat Anti-Human IgG, Fc Gamma-specific (HC-PE), using two staining protocols: less or enhanced dissociation phase. (**C**) Comparison of relative binding affinities determined by Octet® analysis and flow cytometry of cells stained with a less or enhanced dissociation phase. The relative binding affinity was normalized to the high-affinity antibody measured under the same method and conditions. Octet® (■), RRKRAPNFS (▒), and RRKRATNFP (□).

Enhancing the dissociation phase in the staining protocol improved the ability to differentiate cell populations (Figure 5B, Enhanced Dissociation). For high-display pools, single-staining could distinguish cells displaying low-binding antibodies from those with high or medium-binding antibodies, though high- and medium-binding populations remained indistinguishable. For low-display pools, discrimination between high- and medium-affinity populations improved slightly, though significant overlap persisted. Under double-staining conditions, high-display pools were still unable to distinguish between high- and medium-binding affinities. In contrast, low-display pools clearly separated cell populations expressing high, medium, and low binding affinities. Supporting this observation, the relative binding affinity of cells stained using the enhanced dissociation protocol for low-display pools closely reflected actual antibody binding kinetics as measured by Octet® (Figure 5C, Enhanced Dissociation). Meanwhile, high-display pools continued to overestimate binding affinities, yielding values several-fold higher than the actual measurements.

3.5. Evaluation of Simultaneous Display and Secretion Platform for Sorting Sensitivity

To assess the sensitivity of the simultaneous display and secretion platform to isolate cells expressing high-affinity antibodies from a mixed population via fluorescence-activated cell sorting (FACS), the three low-display vector-generated pools expressing high-, medium-, and low-affinity antibodies were mixed in a 1:1:1 ratio. The mixed cell pool was double-stained with HER2-FITC and HC-PE using the modified staining protocol, which enhanced the dissociation step. The low-display pool expressing the high-affinity antibody was included as a positive control to establish gating parameters. Flow cytometry analysis of the mixed pool revealed three clearly separated populations with similar ratios. Cells within the top 1% gate were sorted and double-stained, showing fluorescence intensities comparable to those of the high-affinity control pool. Sequencing of the LC and HC cDNA from cells in the sorted pools confirmed they were expressing the high-affinity antibody (Supplementary Figure S2).

To further evaluate the platform's capability for screening antibody libraries to identify high-affinity binders, a humanized antibody library was constructed. This library was generated by shuffling the human antibody germline frameworks and complementarity-determining regions (CDRs) of the mouse anti-HER2 4D5 antibody, following a previously described method [61]. These humanized antibodies were cloned into the low-display RRKRATNFP vector and co-transfected with an FLPe-expressing helper vector into the CHO master clone, followed by puromycin selection to create a stably transfected cell pool. Cells from the high-affinity control pool were spiked into this newly generated CHO cell pool expressing the humanized antibody library at ratios of 0.01% and 0.0001%. The mixed cell pools were double-stained using the same above-described protocol and subjected to repeated rounds of FACS, selecting the top 1% of high HER2 binders in each round.

Under the 0.01% spiking condition, cells expressing antibodies with binding affinities similar to the high-affinity control pool were enriched after three rounds of sorting. For the 0.0001% spiking condition, four rounds of sorting were required to obtain an enriched pool expressing antibodies with affinities similar to the control pool (Figure 6). Sequencing of the LC and HC cDNA in cells from the enriched pools confirmed that the enriched pools were expressing the high-affinity antibody (Supplementary Figure S2).

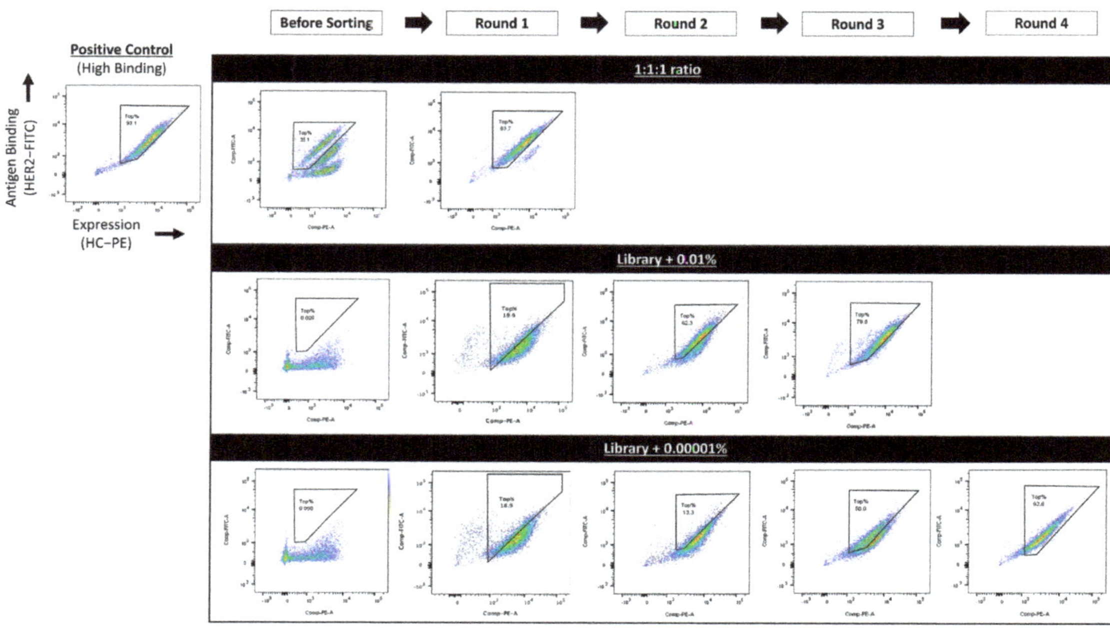

Figure 6. Sorting of stable cell pools spiked with varying percentages of cells expressing high-binding antibodies. Fluorescence-activated cell sorting (FACS) panels were configured with two parameter axes: (1) antigen binding using FITC-labeled Human Her2 (HER2-FITC) and (2) surface expression using R-PE AffiniPure F(ab′)$_2$ Fragment Goat Anti-Human IgG, Fc Gamma-specific (HC-PE). Each dot—colored according to the heatmap—represents an individual cell (or 'event'), with yel-low/red indicating high density and blue/green indicating lower density. A positive control gate, denoted as "Top%", was established based on the stable pool expressing high-binding antibodies. This gate was consistently applied across repeated rounds of FACS to ensure uniformity in sorting criteria.

4. Discussion

Previous studies utilized either mFCS or 2A peptide alone between HC and a membrane anchor to achieve simultaneous antibody display and secretion [43,44]. However, engineering mFCS to achieve different display-to-secretion ratios is challenging due to its highly conserved sequences. Meanwhile, the use of 2A peptides resulted in secreted antibodies with attached 2A residues. To address this limitation, we explored the mFCS-2A peptide combinations between HC and a GPI membrane anchor to enable simultaneous antibody display and secretion. The combinations of mFCS (RRKR) with four 2A peptides—P2A, F2A, E2A, and T2A—were evaluated. As anticipated, all vectors linking HC and GPI with these 2A peptides alone produced secreted antibodies containing 2A residues (Figure 2). Notably, the RRKR-P2A vector was the only construct that produced secreted antibodies without attached 2A residues. In contrast, secreted products from RRKR-F2A, RRKR-T2A, and RRKR-E2A vectors contained a mix of HC polypeptides with and without 2A residues (Figure 2 and Supplementary Figure S1).

The 2A peptides facilitate polypeptide cleavage through ribosomal skipping during translation, whereas furin cleavage occurs in the Golgi post-translation. When the gene encoding HC-RRKR-2A-GPI is expressed, the resulting HC polypeptides exist in two forms, HC-RRKR-2A and HC-RRKR-2A-GPI, depending on whether "self-cleavage" occurs at the 2A peptide (Figure 1B). Upon entering the Golgi, cleavage at the RRKR site in both HC-RRKR-2A and HC-RRKR-2A-GPI results in secreted antibodies lacking the 2A and 2A-GPI residues. If furin cleavage does not occur, two antibody forms are produced: one with 2A-GPI attached, displayed on the cell surface, and the other as secreted antibodies with attached 2A residues. The levels of display and the abundance of secreted antibodies, with or without 2A residues, depend on the cleavage efficiency of furin at the RRKR site and the "self-cleavage" efficiency of the 2A peptides.

Among the four RRKR-2A peptide combinations, previous studies have demonstrated that P2A has the lowest cleavage efficiency, E2A and F2A exhibit moderate cleavage efficiencies, and T2A has the highest [47]. In relation to furin cleavage efficiency, the conserved cleavage sequence is described as RP6–XP5–RP4–XP3–(K/R)P2–RP1–XP1′–XP2′–XP3′–XP4′–XP5′. Although not highly conserved, the five amino acids downstream of the core sequence, R-X-K/R-R, also significantly influence furin cleavage efficiency. S, A, F and G at P1′; L, V, A, and S at P2′; S, D, G, P, and E at P3′; A, Q, S, and T at P4′; and A, G, E and L at P5′ are highly prevalent residues in natural FCS [45,46]. The first five amino acids of P2A, F2A, T2A, and E2A are ATNFS, APVKQ, EGRGS, and QCTNY, respectively. Our data (Figure 4), combined with the prevalence of specific amino acids at positions P1′ to P5′ as reported in previous studies [45,46], suggest that positioning ATNFS downstream of RRKR enhances furin cleavage efficiency, while sequences such as APVKQ, EGRGS, and QCTNY may inhibit it. As a result, the dominant HC form expressed from the RRKR-P2A vector in the endoplasmic reticulum is HC-RRKR-P2A-GPI, with HC-RRKR-P2A as the minor form due to P2A's low cleavage efficiency. Upon Golgi entry, efficient cleavage at RRKR in both HC-RRKR-P2A and HC-RRKR-P2A-GPI results in high levels of secreted antibodies without 2A residues and low levels of cell surface display. For the RRKR-F2A and RRKR-E2A vectors, HC-RRKR-2A and HC-RRKR-2A-GPI are both dominant due to moderate 2A cleavage efficiency. Low furin cleavage efficiency at RRKR in the Golgi results in incomplete processing of both HC-RRKR-2A and HC-RRKR-2A-GPI, leading to the secretion of antibodies in two forms, with and without 2A residues, and relatively higher display levels compared to the RRKR-P2A vector. In the RRKR-T2A vector, HC-RRKR-2A is the dominant form due to T2A's high cleavage efficiency while HC-RRKR-T2A-GPI is the minor form. However, low furin cleavage efficiency in the Golgi leads to high secretion levels of antibodies in both forms, with and without 2A residues, and lower display levels. Further studies showed that mutating P2A in the mFCS-P2A vector increased the antibody display-to-secretion ratio by inhibiting furin cleavage, resulting in incomplete processing of HC-RRKR-P2A-GPI. However, this also led to the secretion of antibodies with 2A residues attached. These findings indicate that achieving both high antibody display levels and secretion of only the correct product is not feasible with the RRKR-2A strategy.

A previous study demonstrated simultaneous antibody display and secretion by linking the HC to the PDGFR membrane anchor via an mFCS sequence, RIRR [43]. However, in our study, when HC was linked to a GPI membrane anchor using a similar mFCS sequence, RRKR, antibodies were expressed exclusively in the membrane-bound form, with no detectable secretion (Figure 1). Given that RIRR and RRKR are expected to have comparable furin cleavage efficiencies, the absence of secretion in our study is likely attributable to the use of different membrane anchors rather than differences in the mFCS sequences. The first five amino acids of GPI are P, N, K, G, and S, whereas those of PDGFR are A, V, G, Q and D. Notably, P at P1′ and N at P2′ are rarely observed in natural

FCS, whereas A at P1′ and V at P2′ are highly prevalent in FCS with high furin cleavage efficiency [45,46]. This difference likely explains the lack of secretion in the RRKR vector, as furin cleavage is inhibited by the P and/or N residues at P1′ and P2′ in GPI. This hypothesis is further supported by data showing that the RRKR-P2A vector exhibited very low display levels, while mutating the first amino acid of P2A from A to P significantly increased display levels (Figure 3). Similarly, the RRKRATNFS variant showed low display levels, whereas the RRKRPTNFS variant exhibited high display levels (Figure 4). These results suggest that P at P1′ strongly inhibits furin cleavage efficiency.

Further analysis of the ten nine-amino-acid FCS variants, incorporating the first five amino acids from the N-terminus of the wild-type and mutated P2A sequences downstream of RRKR, provided deeper insights into how specific amino acids and their positions affect furin cleavage efficiency (Figure 4). For instance, P at positions P2′ to P5′ inhibited furin cleavage, although the effect was less pronounced than at P1′. G at P2′ strongly inhibited cleavage, while its effect at P1′ was weaker. A at P3′ to P4′ also inhibited cleavage but had no impact when located elsewhere. The amino acid composition at specific positions in FCS variants with high furin cleavage efficiency aligned with the prevalence of highly conserved amino acids at positions P1′ to P5′, as reported in previous studies [45,46]. This suggests that incorporating highly conserved amino acids at each position is beneficial for enhancing furin cleavage efficiency. These findings are pivotal for designing optimized FCS variants, enabling precise control over antibody display-to-secretion ratios and supporting diverse applications.

The ten nine-amino-acid FCS variants identified in this study modulated antibody display-to-secretion ratios ranging from 3.1% to 61.6%. Since the total amount of antibodies expressed by a single cell is fixed, increasing the amount of antibodies displayed on the cell surface will lead to lower secretion levels (Figure 4). Higher secretion levels are preferable for reducing cost-of-goods, making the optimal display-to-secretion ratio one that minimizes display levels while still enabling effective antibody library screening. In addition, excessively high display levels can saturate the cell surface, diminishing the platform's ability to differentiate between cells expressing antibodies with similar binding affinities [38]. By comparing two targeting vectors with display levels of 9% and 39% relative to a control display vector, we demonstrated that lower display levels more effectively distinguish cells expressing antibodies with high and closely related binding affinities. While further reducing display levels could enhance the ability to discriminate between antibodies with different affinities, excessively low display levels can result in poor signal-to-noise ratios, negatively impacting platform performance. Low display levels may not be ideal for all applications. For example, in antibody discovery using large libraries, where positive antibodies are rare and often exhibit low binding affinities, higher display levels can improve surface staining signals and increase sensitivity for detecting positive hits.

Our simultaneous antibody display and secretion platform demonstrated high sensitivity for isolating high-affinity antibody-expressing cells, while achieving antibody production levels exceeding 200 mg/L in fed-batch cultures. We validated its sensitivity by successfully isolating cells expressing a known high-affinity antibody from a pool of humanized antibody library variants, with the cells expressing the known antibody spiked at only 0.0001%. However, no antibodies with higher binding affinity than the control antibody were identified. This outcome is likely due to the limited size of the humanized antibody library integrated into the CHO master cells. The humanized antibody library we constructed contains approximately 10 million variants, but the FLP recombinase system used has an integration efficiency of around 1% (unpublished data). As a result, the transfection of 10 million CHO master cells results in the expression of only about 1×10^5 unique antibodies. This limitation in library size is a common drawback of mammalian cell display systems. Enhancing targeted integration efficiency through more effective recombinase systems, such as BxB1, could address this issue [64–67]. Additional strategies include enriching libraries via phage display before transitioning to mammalian display, utilizing immunized libraries, or designing high-quality libraries using artificial intelligence [6,55].

Another critical parameter for our platform is secretion level. While the achieved production level of over 200 mg/L is sufficient for many preclinical developability and functional studies—such as assessing antibody stability, aggregation, and some animal studies—applications like toxicity studies in animals and clinical trials typically require titers in the range of several grams per liter. To expand the platform's applications, strategies for further enhancing expression levels are necessary. Genome-wide screening to identify more active genomic sites could significantly boost expression. Optimizing the targeting vector is another approach. The current vector employs EMCV IRES to drive HC expression, which has lower translation efficiency compared to cap-dependent translation. Replacing EMCV IRES with 2A peptides for co-expression of LC and HC could substantially enhance antibody expression levels [68].

Supplementary Materials: The following supporting information can be downloaded at https://www.mdpi.com/article/10.3390/antib14020038/s1, Supplementary Figure S1: Peptide mapping analysis of HC polypeptides expressed from various targeting vectors. Supplementary Figure S2: Validation of sorted clones. Supplementary Table S1: PCR primer pairs for HC and LC.

Author Contributions: J.P.Z.N. designed and conducted experiments, analyzed data, and wrote the manuscript. M.M. conducted experiments and data analysis. J.B. conducted experiments and data analysis. M.W.C. conceived the project and revised the manuscript. Y.Y. conceived and designed the project and wrote the manuscript. All authors have read and agreed to the published version of the manuscript.

Funding: This work was supported by the Biomedical Research Council (BMRC) of the Agency for Science, Technology and Research (A*STAR), Singapore, and the Bioprocessing Technology Institute, A*STAR, Singapore.

Institutional Review Board Statement: Not applicable.

Informed Consent Statement: Not applicable.

Data Availability Statement: All data generated or analyzed during this study are included in this published article and Supplementary Materials.

Conflicts of Interest: The authors declare no conflicts of interest.

Abbreviations

The following abbreviations are used in this manuscript:

CHO	Chinese hamster ovary
ZZ-PDGFR	Dual Z-domain with PDGFR transmembrane domain
EMCV	Encephalomyocarditis virus
ECSTASY	Enzyme-cleavable surface-tethered all-purpose screening system
E2A	Equine rhinitis A virus 2A peptide
FACS	Fluorescence-activated cell sorting
Fab	Fragment antigen binding
FITC	Fluorescein isothiocyanate
FITC-HC	FITC-conjugated IgG specific to the heavy chain
F2A	Foot-and-mouth disease virus 2A peptide
GPI	Glycophospholipid
HC	Heavy chain

HC-PE	F(ab′)$_2$ Fragment Goat Anti-Human IgG, Fc Gamma specific phycoerythrin
HER2	Human epidermal growth factor receptor 2
HER2-FITC	HER2-Fluorescein isothiocyanate
HYGR	Hygromycin resistance
IgG	Immunoglobulin G
IVCD	Integrated viable cell density
IRES	Internal ribosomal entry site
LC	Light chain
MFI	Mean fluorescence intensity
mFCS	Minimal furin cleavage sequence
F3	Mutated FRT3
PFA	Peptide feature area
Pcd	Pg cell^{-1} day^{-1}
PDGFR	Platelet-derived growth factor receptor
pA	Polyadenylation signals
P2A	Porcine teschovirus-1 2A peptid
RMCE	Recombinase-mediated cassette exchange
R-PE	R-phycoerythrin
scFv	Single-chain variable fragment
qMab	Specific productivity
(ATG-)Puro	Start codon-deficient puromycin resistance gene
T2A	Thosea asigna virus 2A peptide
VH	Variable Heavy
VL	Variable Light
F	Wild-type FRT

References

1. Lu, R.-M.; Hwang, Y.-C.; Liu, I.J.; Lee, C.-C.; Tsai, H.-Z.; Li, H.-J.; Wu, H.-C. Development of therapeutic antibodies for the treatment of diseases. *J. Biomed. Sci.* **2020**, *27*, 1. [CrossRef] [PubMed]
2. Mitra, S.; Tomar, P.C. Hybridoma technology; advancements, clinical significance, and future aspects. *J. Genet. Eng. Biotechnol.* **2021**, *19*, 1–12. [CrossRef]
3. Valldorf, B.; Hinz, S.C.; Russo, G.; Pekar, L.; Mohr, L.; Klemm, J.; Doerner, A.; Krah, S.; Hust, M.; Zielonka, S. Antibody display technologies: Selecting the cream of the crop. *Biol. Chem.* **2022**, *403*, 455–477. [CrossRef]
4. Wang, S.; Li, Y.; Mei, J.; Wu, S.; Ying, G.; Yi, Y. Precision engineering of antibodies: A review of modification and design in the Fab region. *Int. J. Biol. Macromol.* **2024**, 133730. [CrossRef]
5. Simons, J.F.; Lim, Y.W.; Carter, K.P.; Wagner, E.K.; Wayham, N.; Adler, A.S.; Johnson, D.S. Affinity maturation of antibodies by combinatorial codon mutagenesis versus error-prone PCR. *Mabs* **2020**, *12*, 1803646. [CrossRef]
6. Slavny, P.; Hegde, M.; Doerner, A.; Parthiban, K.; McCafferty, J.; Zielonka, S.; Hoet, R. Advancements in mammalian display technology for therapeutic antibody development and beyond: Current landscape, challenges, and future prospects. *Front. Immunol.* **2024**, *15*, 1469329. [CrossRef] [PubMed]
7. Gaa, R.; Mayer, H.M.; Noack, D.; Kumari, K.; Guenther, R.; Tsai, S.-P.; Ji, Q.; Doerner, A. Mammalian display to secretion switchable libraries for antibody preselection and high throughput functional screening. In *Mabs*; Taylor & Francis: Abingdon, UK, 2023; Volume 15, p. 2251190.
8. King, D.J.; Bowers, P.M.; Kehry, M.R.; Horlick, R.A. Mammalian cell display and somatic hypermutation in vitro for human antibody discovery. *Curr. Drug Discov. Technol.* **2014**, *11*, 56–64. [CrossRef] [PubMed]
9. Fernández-Quintero, M.L.; Ljungars, A.; Waibl, F.; Greiff, V.; Andersen, J.T.; Gjølberg, T.T.; Jenkins, T.P.; Voldborg, B.G.; Grav, L.M.; Kumar, S.; et al. Assessing developability early in the discovery process for novel biologics. *mAbs* **2023**, *15*, 2171248. [CrossRef]
10. Wang, Y.; Zhang, K.; Zhao, Y.; Li, Y.; Su, W.; Li, S. Construction and applications of mammalian cell-based DNA-encoded peptide/protein libraries. *ACS Synth. Biol.* **2023**, *12*, 1874–1888. [CrossRef]
11. Ho, M.; Nagata, S.; Pastan, I. Isolation of anti-CD22 Fv with high affinity by Fv display on human cells. *Proc. Natl. Acad. Sci. USA* **2006**, *103*, 9637–9642. [CrossRef]
12. Beerli, R.R.; Bauer, M.; Buser, R.B.; Gwerder, M.; Muntwiler, S.; Maurer, P.; Saudan, P.; Bachmann, M.F. Isolation of human monoclonal antibodies by mammalian cell display. *Proc. Natl. Acad. Sci. USA* **2008**, *105*, 14336–14341. [CrossRef] [PubMed]

13. Zhou, C.; Jacobsen, F.W.; Cai, L.; Chen, Q.; Shen, D. Development of a novel mammalian cell surface antibody display platform. In *Mabs*; Taylor & Francis: Abingdon, UK, 2010; Volume 2, pp. 508–518.
14. Moran, P.; Raab, H.; Kohr, W.J.; Caras, I.W. Glycophospholipid membrane anchor attachment. Molecular analysis of the cleavage/attachment site. *J. Biol. Chem.* **1991**, *266*, 1250–1257. [CrossRef]
15. Akamatsu, Y.; Pakabunto, K.; Xu, Z.; Zhang, Y.; Tsurushita, N. Whole IgG surface display on mammalian cells: Application to isolation of neutralizing chicken monoclonal anti-IL-12 antibodies. *J. Immunol. Methods* **2007**, *327*, 40–52. [CrossRef]
16. Medof, M.E.; Walter, E.I.; Roberts, W.L.; Haas, R.; Rosenberry, T.L. Decay accelerating factor of complement is anchored to cells by a C-terminal glycolipid. *Biochemistry* **1986**, *25*, 6740–6747. [CrossRef]
17. Dyson, M.R.; Masters, E.; Pazeraitis, D.; Perera, R.L.; Syrjanen, J.L.; Surade, S.; Thorsteinson, N.; Parthiban, K.; Jones, P.C.; Sattar, M.; et al. Beyond affinity: Selection of antibody variants with optimal biophysical properties and reduced immunogenicity from mammalian display libraries. *Mabs* **2020**, *12*, 1829335. [CrossRef] [PubMed]
18. Huhtinen, O.; Salbo, R.; Lamminmäki, U.; Prince, S. Selection of biophysically favorable antibody variants using a modified Flp-In CHO mammalian display platform. *Front. Bioeng. Biotechnol.* **2023**, *11*, 1170081. [CrossRef]
19. Luo, R.; Qu, B.; An, L.; Zhao, Y.; Cao, Y.; Ren, P.; Hang, H. Simultaneous maturation of single chain antibody stability and affinity by CHO cell display. *Bioengineering* **2022**, *9*, 360. [CrossRef] [PubMed]
20. Robertson, N.; Lopez-Anton, N.; Gurjar, S.A.; Khalique, H.; Khalaf, Z.; Clerkin, S.; Leydon, V.R.; Parker-Manuel, R.; Raeside, A.; Payne, T. Development of a novel mammalian display system for selection of antibodies against membrane proteins. *J. Biol. Chem.* **2020**, *295*, 18436–18448. [CrossRef]
21. Zhang, H.; Yea, K.; Xie, J.; Ruiz, D.; Wilson, I.A.; Lerner, R.A. Selecting agonists from single cells infected with combinatorial antibody libraries. *Chem. Biol.* **2013**, *20*, 734–741. [CrossRef]
22. Waldmeier, L.; Hellmann, I.; Gutknecht, C.K.; Wolter, F.I.; Cook, S.C.; Reddy, S.T.; Grawunder, U.; Beerli, R.R. Transpo-mAb display: Transposition-mediated B cell display and functional screening of full-length IgG antibody libraries. *Mabs* **2016**, *8*, 726–740. [CrossRef]
23. Hellmann, I.; Waldmeier, L.; Bannwarth-Escher, M.-C.; Maslova, K.; Wolter, F.I.; Grawunder, U.; Beerli, R.R. Novel antibody drug conjugates targeting tumor-associated receptor tyrosine kinase ROR2 by functional screening of fully human antibody libraries using Transpo-mAb display on progenitor B cells. *Front. Immunol.* **2018**, *9*, 2490. [CrossRef] [PubMed]
24. Wang, Y.; Jin, R.; Shen, B.; Li, N.; Zhou, H.; Wang, W.; Zhao, Y.; Huang, M.; Fang, P.; Wang, S. High-throughput functional screening for next-generation cancer immunotherapy using droplet-based microfluidics. *Sci. Adv.* **2021**, *7*, eabe3839. [CrossRef]
25. Segaliny, A.I.; Jayaraman, J.; Chen, X.; Chong, J.; Luxon, R.; Fung, A.; Fu, Q. Jiang, X.; Rivera, R.; Ma, X. A high throughput bispecific antibody discovery pipeline. *Commun. Biol.* **2023**, *6*, 380. [CrossRef]
26. Josephides, D.; Davoli, S.; Whitley, W.; Ruis, R.; Salter, R.; Gokkaya, S.; Vallet, M.; Matthews, D.; Benazzi, G.; Shvets, E. Cyto-mine: An integrated, picodroplet system for high-throughput single-cell analysis, sorting, dispensing, and monoclonality assurance. *SLAS Technol. Transl. Life Sci. Innov.* **2020**, *25*, 177–189. [CrossRef] [PubMed]
27. Fischer, K.; Lulla, A.; So, T.Y.; Pereyra-Gerber, P.; Raybould, M.I.J.; Kohler, T.N.; Yam-Puc, J.C.; Kaminski, T.S.; Hughes, R.; Pyeatt, G.L. Rapid discovery of monoclonal antibodies by microfluidics-enabled FACS of single pathogen-specific antibody-secreting cells. *Nat. Biotechnol.* **2024**, 1–11. [CrossRef] [PubMed]
28. Adler, A.S.; Mizrahi, R.A.; Spindler, M.J.; Adams, M.S.; Asensio, M.A.; Edgar, R.C.; Leong, J.; Leong, R.; Johnson, D.S. Rare, high-affinity mouse anti-PD-1 antibodies that function in checkpoint blockade, discovered using microfluidics and molecular genomics. *Mabs* **2017**, *9*, 1270–1281. [CrossRef]
29. Winters, A.; McFadden, K.; Bergen, J.; Landas, J.; Berry, K.A.; Gonzalez, A.; Salimi-Moosavi, H.; Murawsky, C.M.; Tagari, P.; King, C.T. Rapid single B cell antibody discovery using nanopens and structured light. In *Mabs*; Taylor & Francis: Abingdon, UK, 2019; Volume 11, pp. 1025–1035.
30. Fitzgerald, V.; Manning, B.; O'Donnell, B.; O'Reilly, B.; O'Sullivan, D.; O'Kennedy, R.; Leonard, P. Exploiting highly ordered subnanoliter volume microcapillaries as microtools for the analysis of antibody producing cells. *Anal. Chem.* **2015**, *87*, 997–1003. [CrossRef] [PubMed]
31. Chakrabarti, L.; Zhuang, L.; Roy, G.; Bowen, M.A.; Dall'Acqua, W.F.; Hawley-Nelson, P.; Marelli, M. Amber suppression coupled with inducible surface display identifies cells with high recombinant protein productivity. *Biotechnol. Bioeng.* **2019**, *116*, 793–804. [CrossRef]
32. Chuang, K.H.; Hsieh, Y.C.; Chiang, I.S.; Chuang, C.H.; Kao, C.H.; Cheng, T.C.; Wang, Y.T.; Lin, W.W.; Chen, B.M.; Roffler, S.R.; et al. High-throughput sorting of the highest producing cell via a transiently protein-anchored system. *PLoS ONE* **2014**, *9*, e102569. [CrossRef]
33. Bouquin, T.; Rasmussen, P.B.; Bertilsson, P.-A.; Okkels, J.S. Regulated readthrough: A new method for the alternative tagging and targeting of recombinant proteins. *J. Biotechnol.* **2006**, *125*, 516–528. [CrossRef]

34. Chen, C.-P.; Hsieh, Y.-T.; Prijovich, Z.M.; Chuang, H.-Y.; Chen, K.-C.; Lu, W.-C.; Tseng, Q.; Leu, Y.-L.; Cheng, T.-L.; Roffler, S.R. ECSTASY, an adjustable membrane-tethered/soluble protein expression system for the directed evolution of mammalian proteins. *Protein Eng. Des. Sel.* **2012**, *25*, 367–375. [CrossRef] [PubMed]
35. Yu, B.; Wages, J.M.; Larrick, J.W. Antibody-membrane switch (AMS) technology for facile cell line development. *Protein Eng. Des. Sel.* **2014**, *27*, 309–315. [CrossRef] [PubMed]
36. Breous-Nystrom, E.; Schultze, K.; Meier, M.; Flueck, L.; Holzer, C.; Boll, M.; Seibert, V.; Schuster, A.; Blanusa, M.; Schaefer, V.; et al. Retrocyte Display® technology: Generation and screening of a high diversity cellular antibody library. *Methods* **2014**, *65*, 57–67. [CrossRef] [PubMed]
37. Tomimatsu, K.; Matsumoto, S.; Tanaka, H.; Yamashita, M.; Nakanishi, H.; Teruya, K.; Kazuno, S.; Kinjo, T.; Hamasaki, T.; Kusumoto, K. A rapid screening and production method using a novel mammalian cell display to isolate human monoclonal antibodies. *Biochem. Biophys. Res. Commun.* **2013**, *441*, 59–64. [CrossRef]
38. Nguyen, A.W.; Le, K.C.; Maynard, J.A. Identification of high affinity HER2 binding antibodies using CHO Fab surface display. *Protein Eng. Des. Sel.* **2018**, *31*, 91–101. [CrossRef]
39. Horlick, R.A.; Macomber, J.L.; Bowers, P.M.; Neben, T.Y.; Tomlinson, G.L.; Krapf, I.P.; Dalton, J.L.; Verdino, P.; King, D.J. Simultaneous surface display and secretion of proteins from mammalian cells facilitate efficient in vitro selection and maturation of antibodies. *J. Biol. Chem.* **2013**, *288*, 19861–19869. [CrossRef]
40. Bowers, P.M.; Horlick, R.A.; Kehry, M.R.; Neben, T.Y.; Tomlinson, G.L.; Altobell, L.; Zhang, X.; Macomber, J.L.; Krapf, I.P.; Wu, B.F. Mammalian cell display for the discovery and optimization of antibody therapeutics. *Methods* **2014**, *65*, 44–56. [CrossRef]
41. Aebischer-Gumy, C.; Moretti, P.; Ollier, R.; Ries Fecourt, C.; Rousseau, F.; Bertschinger, M. SPLICELECT™: An adaptable cell surface display technology based on alternative splicing allowing the qualitative and quantitative prediction of secreted product at a single-cell level. *Mabs* **2020**, *12*, 1709333. [CrossRef]
42. Lang, S.; Drewello, D.; Wichter, J.; Nommay, A.; Wilms, B.; Knopf, H.P.; Jostock, T. Surface display vectors for selective detection and isolation of high level antibody producing cells. *Biotechnol. Bioeng.* **2016**, *113*, 2386–2393. [CrossRef]
43. Zhou, Y.; Wang, J.; Zhou, I.; Lou, H.; Li, C.Z.; Chen, Z.R.; Zhang, Z.H.; Liu, S.; Wu, S.; Tan, W.; et al. Simultaneous expression of displayed and secreted antibodies for antibody screen. *PLoS ONE* **2013**, *8*, e80005. [CrossRef]
44. Yu, K.K.; Aguilar, K.; Tsai, J.; Galimidi, R.; Gnanapragasam, P.; Yang, L.; Baltimore, D. Use of mutated self-cleaving 2A peptides as a molecular rheostat to direct simultaneous formation of membrane and secreted anti-HIV immunoglobulins. *PLoS ONE* **2012**, *7*, e50438. [CrossRef] [PubMed]
45. Izidoro, M.A.; Gouvea, I.E.; Santos, J.A.; Assis, D.M.; Oliveira, V.; Judice, W.A.; Juliano, M.A.; Lindberg, I.; Juliano, L. A study of human furin specificity using synthetic peptides derived from natural substrates, and effects of potassium ions. *Arch. Biochem. Biophys.* **2009**, *487*, 105–114. [CrossRef]
46. Weldon, J.E.; Skarzynski, M.; Therres, J.A.; Ostovitz, J.R.; Zhou, H.; Kreitman, R.J.; Pastan, I. Designing the furin-cleavable linker in recombinant immunotoxins based on Pseudomonas exotoxin A. *Bioconjug. Chem.* **2015**, *26*, 1120–1128. [CrossRef]
47. Chng, J.; Wang, T.; Nian, R.; Lau, A.; Hoi, K.M.; Ho, S.C.L.; Gagnon, P.; Bi, X.; Yang, Y. Cleavage efficient 2A peptides for high level monoclonal antibody expression in CHO cells. In *Mabs*; Taylor & Francis: Abingdon, UK, 2015; Volume 7, pp. 403–412.
48. Higuchi, K.; Araki, T.; Matsuzaki, O.; Sato, A.; Kanno, K.; Kitaguchi, N.; Ito, H. Cell display library for gene cloning of variable regions of human antibodies to hepatitis B surface antigen. *J. Immunol. Methods* **1997**, *202*, 193–204. [CrossRef] [PubMed]
49. Ho, M.; Pastan, I. Display and selection of scFv antibodies on HEK-293T cells. *Methods Mol. Biol.* **2009**, *562*, 99–113. [CrossRef] [PubMed]
50. Zhang, J.; Zhang, X.; Liu, Q.; Li, M.; Gao, L.; Gao, X.; Xiang, S.; Wu, L.; Fu, J.; Song, H. Mammalian cell display for rapid screening scFv antibody therapy. *Acta Biochim. Biophys. Sin.* **2014**, *46*, 859–866. [CrossRef]
51. Bowers, P.M.; Horlick, R.A.; Neben, T.Y.; Toobian, R.M.; Tomlinson, G.L.; Dalton, J.L.; Jones, H.A.; Chen, A.; Altobell, L.; Zhang, X.; et al. Coupling mammalian cell surface display with somatic hypermutation for the discovery and maturation of human antibodies. *Proc. Natl. Acad. Sci. USA* **2011**, *108*, 20455–20460. [CrossRef]
52. Hillenbrand, M.; Esslinger, C.; Seidenberg, J.; Weber, M.; Zingg, A.; Townsend, C.; Eicher, B.; Rutkauskaite, J.; Riese, P.; Guzman, C.A. Fast-Track Discovery of SARS-CoV-2-Neutralizing Antibodies from Human B Cells by Direct Functional Screening. *Viruses* **2024**, *16*, 339. [CrossRef]
53. Chang, J.; Rader, C.; Peng, H. A mammalian cell display platform based on scFab transposition. *Antib. Ther.* **2023**, *6*, 157–169. [CrossRef]
54. Chen, C.; Li, N.; Zhao, Y.; Hang, H.Y. Coupling recombinase-mediated cassette exchange with somatic hypermutation for antibody affinity maturation in CHO cells. *Biotechnol. Bioeng.* **2016**, *113*, 39–51. [CrossRef]
55. Jin, Y.; Yu, D.; Tian, X.; Li, H.; Zhou, X.; Kong, Y.; Zhang, W.; Zhang, L.; Lei, C.; Yang, Z. A novel and effective approach to generate germline-like monoclonal antibodies by integration of phage and mammalian cell display platforms. *Acta Pharmacol. Sin.* **2022**, *43*, 954–962. [CrossRef] [PubMed]

56. Mason, D.M.; Weber, C.R.; Parola, C.; Meng, S.M.; Greiff, V.; Kelton, W.J.; Reddy, S.T. High-throughput antibody engineering in mammalian cells by CRISPR/Cas9-mediated homology-directed mutagenesis. *Nucleic Acids Res.* **2018**, *46*, 7436–7449. [CrossRef] [PubMed]
57. Parola, C.; Neumeier, D.; Friedensohn, S.; Csepregi, L.; Di Tacchio, M.; Mason, D.M.; Reddy, S.T. Antibody discovery and engineering by enhanced CRISPR-Cas9 integration of variable gene cassette libraries in mammalian cells. In *Mabs*; Taylor & Francis: Abingdon, UK, 2019; Volume 11, pp. 1367–1380.
58. Parthiban, K.; Perera, R.L.; Sattar, M.; Huang, Y.; Mayle, S.; Masters, E.; Griffiths, D.; Surade, S.; Leah, R.; Dyson, M.R.; et al. A comprehensive search of functional sequence space using large mammalian display libraries created by gene editing. *Mabs* **2019**, *11*, 884–898. [CrossRef]
59. Nguyen, N.T.B.; Lin, J.; Tay, S.J.; Mariati Yeo, J.; Nguyen-Khuong, T.; Yang, Y. Multiplexed engineering glycosyltransferase genes in CHO cells via targeted integration for producing antibodies with diverse complex-type N-glycans. *Sci. Rep.* **2021**, *11*, 12969. [CrossRef]
60. Loh, H.P.; Mahfut FBin Chen, S.W.; Huang, Y.; Huo, J.; Zhang, W.; Lam, K.P.; Xu, S.; Yang, Y. Manufacturability and functionality assessment of different formats of T-cell engaging bispecific antibodies. In *Mabs*; Taylor & Francis: Abingdon, UK, 2023; Volume 15, p. 2231129.
61. Dall'Acqua, W.F.; Damschroder, M.M.; Zhang, J.; Woods, R.M.; Widjaja, L.; Yu, J.; Wu, H. Antibody humanization by framework shuffling. *Methods* **2005**, *36*, 43–60. [CrossRef]
62. Sharma, P.; Yan, F.; Doronina, V.A.; Escuin-Ordinas, H.; Ryan, M.D.; Brown, J.D. 2A peptides provide distinct solutions to driving stop-carry on translational recoding. *Nucleic Acids Res.* **2012**, *40*, 3143–3151. [CrossRef] [PubMed]
63. Johnstone, R.W.; Andrew, S.M.; Hogarth, M.P.; Pietersz, G.A.; McKenzie, I.F.C. The effect of temperature on the binding kinetics and equilibrium constants of monoclonal antibodies to cell surface antigens. *Mol Immunol.* **1990**, *27*, 327–333. [CrossRef]
64. Chi, X.; Zheng, Q.; Jiang, R.; Chen-Tsai, R.Y.; Kong, L.-J. A system for site-specific integration of transgenes in mammalian cells. *PLoS ONE* **2019**, *14*, e0219842. [CrossRef]
65. Jusiak, B.; Jagtap, K.; Gaidukov, L.; Duportet, X.; Bandara, K.; Chu, J.; Zhang, L.; Weiss, R.; Lu, T.K. Comparison of integrases identifies Bxb1-GA mutant as the most efficient site-specific integrase system in mammalian cells. *ACS Synth. Biol.* **2019**, *8*, 16–24. [CrossRef]
66. Pandey, S.; Gao, X.D.; Krasnow, N.A.; McElroy, A.; Tao, Y.A.; Duby, J.E.; Steinbeck, B.J.; McCreary, J.; Pierce, S.E.; Tolar, J. Efficient site-specific integration of large genes in mammalian cells via continuously evolved recombinases and prime editing. *Nat. Biomed. Eng.* **2024**, 1–18. [CrossRef]
67. Hew, B.E.; Gupta, S.; Sato, R.; Waller, D.F.; Stoytchev, I.; Short, J.E.; Sharek, L.; Tran, C.T.; Badran, A.H.; Owens, J.B. Directed evolution of hyperactive integrases for site specific insertion of transgenes. *Nucleic Acids Res.* **2024**, *52*, e64. [CrossRef] [PubMed]
68. Ho, S.C.L.; Bardor, M.; Li, B.; Lee, J.J.; Song, Z.; Tong, Y.W.; Goh, L.-T.; Yang, Y. Comparison of internal ribosome entry site (IRES) and Furin-2A (F2A) for monoclonal antibody expression level and quality in CHO cells. *PLoS ONE* **2013**, *8*, e63247. [CrossRef] [PubMed]

antibodies

MDPI

Article

CEA-4-1BBL: CEACAM5-Targeted 4-1BB Ligand Fusion Proteins for Cis Co-Stimulation with CEA-TCB

Christina Claus [1], Claudia Ferrara-Koller [1], Johannes Sam [1], Sabine Lang [1], Rosmarie Albrecht [1], Regula B. Buser [1], Esther Bommer [1], Grégory La Sala [1], Valeria G. Nicolini [1,*], Sara Colombetti [2], Marina Bacac [1], Pablo Umaña [1] and Christian Klein [1,3]

[1] Roche Innovation Center Zurich, Roche Pharma Research and Early Development (pRED), Wagistrasse 10, 8952 Schlieren, Switzerland

[2] Roche Innovation Center Basel, Roche Pharma Research and Early Development (pRED), Grenzacherstrasse 124, 4070 Basel, Switzerland

[3] Department of Biochemistry, Faculty of Chemistry and Pharmacy, Ludwig Maximilians University of Munich, 81377 Munich, Germany

* Correspondence: valeria.nicolini@roche.com

Abstract: Background/Objectives: T cell bispecific antibodies (TCBs) result in the activation of T cell receptor signaling upon binding to tumor antigens providing signal 1 to T cells. To enhance and sustain their activity, a co-stimulatory signal 2 is required. Here CEACAM5-targeted 4-1BBL antibody fusion proteins for combination with CEA-TCB (cibisatamab, RG7802) are described in an investigation of the relationship between the CEACAM5 epitope and T cell activity. Methods: CEACAM5-targeted bispecific 4-1BBL antibody fusion proteins (CEA-4-1BBLs) were generated based on different CEACAM5 antibodies and characterized in vitro in Jurkat-4-1BB reporter and PBMC cell assays. The impact of shed CEA on in vitro activity and cynomolgus cross-reactivity was studied. In vivo efficacy was assessed in human stem cell humanized NSG mice xenograft models bearing MKN-45 and HPAFII tumors. Results: MFE23-4-1BBL and Sm9b-4-1BBL showed superior functional activity in Jurkat-4-1BB reporter and primary T cell assays when combined with the CD3 antibody V9, whereas T84.66-LCHA-4-1BBL and A5B7-4-1BBL performed better when combined with CEA-TCB. In humanized NSG mice MKN-45 and HPAFII xenograft models, T84.66-LCHA-4-1BBL mediated the best anti-tumor efficacy. Conclusions: For the assessment of the combination of CEA-TCB with CEA-4-1BBL, co-stimulatory antibody fusion protein in vitro assays are not sufficient to fully capture the complex relationships affecting efficacy. Thus, screening with different cell assays and in vivo efficacy studies in combination with CEA-TCB are essential to select the best candidate. Based on the totality of data on the T84.66-LCHA-4-1BBL antibody fusion protein comprising the CEACAM5 antibody, T84.66-LCHA was selected as the optimal combination partner for CEA-TCB.

Keywords: CEACAM5; 4-1BB; TNFRSF9; CD137; membrane-proximal; membrane-distal; epitope

1. Introduction

Checkpoint-inhibitory and bispecific antibodies have shaped cancer immunotherapy strategies during the last decade [1–4]. Tumor-infiltrating T cells can either receive a "natural" activation signal 1 through a T cell receptor (TCR) due to its interaction with peptide–MHC complexes (endogenous immunity) or receive one "artificially" by a T cell bispecific antibody (TCB) simultaneously bound to a tumor cell surface antigen and the

CD3e chain of the TCR complex leading to redirected T cell-mediated tumor cell killing (synthetic immunity) [2,5–9]. To enhance and sustain the activity of T cells, a co-stimulatory signal 2 is required, and various approaches have been described to provide this signal 2 co-stimulatory signal by agonizing CD28 or 4-1BB/CD137 with bispecific antibodies or fusion proteins [10–14]. In the clinic, promising data have been reported for several bispecific 4-1BB agonists, both as single agents and in combination with checkpoint inhibition [15–20], while clinical data for the combination with TCBs are eagerly awaited.

Carcinoembryonic antigen cell adhesion molecule 5 (CEACAM5), also known as CEA or CD66e, is a highly glycosylated cell surface protein anchored in the membrane via a glycosylphosphatidylinositol (GPI) anchor and normally functions as an adhesion molecule that can interact with CEACAM6 and CEACAM1 and plays a pivotal role in cancer biology [21,22]. CEACAM5 expression has been shown to be important for metastasis in colon cancer, but also for immune evasion of tumor cells [21,23]. CEACAM5 is overexpressed in various epithelial malignancies, including gastric, colorectal, pancreatic, breast and non-small cell lung cancer, making it an attractive drug target. Furthermore, in colorectal cancer, shed CEA is an FDA-approved diagnostic tumor marker [21,23].

We have previously described tumor stroma-targeted FAP-4-1BBL and tumor-targeted CD19-4-1BBL antibody fusion proteins that can bind simultaneously to fibroblast activation protein (FAP) or CD19, respectively, and the co-stimulatory receptor 4-1BB/CD137 [24,25]. These 4-1BBL antibody fusion proteins provide signal 2 to 4-1BB-expressing T cells, either in cis by binding to CD19-expressing malignant B cells in combination with the CD20-TCB glofitamab [26] or in trans by binding to FAP-expressing tumor fibroblasts in combination with the CEA-TCB cibisatamab [27,28]. Likewise, the combination of a CEACAM5-targeted CEA-4-1BBL fusion protein in combination with CEA-TCB was shown to strongly enhance T cell infiltration in the MKN-45 xenograft model in humanized NSG mice using the CD8-specific PET-Tracer 89Zr-Df-IAB22M2C [29]. While we have not seen major differences between 4-1BB co-stimulation in cis versus in trans, Otano et al. found that co-stimulation may be more potent when provided in cis as compared to in trans [30]. Another recent approach for 4-1BB co-stimulation in cis has been the generation of trispecific T cell engagers with integrated co-stimulation through 4-1BB binding moieties [31–33].

Here the design, optimization and characterization of novel CEACAM5-targeted CEA-4-1BBL antibody fusion proteins with the goal of providing signal 2 to T cells receiving their signal 1 from CEA-TCB in cis—e.g., independently of FAP expression on fibroblasts—is described. Several CEA-4-1BBL antibody fusion proteins directed against different domains of CEACAM5 were generated and tested for functionality in vitro and in vivo. The CEACAM5 antibodies chosen for the construction of the respective CEA-4-1BBL antibody fusion proteins recognize membrane-distal or membrane-proximal epitopes. Reporter T cell assays were applied for a first ranking of functionality. Targeting membrane-distal CEACAM5 epitopes mediated better T cell activation in reporter cells compared to targeting membrane-proximal epitopes when tested as a monotherapy. In contrast to the reporter cell line assays, two different PBMC-based in vitro activation assays in combination with different signal 1 providers were not able to provide a clear differentiation between the different CEACAM5 targeting antibodies. However, in MKN45 and HPAFII in vivo xenograft models in humanized NSG mice, the CEA-4-1BBL antibody fusion protein constructed with the CEACAM5 antibody T84.66-LCHA targeting a membrane-proximal CEACAM5 epitope showed the best tumor control overall in combination with CEA-TCB. Based on the totality of data, the T84.66-LCHA-4-1BBL antibody fusion protein was selected as the optimal co-stimulatory antibody fusion protein for combination with CEA-TCB.

Furthermore, the preclinical data presented indicate that reporter cell line assays can be limited in their translatability and that in order to select the best candidate screening with different cell assays, in vivo efficacy studies together with potential combination partners are essential.

2. Materials and Methods

Reagents: CEA–4-1BBL molecules, non-binding control DP47–4-1BBL, CEA-TCB (cibisatamab, RG7802) and huCEACAM5 peptides (NCBI EDL26990.1) were generated by Roche Innovation Centers, Zurich or Munich. CEA–4-1BBL and DP47-4-1BBL were assembled using the following polypeptide chains: (a) two ectodomains of 4-1BBL (UniProtKB P41273) fused to CL followed by a human IgG1 Fc-knob chain, (b) one ectodomain of 4-1BBL fused to CH1, (c) a CEACAM5 or DP47 VH-CH1 fused to a human IgG1 Fc-hole chain, and (d) a CEACAM5 or DP47 VL-CL. As previously described, correct chain association was enabled using knob-into-hole technology [34] and CH1-CL crossover with charge mutations [35,36], and Fc silencing was achieved with P329G LALA mutations [37]. Molecules were produced in HEK293-EBNA or CHO cells and purified by affinity chromatography followed by size exclusion chromatography as previously described [24].

The monomer content and integrity of the molecules were determined, respectively, by size exclusion on HPLC using a TSKgel UP-SW3000 column (Tosoh Biosciences LLC, King of Prussia, PA, USA) (<95%) and by CE-SDS (LabChip GXII Touch HT, Revvity, Zurich, Switzerland) (<95%). The mass of the purified molecules was confirmed by LC-MS analysis in a non-reduced state, and endotoxin content was determined. All constructs fulfilled the internal standards and were released for use in vivo. CEACAM5 Fabs were prepared by plasmin (Roche, Mannheim, Germany 10602361001) digestion of the respective human IgG1s as recently described [24]. The NABA constructs were used as antigens for the CEACAM5 Fabs. Since the CEACAM5 binders recognize different CEACAM5 domains, the following NABA constructs, containing N, A or B domains of human CEACAM1 and CEACAM5, were prepared. The constructs were purified by standard chromatographic methods (chelating chromatography and preparative SEC).

Surface plasmon resonance: All SPR experiments were performed on a Biacore T200 (Cytiva, Freiburg, Germany) at 25 °C with HBS-EP as a running buffer (0.01 M HEPES (pH 7.4), 0.15 M NaCl, 3 mM EDTA, 0.005% surfactant P20; Cytiva, Freiburg, Germany). His-specific antibody (Qiagen, Hilden, Germany, 34660) was coupled on a CM3 chip at pH 5.0 using the standard amine coupling kit (Cytiva, Freiburg, Germany). The NABA constructs were captured with a flow rate of 10 μL/minute for 30 or 40 s on flow cell 2 (Supplementary Table S1). Dilution series of CEACAM5 Fabs were passed on both flow cells at 30 μL/minute for 150 s to record the association phase. The dissociation phase was monitored for 300 s and triggered by switching from the sample solution to HBS-EP. The chip surface was regenerated after every cycle using a double injection of 60 s Glycine (pH 2.0). Bulk refractive index differences were corrected for by subtracting the response obtained on the reference flow cell, where no NABA construct was captured. The affinity constants were derived from the rate constants by fitting to a 1:1 Langmuir binding curve using the Biacore T200 Evaluation Version 3.2.2 (Cytiva, Freiburg, Germany). The results are the average of three experiments.

Cell lines: Human gastric cancer cell line MKN-45 (DSMZ ACC 409) and human colon adenocarcinoma cell lines CX-1 (DSMZ ACC 129) and HT-29 (ATCC HTB-38) were cultured in RPMI 1640 (Gibco, Thermo Fisher Scientific, Reinach, Switzerland, 42401) supplemented with 10% FBS (Gibco, Thermo Fisher Scientific, Reinach, Switzerland, 16140 or Sigma-Aldrich, Buchs, Switzerland, F4135) and 2 mM GlutaMAX (Gibco, Thermo Fisher Scientific, Reinach, Switzerland, 35050). Human pancreatic adenocarcinoma cell line HPAF-II (ATCC CRL-1997) was cultured in MEM/EBSS (Gibco, Thermo Fisher Scientific, Reinach, Switzerland, 21090) supplemented with 10% FBS, 2 mM GlutaMAX, 1% MEM-Nonessential amino acids (Sigma-Aldrich, Buchs, Switzerland, M7145) and 1 mM Sodium-Pyruvate (Sigma-Aldrich, Buchs, Switzerland, S8636). Human colorectal adenocarcinoma cell line LS-180 (ATCC CL-187) was cultured in MEM/EBSS supplemented with 10% FBS and 2 mM GlutaMAX. LoVo (ATCC CCL-229) was cultured in DMEM/F12 (Gibco, Thermo Fisher Scientific, Reinach, Switzerland, 31331) supplemented with 10% FBS and 2 mM GlutaMAX. All tumor cell lines grew adherently and were split 2-3 times per week, keeping a confluence between 20 and 80%. Jurkat-human-4-1BB-NFkB-luc2 (Promega CS196004) reporter cell line was cultured in RPMI 1640 supplemented with 10% FBS, 2 mM GlutaMAX, 1 mM Sodium-Pyruvate, 1% (*v*/*v*) MEM-nonessential Amino acid solution, 25 mM HEPES (Sigma-Aldrich, Buchs, Switzerland, H0887), 600 µg/mL G-418 (Roche, Mannheim, Germany 04727894001) and 400 µg/mL Hygromycin B (Roche, Mannheim, Germany 10843555001). The suspension cells were split twice per week, keeping a cell density of 0.2–1.2 × 106 cells/mL. CHO-k1 (ATCC CCL-61) cells were transfected with transfection plasmid pETR19312 to express cynomolgus monkey CEACAM5, and CHO-k1-cynoCEACAM5 clone 8 was selected. CHO-k1-cynoCEACAM5 clone 8 cell line was cultured in DMEM/F-12 (Gibco, Thermo Fisher Scientific, Reinach, Switzerland, 31330) supplemented with with 10% FBS and 6 µg/mL Puromycin (Gibco, Thermo Fisher Scientific, Reinach, Switzerland, A1113802) and split 3 times per week, keeping a confluence between 20 and 80%. To generate CHO-k1CEACAM5 clone 11 and CHO-k1-huCEACAM5 clone 12 cell lines CHO-k1 (ATCC CCL-61), cells were transfected with transfection plasmid pETR14582 to express human CEACAM5. Cells were cultured in DMEM/F-12 (Gibco, Thermo Fisher Scientific, Reinach, Switzerland, 31330) supplemented with 10% FBS and 500 µg/mL Zeocin (Gibco R25001) and split 3 times per week, keeping a confluence between 20 and 80%.

Cell binding assay: Tumor cell lines or CEACAM5-transfected CHO-k1 cells were harvested using 0.05% Trypsin-EDTA (Gibco, Thermo Fisher Scientific, Reinach, Switzerland, 25300) for 10 min, washed and resuspended in DPBS (Gibco, Thermo Fisher Scientific, Reinach, Switzerland, 14190). Then, 2 × 105 cells/well or 3 × 103 cells/well were seeded on a round-bottom 96-well plate (Cellstar Sigma-Aldrich, Buchs, Switzerland, 650185) or a round-bottom 384-well plate (Corning, New York, NY, USA, 3830), respectively. Cells were stained with fixable viability stain eF450 (eBioscience, Thermo Fisher Scientific, Reinach, Switzerland, 65-0863-18) according to the manufacturer's recommendations. Cells were washed with DPBS and incubated for 1 h at 4 °C in flow cytometry staining buffer containing different concentrations of CEA-4-1BBL molecules (starting from a concentration of 300 nM, titrated in 1:5 or 1:4 dilution steps). Cells were washed with DPBS and incubated for 30 min in flow cytometry staining buffer containing 2.5 µg/mL PE-conjugated secondary antibody (Jackson ImmunoResearch, Cambridge, UK, 109-116-098). Cells were washed and fixed in PBS with 1% Paraformaldehyde for 15 min. Cells were acquired using the MACS Quant X (Miltenyi Biotec, Bergisch Gladbach, Germany) coupled to a Cytomat line 2, a 4 °C fridge and a plate handling system (Thermo Fisher Scientific, Reinach, Switzerland). Data were analyzed using FlowJo Version 10 to gate on living cells, and the median fluorescence intensity of the PE-conjugated secondary antibody was evaluated.

Reporter cell line assay: NFkB-downstream signaling after 4-1BB activation was measured by co-culturing 2 × 10^3 Jurkat-human-4-1BB-NFkB-luc2 reporter cells with 1 × 104 tumor cells or CEACAM5-transfected CHO-k1 cells in 384-well flat-bottom white plates (Corning, New York, NY, USA 3826). Titrated CEA-4-1BBL molecules were added, and the plates were incubated for 5 h at 5% CO2 and 37 °C. For one experiment, a fixed dose of 1 nM CEA-4-1BBL was used and different concentrations of soluble CEACAM5 (BioRad, Basel, Switzerland, PHP282) were added (starting dose: 190 nM, titrated in 1:5 dilution steps). Afterwards, One-Glo Luciferase reagent (Promega, Walldorf, Germany E6110) was added, and luminescence was measured for 0.5 s/well using a Tecan Spark plate reader.

Human PBMC activation assay: Buffy coats were obtained from the Zurich blood donation center in accordance with the Declaration of Helsinki. Donors signed a written informed consent before sample collection. Peripheral blood mononuclear cells (PBMCs) were isolated by Ficoll density centrifugation as follows: buffy coat was mixed 1:1 with sterile DPBS (Gibco, Thermo Fisher Scientific, Reinach, Switzerland, 14190-136), and 35 mL of this mixture was layered over 15 mL Histopaque 1077 (Sigma-Aldrich, Buchs, Switzerland, 10771, density: 1.077 g/mL) in 50 mL Falcon containers. The containers were centrifuged for 30 min at 450× *g* at room temperature without a break. The lymphocyte fraction was harvested, pooled and washed several times with DPBS. PBMCs were frozen in 80% FBS and 10% DMSO (Sigma-Aldrich, Buchs, Switzerland, D2650) and stored in vapor phase. PBMCs were thawed, washed and directly seeded. Then, 7.5 × 10^4 PBMCs, 2 × 10,445 Gy-irradiated MKN-45 tumor cells, 2nM agonistic human CD3 human IgG1 (clone V9) or titrated concentrations of CEA-TCB in combination with different concentrations of CEA-4-1BBL or untargeted controls were combined in 200 μL/well of assay medium in a 96-well round-bottom plate (TTP 92097). As assay medium, RPMI 1640 (GIBCO, Thermo Fisher Scientific, Reinach, Switzerland, 42401) supplemented with 10% (*v*/*v*) FBS, 2 mM GlutaMAX (Gibco, Thermo Fisher Scientific, Reinach, Switzerland, 35050), 1 mM Sodium-Pyruvat (Sigma-Aldrich, Buchs, Switzerland S8636), 1% (*v*/*v*) MEM-nonessential Amino acid solution (Sigma-Aldrich, Buchs, Switzerland, M7145) and 50 μM beta-mercaptoethanol was used. Cells were incubated under humidity at 5% CO_2 and 37 °C. At day 4, cells were harvested, washed and stained with LIVE/DEAD Fixable Aqua Dead Cell Stain (Molecular Technology, Berlin, Germany L34957) according to the manufacturer's recommendations. Afterwards, cells were stained in 50 μL/well flow cytometry staining buffer containing 0.3 μg/mL human CD25-APC (BioLegend, San Diego, CA, USA, 301016, clone BC96), 0.67 μg/mL human CD4-BV421 (BioLegend, San Diego, CA, USA, 300532, clone RPA-T4) and 0.67 μg/mL human CD8-APC/Cy7 (BioLegend, San Diego, CA, USA, 301016, clone RPA-T8) for 30 min at 4 °C. Cells were washed, fixed in PBS with 1% paraformaldehyde for 15 min and acquired using the Canto II (BD). Data were analyzed using FlowJo Version 10 to gate on living CD4 and CD8 T cells to determine the frequency of CD137+ or CD25+ cells of parental populations.

In vivo studies in humanized mice: The experimental study protocol was reviewed and approved by local government (code: P2011/128, date: 23 September 2016; code: ZH193-2014, date: 26 May 2020; code: ZH223/17 date: 14 May 2020). NOD.Cg-PrkdcscidIL2rgtm1Wjl/SzJ (NSG) was supplied by The Jackson Laboratory. Female NSG mice were bred by Charles River Laboratories (Lyon, France). After arrival, the animals were maintained for one week to get accustomed to the new environment and for observation. Continuous health monitoring was carried out on a regular basis. NSG mice were at the age of 4–5 weeks at the start of human stem cell engraftment and were maintained under specific-pathogen-free conditions with daily cycles of 12 h light/12 h darkness according to committed guidelines (GV-Solas; Felasa; TierschG). To generate human stem cell-engrafted (HSC)-NSG mice, NSG mice were administered 15 mg/kg busulfan (Busilvex, Pierre Fabre) intraperitoneally (i.p.). Twenty-four hours later, each mouse received an intravenous (i.v.) injection of 1 × 10^5 human CD34+ cord blood cells (purchased from STEMCELL Technologies). At 15–17 weeks post-engraftment, the mice were screened and only mice with more than 25% human CD45+ cells and a T cell count greater than 80 cells/μL were enrolled in the studies. For tumor inoculation, 1 × 10^6 tumor cells were resuspended in a 1:1 mixture of RPMI 1640 (Gibco, Thermo Fisher Scientific, Reinach, Switzerland, 42401) and GFR Matrigel (Corning, New York, NY, USA, 734-0269) to a total volume of 100 μL and injected subcutaneously into the flank. Tumor volume was calculated using the formula: V = length × width^2 × 0.5, and it was used to randomize groups one day before treatment.

Tumor-bearing mice (average tumor size: 200 mm^3, randomized groups) were treated intravenously with vehicle or 2.5 mg/kg CEA-TCB alone or in combination with 1, 3 or 10 mg/kg CEA–4-1BBL molecules. The dose for CEA-TCB was determined in previous experiments as optimal for xenograft studies in (HSC)-NSG mice and translatable. Tumor volume was calculated using the formula: $V = \text{length} \times \text{width}^2 \times 0.5$.

Flow cytometry measurement ex vivo: For ex vivo flow cytometry analysis, tumor single-cell suspensions were prepared using the gentleMACS Octo Dissociator (Miltenyi Biotec, Germany) according to the manufacturer's protocol and subsequently analyzed by flow cytometry. Formaldehyde-fixed tumor tissues were utilized for immunohistochemistry. Digested tumor single-cell suspensions were stained with 1:500 diluted live/death fixable blue dye (molecular probes: L-23105) according to the manufacturer's recommendations. Cells were washed and stained in flow cytometry staining buffer containing mouse CD45-AF700 (BioLegend, San Diego, CA, USA, 103128, clone 30-F11), human CD3-BV605 (BD 750985) and human CD4-BV421 (BioLegend, San Diego, CA, USA, 317434, clone OKT4), all diluted 1:300 and incubated for 30 min at 4 °C in the dark. Cells were washed, fixed and permeabilized using Permeablization buffer 10× from the FoxP3 Transcription Buffer staining kit (eBioscience, Thermo Fisher Scientific, Reinach, Switzerland, 00-5523-00). Cells were stained in perm-buffer containing 1:300 diluted human FoxP3-PE and incubated for 30 min at 4 °C in the dark. Cells were acquired with a 5-laser BD Fortessa flow cytometer and analyzed.

Immunohistology ex vivo: Immunohistochemical staining was performed in formalin-fixed paraffin-embedded tissue (FFPET) as described in [3,4] with human CD8 (Cell Marque Tissue Diagnostic, Sigma-Aldrich, Buchs, Switzerland, clone SP16) and human CD3 (Thermo Fisher Scientific, Reinach, Switzerland, MA5-13473, clone C8/144B). Human CD3+ and human CD8+ cells were quantified with Definiens Tissue Studio 3.0 (Definiens, Gilde Healthcare, Munich, Germany).

CEACAM5 modelling: Inter-domain centroid and center-of-mass distance calculations: In order to calculate typical distances between domains of the full-length human CEACAM5 protein, the alphafold2 model AF-P06731-F1-v6 was applied [38,39], which was aligned with available published CEACAM5 CryoEM structures of the N-terminal domain (PDB: 2QSQ, 2QST, 2VER) [40] and A3-B3 domain (PDB: 8BW0) [41]. All structures were analyzed using MOE (Molecular Operating Environment (MOE); 2024.0601 Chemical Computing Group ULC, 910-1010 Sherbrooke St. W., Montreal, QC H3A 2R7, Canada, 2025). All solved individual domains were initially superimposed based on structure and sequence similarities to evaluate the folding reliability of the AlphaFold model. Sequences corresponding to the N-term (A.A. 1–34) and C-term tail (A.A. 677–702) of the CEACAM5 AlphaFold model ranked with very low confidence (per residue confidence score pLDDT < 50, as disclosed in the model description file) were removed from the structure for further analysis. Next, structures were prepared using the protonate3D function and refined using an RMS gradient of 0.1 kcal/mol/A2. Domains were annotated and color-coded according to the figure legends. In order to measure a meaningful inter-domain distance, a centroid was defined as the geometric center corresponding to the average position of all atoms in each domain. Similarly, mass-weighted positions of all atoms in the domain were defined as centers of mass. Inter-centroid and inter-center-of-mass distances were next measured and compared to the maximal paratopic distance between the CDRL1 and CDRH2 of the solved CryoEM structure of tusamitamab bound to CEACAM5 (PDB:8BW0) [41].

Images: The pictograms shown below were designed with Biorender.com.

Statistics: Binding and activation values were baseline-corrected. Curves were analyzed by calculating the area under the curve (AUC), and AUC values were then used to calculate significant differences using an unpaired one-way ANOVA test. For in vivo efficacy studies, statistical analysis was performed using 1-way analysis of variance (ANOVA), and significant p values are indicated as follows: ns: not significant; * $p < 0.05$; ** $p < 0.01$; *** $p < 0.001$; **** $p < 0.0001$.

3. Results

3.1. CEA-4-1BBL Antibody Fusion Protein Design

Following the recently described 4-1BBL antibody fusion protein design [24], different CEA-4-1BBL antibody fusion proteins were generated using five different human CEACAM5-specific antibodies for tumor targeting (Figure 1A). As a non-tumor-targeted control, the previously published DP47-4-1BBL fusion protein was applied [24]. This construct does not bind to the cell surface of cancer cells. CEA-4-1BBL was specifically designed for combination with CEA-TCB, which comprises the human CEA-specific antibody CH1A1A98/99x2F1 and targets the membrane-proximal CEACAM5 domain B3 [27,42]. Thus, to allow the combination of CEA-TCB with CEA-4-1BBL, only CEACAM5-specific antibodies, which do not share the same CEACAM5 epitope and do not compete for binding with CEA-TCB, were selected (Figures 1B and S1). The following antibodies known from the literature were chosen: (1) the humanized antibody T84.66-LCHA binding to CEACAM5 domain A3 [43–45], (2) the parental antibody A5B7 and its humanized version huA5B7 binding to the membrane-proximal CEACAM5 domain A2 [46,47], and (3) the parental antibody MFE23 [48,49] and its humanized, stabilized and affinity-matured version Sm9b binding to CEACAM5 domain A1 [50]. All five of these CEACAM5 antibodies recognize soluble CEACAM5, a product induced by glycosylphosphatidylinositol-phospholipase D cleavage activity [51], as their epitopes are above the shedding site (Figure 1B). In contrast, the epitope of CH1A1A98/99x2F1 applied in CEA-TCB is close to the shedding site, and binding to soluble CEACAM5 as well as inhibition by soluble CEACAM5 is considered negligible, as it has been shown for its parental antibody, PR1A3 [52]. To confirm that simultaneous multiple Fab binding to CEACAM5 domains is possible from a structural point of view, inter-domain distances relative to the B3 domain of the AlphaFold model of human CEACAM5 were calculated (Supplementary Figure S1A,B). Measurements show that a typical Fv paratope length scores shorter than inter-domain distances (Supplementary Figure S1C), further supporting that simultaneous binding to individual domains occurs.

3.2. CEA-4-1BBL In Vitro Characterization

The affinity determination of the CEACAM5-targeting Fab fragments for CEACAM5 NABA constructs covering the N-A1-B2-A2 domains (Supplementary Table S1) via surface plasmon resonance (SPR) demonstrated that T84.66-LCHA and A5B7 have the highest and MFE23 and Sm9b the lowest affinities (Table 1). To test the cellular binding of the different CEA-4-1BBL antibody fusion proteins, six different human tumor cell lines were chosen to be representative for high to low CEACAM5 surface expression, as described previously [27,42]. The ability of the different CEA-4-1BBL fusion proteins to bind these CEACAM5+ human cancer cell lines was assessed by flow cytometry using a binding assay, as described in Figure 1C. The measured binding as median fluorescence intensity (MdFI) showed that Sm9b-4-1BBL and its parental variant, MFE23-4-1BBL, displayed similar MdFI values, which were higher than for A5B7-4-1BBL or T84.66-LCHA-4-1BBL (Figure 1E, Supplementary Table S2). The discrepancy between affinity determined by surface plasmon resonance and cell binding is likely caused by a different epitope accessibility on cells and/or differences in the glycosylation profiles between recombinant NABA-CEACAM5 and CEACAM5 expressed on tumor cell surfaces. To test if these differences in tumor cell binding translate into differences in functionality, the Jurkat-human-4-1BB-NFkB-luc2 reporter assay illustrated in Figure 1D was applied using the same tumor cell line panel as applied in the cell binding assay (Figure 1F, Supplementary Table S3). MFE23-4-1BBL and Sm9b-4-1BBL displayed comparable functionality similar to the cell binding results. On the other hand, A5B7-4-1BBL and T84.66-LCHA-4-1BBL, which showed lower overall cell binding, displayed lower activity in the reporter cell assay, with T84.66-LCHA-4-1BBL showing the lowest activity, while A5B7-4-1BBL performed more similarly to MFE23-4-1BBL and Sm9b-4-1BBL (Figure 1F).

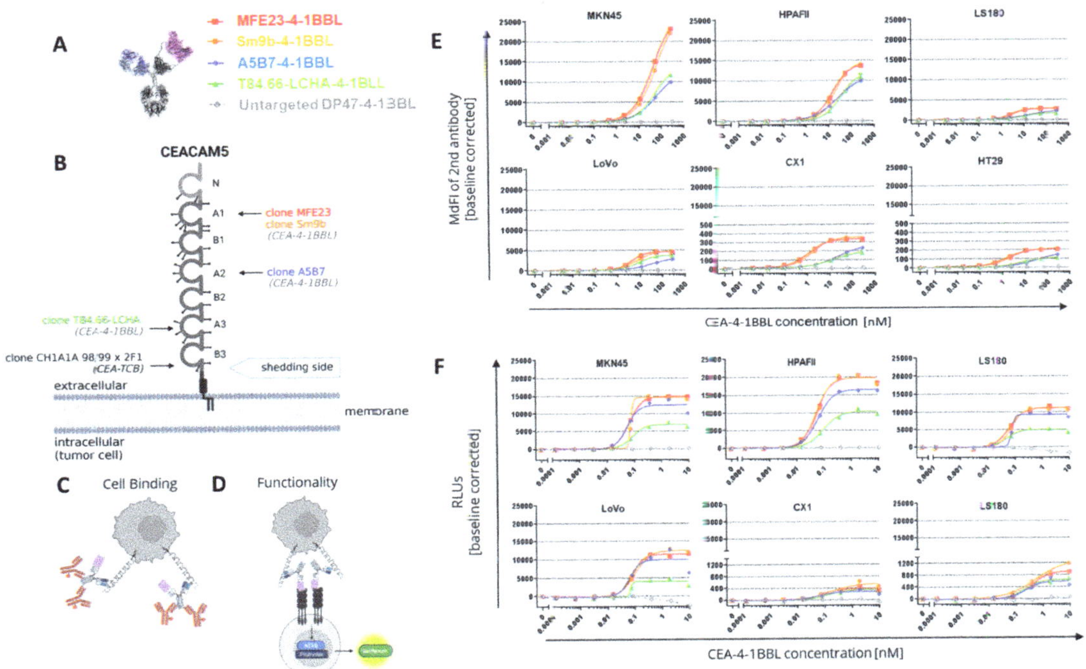

Figure 1. Evaluating binding to CEACAM5-expressing tumor cells and crosslinking functionality in an analogous reporter cell line assay. (**A**) Tested CEA-4-1BBL bispecific molecules. The CEACAM5 binding Fab is indicated in blue, the split trimeric 4-1BB ligand in purple and the Fc portion in gray. (**B**) Different CEACAM5 epitopes and binders. (**C**) Schematic illustration of binding assay. (**D**) Schematic illustration of functional reporter cell line assay. (**E**) CEA-4-1BBL binding to different CEACAM5-expressing tumor cells. CEA-4-1BBL molecules were incubated with the indicated tumor cell lines. Specific binding was detected using a PE-conjugated secondary antibody, and median of fluorescence intensity (MdFI) was measured by flow cytometry and baseline-corrected. Each point is the mean of technical duplicates. One of three similar experiments is shown. (**F**) CEA-4-1BBL molecules were titrated and incubated for 5 h with the reporter cell line Jurkat-human-4-1BB-NFkB-luc and different tumor cells in a 1:1 ratio. Emitting luminescence after substrate addition was measured for 0.5 s/well as relative light units (RLUs). Each point is the mean of technical duplicates. One of three independent and similar experiments is shown.

In subsequent experiments it was tested whether these differences in functional activity can also be observed in primary T cell assays. Human PBMCs obtained from different healthy donors were activated in the presence of CEACAM5-expressing MKN-45 tumor cells in the presence of 2 nM of the agonistic human CD3 IgG1 antibody V9 (Figure 2B) or a titration of CEA-TCB (Figure 2C). This activation step was introduced to induce sufficient 4-1BB expression, as resting T cells do not express 4-1BB constitutively [24]. The timing of this assay was optimized in pre-experiments, and four days was found to be the optimal time point. Interestingly, in both assays, in the presence of TCR signal 1, no major differences between the tested CEA-4-1BBL fusion proteins were observed, as were seen in the Jurkat-human-4-1BB-NFkB-luc2 reporter cell line assays with CEA-4-1BBL as a monotherapy (Figure 2D,E, Supplementary Tables S4 and S5). In the primary T cell assays, only minimal differences were observed. However, there was a trend suggesting that MFE23-4-1BBL and Sm9b-4-1BBL showed better performance when combined with the agonistic human CD3 antibody V9 (Figure 2D), whereas T84.66-LCHA-4-1BBL and A5B7-4-1BBL appeared to perform better when combined with CEA-TCB (Figure 2F). This demonstrates that the resolution of PBMC activation assays is smaller and different to the reporter cell assay. Furthermore, it demonstrated that the signal 1 provider, CD3 antibody versus TCB, may have an impact on the activity of the CEA-4-1BBL fusion proteins.

Table 1. Affinity constants of CEACAM5 Fabs for human CEACAM5 NABA constructs. The results are the average of three experiments.

Anti-CEACAM5 Fab	Human CEACAM5 NABA Construct	ka (1/Ms)	kd (1/s)	K_D (M)	SD
MFE23	huCEACAM1(A2,B1)-huCEACAM5(N, A1) avi His	9.2×10^4	1.1×10^{-3}	1.2×10^{-8}	1.3×10^{-9}
Sm9b	huCEACAM1(A2,B1)-huCEACAM5(N, A1) avi His	8.1×10^4	9.0×10^{-4}	1.1×10^{-8}	1.3×10^{-9}
A5B7	huCEACAM1(N,A2)-huCEACAM5(A2,B2) avi His	3.6×10^5	3.1×10^{-4}	8.6×10^{-10}	5×10^{-11}
huA5B7	huCEACAM1(N,A2)-huCEACAM5(A2,B2) avi His	1.9×10^5	4.6×10^{-4}	2.5×10^{-9}	5.5×10^{-11}
T84.66-LCHA	huCEACAM1(N,A2)-huCEACAM5(A3,B3) avi His	8.9×10^5	1.1×10^{-4}	1.2×10^{-10}	3.0×10^{-11}
CH1A1A 98/99 × 2F1	huCEACAM1(N,A2)-huCEACAM5(A3,B3) avi His	9.0×10^5	8.1×10^{-3}	9.1×10^{-9}	7.9×10^{-10}

As the PBMC-based assays had a longer incubation time of four days than the reporter cell line assay with an incubation time of five hours, subsequent experiments investigated whether the accumulation of soluble CEACAM5 may result in a reduction in the binding to cell-bound CEACAM5 and could provide an explanation for the findings. Indeed, overexpression of CEACAM5 in colorectal cancer correlates with soluble CEACAM5 concentrations in serum, and in cancer patients concentrations from 0.002 to 5.187 ug/mL have been reported, with soluble CEACAM5 values generally correlating with advanced cancer stages [53]. To test whether soluble human CEACAM5 can lead to inhibition of the functionality of the different CEA-4-1BBL fusion proteins (Figure 3A), a fixed concentration of 1 nM CEA-4-1BBL was chosen (Figure 3B). After adding increasing concentrations of soluble human CEACAM5 in the Jurkat-human-4-1BB-NFkB-luc2 reporter cell line assay, the IC50 values were evaluated. In these experiments the strongest inhibition was seen for T84.66-LCHA-4-1BBL and the lowest for A5B7-4-1BBL and MF23-4-1BBL (Figure 3B). Therefore, the functional differences in the different assays could not be explained by the sole impact of soluble CEACAM5.

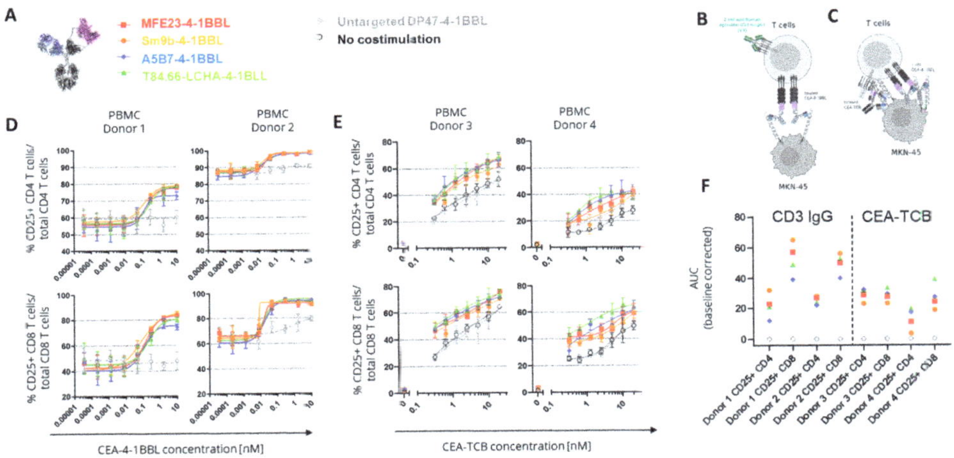

Figure 2. Co-stimulatory activity of CEA-4-1BBL molecules on primary human T cells. (**A**) Tested CEA-4-1BBL bispecific molecules. (**B**) Schematic display of the assay with CD3 IgG1. (**C**) Schematic display of the assay with CEA-TCB. (**D**) CEA-4-1BBL molecules were titrated and incubated with human PBMCs from two different donors, 45 Gy-irradiated MKN-45 tumor and 2 nM agonistic CD3 human IgG1 V9 antibody delivering signal 1, as indicated. After 4 days, T cells were analyzed by flow cytometry. The frequency of CD25 upregulation in the parental CD4 or CD8 T cell population was determined as indicated. Each point is the mean of technical triplicates. Error bars indicate the standard deviation. Only one experiment was performed. (**E**) Human PBMCs from two different donors were incubated with 45 Gy-irradiated MKN-45 tumor cells, titrated CEA-TCB and 1 nM CEA-4-1BBL molecules, as indicated. After 4 days, T cells were analyzed by flow cytometry. The frequency of CD25 upregulation in the parental CD4 or CD8 T cell population was determined. Each point is the mean of technical triplicates. Error bars indicate the standard deviation. Only one experiment was performed. (**F**) Area under the curve (AUC) was calculated from both assays and baseline-corrected by subtracting the AUC value from the DP47-4-1BBL control. The mean is shown.

3.3. *Cynomolgus Cross-Reactivity*

To select the best molecule for further development, the cynomolgus cross-reactivity of the different CEA-4-1BBL molecules was determined. Cynomolgus cross-reactivity has already been shown for human 4-1BBL [24], so only the cynomolgus cross-reactivity of the different CEACAM5 antibodies was determined. For this purpose, CHO-k1 cell lines stably expressing either human or cynomolgus monkey CEACAM5 were generated. To compare the cynoCEACAM5 expression levels of the different chosen CHO-k1 clones, a human/cynomolgus cross-reactive CEACAM5 (CD66e) APC-conjugated detection antibody was chosen. Based on the expression data, CHO-k1-cynoCEACAM5 clone 8 with good cynoCEACAM5 expression was selected for subsequent experiments. Furthermore, CHO-k1-huCEACAM5 clone 12 and clone 11 were selected as controls (Figure 3C). With these three cell lines, cell binding assays were performed (Figure 3D, Supplementary Table S6). Only A5B7-4-1BBL showed cyno-cross-reactivity, although with reduced binding to cynoCEACAM5. A5B7-4-1BBL bound much more weakly to CHO-k1-cynoCEACAM5 clone 8 (EC50 = 41.8 nM) than to CHO-k1-huCEACAM5 clone 12 (EC50 = 21.69 nM), although both clones displayed similar CEACAM5 expression levels (Figure 3C). As A5B7 was the only cyno-cross-reactive clone, it was humanized and converted into the CEA-4-1BBL format (huA5B7-4-1BBL). The humanization led to slightly reduced binding to CHO-k1-huCEACAM5 cells; however, unfortunately, the binding to CHO-k1-cynoCEACAM5 clone 8 decreased dramatically as a result of the humanization (Figure 3D, Supplementary Table S6). To test if the low binding to cynoCECAM5 had an impact on functionality, the same transgenic CHO-k1 cell lines were used in the functional Jurkat-human-4-1BB-NFkB-luc2 reporter assay (Figure 3E, Supplementary Table S7). As expected, only A5B7-4-1BBL and huA5B7-4-1BBL induced reporter cell line activation in the presence of CHO-k1-cynoCEACAM5. However, both antibody fusion proteins showed a lower activity with CHO-k1-cynoCEACAM5 than with CHO-k1-humanCEACAM5 controls. Interestingly, compared to A5B7-4-1BBL, huA5B7-4-1BBL functionality was not reduced as much as the binding to CHO-k1-cynoCEACAM5 would have suggested. Furthermore, no significant differences between CHO-k1-huCEACAM5 clone 11 and CHO-k1-huCEACAM5 clone 12 were observed in the reporter assay (Figure 3E), although both cell lines have different CEACAM5 expression (Figure 3C). It is important to mention that the CHO-k1 cell lines express high levels of CEACAM5 compared to tumor cell lines (Figure 1D) and that saturation of the reporter cell line activity with CHO-k1-huCEACAM5 clone 11 might have been reached.

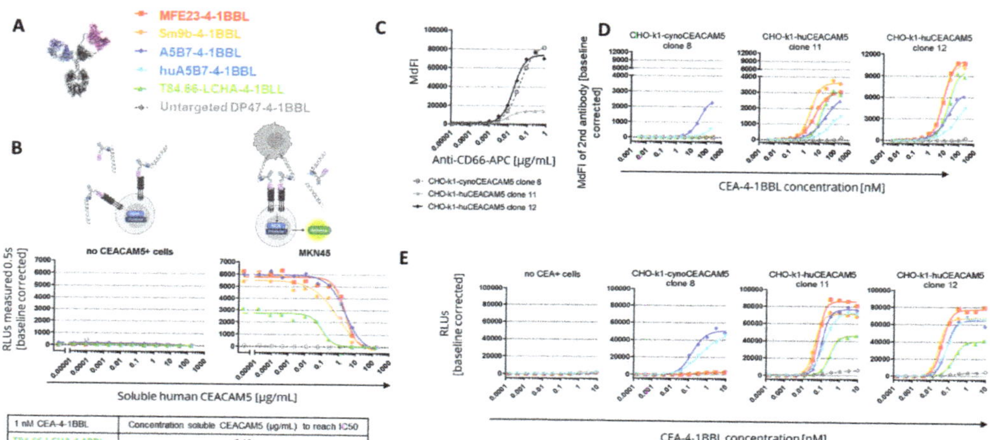

1 nM CEA-4-1BBL	Concentration soluble CEACAM5 (µg/mL) to reach IC50
T84.66-LCHA-4-1BBL	0.13
A5B7-4-1BBL	2.67
Sm9b-4-1BBL	1.05
MF23-4-1BBL	2.74

Figure 3. Inhibitory capacity of soluble CEACAM5 and cynomolgus monkey cross-reactivity. (**A**) Tested CEA-4-1BBL bispecific molecules. (**B**) Quantities of 1 nM of indicated CEA-4-1BBL molecules were incubated for 5 h with the reporter cell line Jurkat-human-4-1BB-NFkB-luc, with or without MKN-45 tumor cells, in a 1:1 ratio and in the presence of titrated concentrations of soluble human CEACAM5. After substrate addition, emitting luminescence was measured for 0.5 s/well as relative light units (RLUs) and baseline-corrected. Each point is the mean of technical duplicates. IC50 values were calculated and are displayed. Only one experiment was performed. (**C**) CHO-k1 cells were stably transduced with cynomolgus monkey (cynoCEACAM5) or human CEACAM5 (huCEACAM5). To compare expression levels between the different clones, a human/cynomolgus monkey cross-reactive CD66e-APC detection antibody was used for flow cytometry analysis. The concentration of the CD66e-APC detection antibody is displayed against the median fluorescence intensity (MdFI). Each point indicates the mean of technical duplicates. One of five independent and similar experiments is shown. (**D**) CEA-4-1BBL molecules were incubated with the indicated transgenic CHO-k1 cells and detected with a PE-conjugated secondary antibody. The MdFI was measured by flow cytometry, and values were baseline-corrected. Each point is the mean of technical duplicates. One of five independent and similar experiments is shown. (**E**) CEA-4-1BBL molecules were titrated and incubated for 5 h with the reporter cell line Jurkat-human-4-1BB-NFkB-luc and in the absence or presence of transgenic CHO-k1 cells in a 1:1 ratio. After addition of substrate, emitting luminescence was measured for 0.5 s/well as RLUs. Each point is the mean of technical duplicates. One of three independent and similar experiments is shown.

3.4. In Vivo Efficacy Studies

Finally, the potential of the different CEA-4-1BBL antibody fusion proteins as signal 2 providers in combination with CEA-TCB was tested in vivo in human CD34+ stem cell humanized NSG mice bearing subcutaneous human gastric adenocarcinoma MKN-45 or human pancreatic adenocarcinoma HPAFII xenografts. Both models exhibit comparable and high CEACAM5 expression and low CD3 T cell infiltration at baseline (Figure 4A,B), making them xenograft models with good translational potential for colorectal, gastric and pancreatic cancers. All treated mice exhibited reduced tumor growth relative to the vehicle. Notably, only T84.66-LCHA-4-1BBL consistently demonstrated significant tumor growth control across the performed experiments as a combination partner with CEA-TCB when compared with CEA-TCB monotherapy (Figure 4C–E). To establish this activity further and evaluate the dose–response relationship, T84.66-LCHA-4-1BBL was evaluated at three different doses in the MKN45/3T3 co-grafting tumor model in humanized NSG mice [24]. Administration of T84.66-LCHA-4-1BBL at all three tested doses (1, 3 and 10 mg/kg) resulted in the enhancement of CEA-TCB-mediated tumor growth inhibition (Figure 5A), exhibiting a clear dose-dependent trend. Furthermore, treatment with all three doses resulted in comparable immuno-pharmacodynamic effects in the tumor, as evidenced by flow cytometric analysis of human $CD4^+$ and $CD8^+$ T cell infiltration (Figure 5C). Consistently, immunohistochemistry (Figure 5D,E) demonstrated a significant increase in intratumoral human T cell density. Lastly, an increased accumulation of cytokines within the tumor was observed in the combination groups over CEA-TCB monotherapy at the study endpoint on day 44 (Figure 5F).

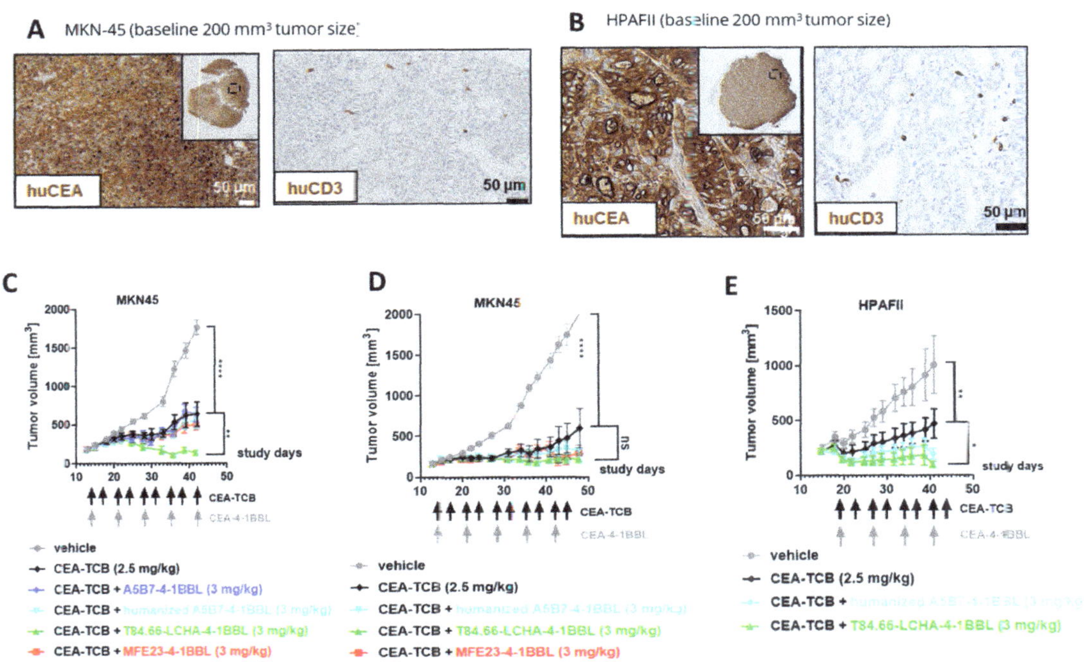

Figure 4. CEA-4-1BBL-mediated tumor growth inhibition in humanized mice (**A**,**B**) The human CEACAM5 expression and human CD3+ cell infiltration measured by IHC on FFPET at a tumor size of 200 mm^3 before treatment of MKN-45 or HPAFII tumors are shown. (**C**) CD34+ human cord blood humanized NOG mice received a subcutaneous injection of MKN-45 tumor cells. When a tumor volume of around 200 mm^3 was reached, the mice were randomized and treated with the vehicle

(gray filled circle) or 2.5 mg/kg CEA-TCB (black filled diamond) twice per week. Four of the six groups received a combination treatment of 2.5 mg/kg CEA-TCB twice per week and 3 mg/kg CEA-4-1BBL once per week (the different colors indicate CEA-4-1BBL molecules implementing different CEACAM5 binders). The schedule of treatment is indicated along the x-axis; tumor growth in mm^3 is indicated along the y-axis. Each point indicates the mean for 10 mice; error bars show the standard error of the mean (SEM). (**D**) A repetition of the experiment shown in C with indicated molecules. (**E**) A repetition of the experiment shown in C but with HPAFII tumor cells instead of MKN45 with indicated molecules. Statistical analysis was performed using 1-way analysis of variance (ANOVA), and significant p values are indicated as follows: ns: not significant; * $p < 0.05$; ** $p < 0.01$; **** $p < 0.0001$.

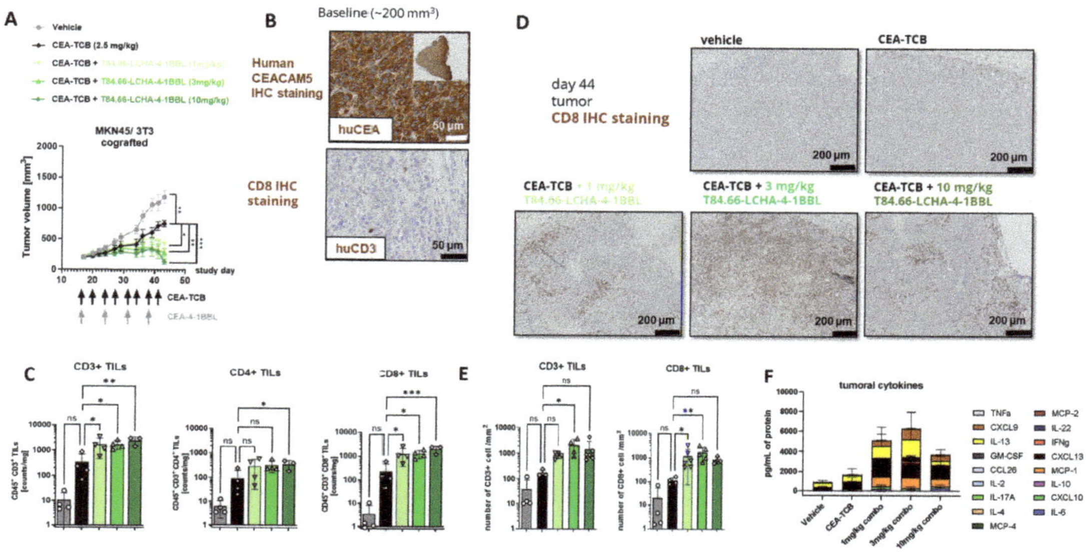

Figure 5. CEA-4-1BBL mediated increased CD8 T cell accumulation in a dose-dependent manner. (**A**) CD34+ human cord blood humanized NOG mice received subcutaneous injection of MKN-45 tumor cells and murine 3T3 fibroblasts. When a tumor volume of around 200 mm^3 was reached, mice were randomized and were treated with vehicle (gray) or 2.5 mg/kg CEA-TCB (black) twice per week. Three of the five groups received a combination treatment of 2.5 mg/kg CEA-TCB twice per week and 1m 3 or 10 mg/kg CEA-4-1BBL once per week (different shades of green). The schedule of treatment is indicated along the x-axis; tumor growth in mm^3 is indicated along the y-axis. Each point indicates the mean of 10 mice; error bars show the standard error of the mean (SEM). (**B**) The human CEACAM5 expression measured by IHC on FFPET at a tumor size of 200 mm^3 before treatment is shown. (**C**) At the endpoint (day 44), tumors were isolated and weighted. Three to four of ten tumors were digested and analyzed by flow cytometry. The counts of CD45+ CD3+ T cells, CD45+ CD3+ CD4+ T cells and CD45+ CD3+ CD8+ T cells per mg tumor are shown. Each symbol indicates an individual mouse; the bars indicate the mean, and the error bars the standard deviation (SD). (**D**) A total of 3-6 tumors per group were harvested at the endpoint (day 44); the tissue was formalin-fixed, paraffin-embedded and stained by immunohistochemistry (IHC) for human CD8 to monitor the tumor infiltration of CD8+ T cells. Tissue sections were scanned, and whole scans were analyzed by Definiens. One representative picture per group is shown. (**E**) The CD3+ and CD8+ T cell counts per square millimeter of immunohistochemistry are shown. Each symbol represents one individual mouse; the bars indicate the mean, and the error bars indicate the standard deviation (SD). Significance was calculated using unpaired one-way ANOVA with Tukey's multiple comparison test. ns: not significant; * = $p \leq 0.05$; ** $p < 0.01$. (**F**) Tumor was digested and centrifuged at endpoint day

44. Human cytokines were measured from supernatant using a 24plex R&D. For vehicle, CEA-TCB and CEA-TCB + 3 mg/kg T84.66-LCHA-4-1BBL, tumors from n = 4 mice were analyzed; for CEA-TCB + 1 mg/kg T84.66-LCHA-4-1BBL, tumors from n = 5 mice were analyzed; and for CEA-TCB + 10 mg/kg T84.66-LCHA-4-1BBL, tumors from n = 3 mice were analyzed. Measured cytokines are shown in different colors as the mean ± SD per treatment group. Statistical analysis was performed using 1-way analysis of variance (ANOVA), and significant p values are indicated as follows: ns: not significant; * $p < 0.05$; ** $p < 0.01$; *** $p < 0.001$.

4. Discussion

CEACAM5 is a well-studied marker for colorectal cancer, and soluble CEA serves as an FDA-approved diagnostic tumor marker [22,23]. Due to its broad and relatively tumor-selective expression, CEACAM5 has a high potential to serve as a target for different antibody-based cancer (immuno-) therapies. Several CEA-targeted TCBs have reached clinical trials, including CEA-TCB (cibisatamab) [27] and MEDI-565/AMG211 [54]. Clinical data have been reported for CEA-TCB as a monotherapy and in combination with the PD-L1 antibody atezolizumab [55]. Recently, promising data were reported for the TopoI payload-based CEACAM5-specific antibody drug conjugate precemtabart tocentecan in a Phase 1 study in colorectal cancer patients [56]. Notably, CEACAM5 has been described as displaying immune cell-inhibitory functions. For example, CEACAM5 expressed by tumor cells can interact with CEACAM1 expressed by NK cells and lead to the inhibition of NK cell-mediated tumor target cell killing [57]. Furthermore, it has been proposed that CEACAM5 can directly interact with CD1d via its B3 domain and with CD8α via its N-domain, leading to the formation of suppressive CD8 regulatory T cells [58]. CEACAM5 expression in pancreatic cancer also correlates with impairment of the tumoricidal function of M1 macrophages and neutrophils, although the mechanism is not fully understood [59].

Here different CEACAM5-specific antibodies directed against different epitopes/domains were tested to see which of them when used as targeting antibodies in CEA-4-1BBL fusion proteins would lead to the best CEA-4-1BBL-mediated tumor control in combination with CEA-TCB. Interestingly, in vitro MFE23-4-1BBL and Sm9b-4-1BBL (targeting membrane-distal CEACAM5 epitopes) performed superiorly in binding to tumor cells and in the functional Jurkat-human-4-1BB-NFkB-luc2 reporter cell assay. Similarly, when combined with a CD3 antibody as a signal 1 provider, both molecules had a tendency to perform better in the PBMC activation assay. In contrast, T84.66-LCHA-4-1BBL targeting a membrane-proximal epitope showed weaker binding on tumor cells and the lowest activity in the reporter cell assay. However, when combined with CEA-TCB as a signal 1 provider, T84.66-LCHA-4-1BBL mediated the best activity overall, both in a human PBMC assay and in vivo in two different xenograft models in human CD34+ stem cell humanized NSG mice.

The location of the tumor-target epitope, whether the binder-specific epitope is proximal or distal to the membrane, can influence T cell bispecific antibody function. Membrane-proximal epitopes are generally considered beneficial for TCBs, as they favor the formation of functional synapses between T and tumor cells. Indeed, Bluemel and colleagues used a MCSP-TCB comparing different MCSP-specific binders which recognized membrane-proximal, central and membrane-distal domains [60]. Membrane-proximal binding led to the strongest T cell-mediated tumor cell killing. Similarly, an EpCAM-TCB was tested, where not the EpCAM antibody but the location of the recognized epitope was modified by introducing spacers to mimic a more distal location of the same EpCAM epitope. In this example, creating a membrane-distal location led to the loss of bispecific-mediated T cell cytotoxicity [60]. Li and colleagues have studied FcRH5-TCB. The membrane-proximal-domain FcRH5 binder demonstrated the strongest in vitro functionality; however, a membrane-distal binder mediated similar tumor cell killing when the epitope was moved to a membrane-proximal position through genetic engineering [61]. Finally, Chen and colleagues investigated BCMA and Flt3 TCBs and their specific lysis capacity depending on the epitope localization, affinity and molecular design of the bispecific antibodies [62]. They showed that molecular design and tumor-target epitope location can decouple cytotoxicity and cytokine release and must be coordinated to achieve the best synapse and tumor killing results [62]. Antigen-binding affinities, on the other hand, appear to be positively correlated with cytotoxicity and cytokine release [62]. It is therefore reasonable to speculate that T84.66-LCHA-4-1BBL, by virtue of binding to the membrane-proximal A3 domain, is the optimal combination partner for CEA-TCB, recognizing the membrane-proximal B3 domain and thus enabling optimal synapse formation (Figures 1B and S1).

In contrast, the improved performance of membrane-distal CEACAM5-binding CEA-4-1BBL bispecific molecules in vitro in the Jurkat-human-4-1BB-NFkB-luc2 reporter cell line assay was surprising, as in general a smaller interspace that better mimics natural synapse formation is considered superior. In order to trigger TCR signaling in Jurkat reporter cells, physical immobilization and crosslinking appear to be more important than the actual synapse formation. Accordingly, the ranking observed in the Jurkat-human-4-1BB-NFkB-luc2 reporter cell line assay was not reflected in the PBMC-based activation assays. In vivo, T84.66-LCHA-4-1BBL displayed the best tumor growth control in combination with CEA-TCB over several in vivo studies and models. Importantly, testing CEA-4-1BBL activity in the in vitro PMBC activation assay or in vivo requires the presence of an active signal 1, whereas the Jurkat reporter cell line works without additional signal 1 and therefore ultimately may not reflect the impact of functional synapse formation in the presence of bispecific molecules. Therefore, the reporter cell line assay in this particular case cannot be considered the assay of choice for the selection of bispecific lead candidates, and broader testing, including TCBs as combination partners and in vivo experiments, is recommended. Based on the totality of the PBMC activation and in vivo data in combination with CEA-TCB, T84.66-LCHA-4-1BBL was chosen as the lead. Due to the discontinuation of the clinical development of CEA-TCB (cibisatamab, RG7802), further preclinical development of T84.66-LCHA-4-1BBL was halted; however, these findings may be of relevance to the development of future CEACAM5-targeted co-stimulatory agents and their combination with T cell engagers.

Supplementary Materials: The following supporting information can be downloaded at https://www.mdpi.com/article/10.3390/antib14040096/s1: Table S1: NABA constructs and conditions used for determining the affinity constants of the CEACAM5 Fabs via surface plasmon resonance (SPR); Table S2: EC50 in nM and area under the curve (AUC) values of binding curves for different CEACAM5-expressing tumor cell lines shown in Figure 1E; Table S3: EC50 in nM and area under the curve (AUC) values of activation curves for the Jurkat-human-4-1BB-NFkB-luc reporter cell line in the presence of different CEACAM5-expressing tumor cell lines shown in Figure 1F; Table S4: EC50 in nM and area under the curve (AUC) values of activation curves for the activation assay for PBMCs from two different donors shown in Figure 2D; Table S5: Area under the curve (AUC) values of activation curves for the activation assay for PBMCs from two different donors shown in Figure 2E; Table S6: EC50 in nM and area under the curve (AUC) values of binding curves for different transgenic cynomolgus monkey or human CEACAM5-expressing CHO-k1 cell lines shown in Figure 3C and 3D; Table S7: EC50 in nM and area under the curve (AUC) values of binding curves for different transgenic cynomolgus monkey or human CEACAM5-expressing CHO-k1 cell lines shown in Figure 3E; Figure S1: Inter-domain distance of human CEACAM5 AlphaFold model. Centroids, defined as geometric centers of individual domains, were calculated and depicted as red dots within each individual domain and used for distance calculation relative to the B3 centroids (depicted as green lines in A and reported in B). Similarly, centers of mass accounting for each domain's atom mass were calculated and used for relative distance measurement (reported in B). Structural overlay of B3-A3 domains with solved CryoEM structure of tusamitamab bound to

CEAMCAM5 (gray, PDB: 8BW0) allowed the calculation of maximal Fv paratope distance between the CDRL1 and CDRH2 of a typical Fv antibody (C).

Author Contributions: Conceptualization, C.C., C.F.-K., J.S., S.C., M.B., P.U. and C.K.; methodology, C.C., C.F.-K., J.S., S.L., R.A., R.B B., E.B. and V.G.N.; writing—original draft preparation, C.C., C.F.-K. and C.K.; writing—review and editing, C.C., C.F.-K., J.S., V.G.N., C.K. and G.L.S.; visualization, C.C., C.F.-K., J.S. and G.L.S.; supervision, C.C., C.F.-K., S.C., M.B., P.U. and C.K. All authors have read and agreed to the published version of the manuscript.

Funding: This research received no external funding.

Institutional Review Board Statement: The experimental in vivo study protocol was reviewed and approved by local government (code: P2011/128, date: 23 September 2016; code: ZH193-2014, date: 26 May 2020; code: ZH223/17, date: 14 May 2020).

Data Availability Statement: Data are available upon reasonable request from the authors.

Acknowledgments: We thank all contributors from Roche Innovation Centers, Basel, Zurich and Munich, for supporting this work.

Conflicts of Interest: All authors declare employment with Roche during the performance of this study. C.C., C.K.-F., J.S., G.L.S., V.G.N., S.C., M.B., P.U. and C.K. declare stock ownership and patents with Roche.

Abbreviations

The following abbreviations are used in this manuscript:

CEA	Carcinoembryonic Antigen 5
CEACAM5	Carcinoembryonic Antigen Cell Adhesion Molecule 5
FAP	Fibroblast Activation Protein
MdFI	Median Fluorescence Intensity
PBMC	Peripheral Blood Mononuclear Cell
PET	Positron Emission Tomography
SPR	Surface Plasmon Resonance
TCB	T Cell Bispecific Antibody
TCR	T Cell Receptor

References

1. Chan, A.C.; Martyn, G.D.; Carter, P.J. Fifty Years of Monoclonals: The Past, Present and Future of Antibody Therapeutics. *Nat. Rev. Immunol.* **2025**, *25*, 745–765. [CrossRef]
2. Klein, C.; Brinkmann, U.; Reichert, J.M.; Kontermann, R.E. The Present and Future of Bispecific Antibodies for Cancer Therapy. *Nat. Rev. Drug Discov.* **2024**, *23*, 301–319. [CrossRef]
3. Surowka, M.; Klein, C. A Pivotal Decade for Bispecific Antibodies? *mAbs* **2024**, *16*, 2321635. [CrossRef]
4. Labrijn, A.F.; Janmaat, M.L.; Reichert, J.M.; Parren, P.W.H.I. Bispecific Antibodies: A Mechanistic Review of the Pipeline. *Nat. Rev. Drug Discov.* **2019**, *18*, 585–608. [CrossRef]
5. Arvedson, T.; Bailis, J.M.; Britten, C.D.; Klinger, M.; Nagorsen, D.; Coxon, A.; Egen, J.G.; Martin, F. Targeting Solid Tumors with Bispecific T Cell Engager Immune Therapy. *Annu. Rev. Cancer Biol.* **2022**, *6*, 17–34. [CrossRef]
6. Baeuerle, P.A.; Wesche, H. T-Cell-Engaging Antibodies for the Treatment of Solid Tumors: Challenges and Opportunities. *Curr. Opin. Oncol.* **2022**, *34*, 552–558. [CrossRef] [PubMed]
7. Middelburg, J.; Kemper, K.; Engelberts, P.; Labrijn, A.F.; Schuurman, J.; van Hall, T. Overcoming Challenges for CD3-Bispecific Antibody Therapy in Solid Tumors. *Cancers* **2021**, *13*, 287. [CrossRef] [PubMed]
8. Cao, L.; Leclercq-Cohen, G.; Klein, C.; Sorrentino, A.; Bacac, M. Mechanistic Insights into Resistance Mechanisms to T Cell Engagers. *Front. Immunol.* **2025**, *16*, 1583044. [CrossRef] [PubMed]
9. Oostindie, S.C.; Lazar, G.A.; Schuurman, J.; Parren, P.W.H.I. Avidity in Antibody Effector Functions and Biotherapeutic Drug Design. *Nat. Rev. Drug Discov.* **2022**, *21*, 715–735. [CrossRef]
10. Claus, C.; Ferrara-Koller, C.; Klein, C. The Emerging Landscape of Novel 4-1BB (CD137) Agonistic Drugs for Cancer Immunotherapy. *mAbs* **2023**, *15*, 2167189. [CrossRef]

11. Lotze, M.T.; Olejniczak, S.H.; Skokos, D. CD28 Co-Stimulation: Novel Insights and Applications in Cancer Immunotherapy. *Nat. Rev. Immunol.* **2024**, *24*, 878–895. [CrossRef]
12. Olejniczak, S.H.; Lotze, M.T.; Skokos, D. Second Signals for Cancer Immunotherapy. *J. Immunother. Cancer* **2025**, *13*, e010530. [CrossRef] [PubMed]
13. Sanmamed, M.F.; Etxeberría, I.; Otano, I.; Melero, I. Twists and Turns to Translating 4-1BB Cancer Immunotherapy. *Sci. Transl. Med.* **2019**, *11*, eaax4738. [CrossRef]
14. Watts, T.H.; Yeung, K.K.M.; Yu, T.; Lee, S.; Eshraghisamani, R. TNF/TNFR Superfamily Members in Costimulation of T Cell Responses-Revisited. *Annu. Rev. Immunol.* **2025**, *43*, 113–142. [CrossRef] [PubMed]
15. Eguren-Santamaría, I.; Sanmamed, M.F.; Molero-Glez, P.; Perez-Gracia, J.L.; Melero, I. Targeting T-Cell Costimulation to the Surface of Tumor Cells. *Clin. Cancer Res.* **2024**, *31*, 231–233. [CrossRef]
16. Piha-Paul, S.; Olwill, S.A.; Hamilton, E.; Tolcher, A.; Pohlmann, P.; Liu, S.V.; Wurzenberger, C.; Hasenkamp, L.-C.; Hansbauer, E.-M.; Shroff, R.; et al. A First-in-Human Study of Cinrebafusp Alfa, a HER2/4-1BB Bispecific Molecule, in Patients with HER2-Positive Advanced Solid Malignancies. *Clin. Cancer Res.* **2025**, *31*, 288–298. [CrossRef] [PubMed]
17. Ku, G.; Shen, L.; Dayyani, F.; Kratz, J.; Liang, X.; Liu, F.; Wang, Z.; Feller, L.; Girda, E.; Pan, H.; et al. A First-in-Human Study of Givastomig, a CLDN18.2 and 4-1BB Bispecific Antibody, as Monotherapy in Patients with CLDN18.2-Positive Advanced or Metastatic Solid Tumors. *Clin. Cancer Res.* **2025**, OF1–OF9. [CrossRef]
18. Muik, A.; Garralda, E.; Altintas, I.; Gieseke, F.; Geva, R.; Ben-Ami, E.; Maurice-Dror, C.; Calvo, E.; LoRusso, P.M.; Alonso, G.; et al. Preclinical Characterization and Phase I Trial Results of a Bispecific Antibody Targeting PD-L1 and 4-1BB (GEN1046) in Patients with Advanced Refractory Solid Tumors. *Cancer Discov.* **2022**, *12*, 1248–1265. [CrossRef]
19. Pretelli, G.; Garralda, E. Rewiring Antitumor Immunity: Targeting CLDN18.2 with Conditional 4-1BB Activation. *Clin. Cancer Res.* **2025**, OF1–OF3. [CrossRef] [PubMed]
20. Melero, I.; Tanos, T.; Bustamante, M.; Sanmamed, M.F.; Calvo, E.; Moreno, I.; Moreno, V.; Hernandez, T.; Martinez Garcia, M.; Rodriguez-Vida, A.; et al. A First-in-Human Study of the Fibroblast Activation Protein-Targeted, 4-1BB Agonist RO7122290 in Patients with Advanced Solid Tumors. *Sci. Transl. Med.* **2023**, *15*, eabp9229. [CrossRef] [PubMed]
21. Benchimol, S.; Fuks, A.; Jothy, S.; Beauchemin, N.; Shirota, K.; Stanners, C.P. Carcinoembryonic Antigen, a Human Tumor Marker, Functions as an Intercellular Adhesion Molecule. *Cell* **1989**, *57*, 327–334. [CrossRef]
22. Thomas, J.; Klebanov, A.; John, S.; Miller, L.S.; Vegesna, A.; Amdur, R.L.; Bhowmick, K.; Mishra, L. CEACAMS 1, 5, and 6 in Disease and Cancer: Interactions with Pathogens. *Genes Cancer* **2023**, *14*, 12–29. [CrossRef] [PubMed]
23. Thirunavukarasu, P.; Sukumar, S.; Sathaiah, M.; Mahan, M.; Pragatheeshwar, K.D.; Pingpank, J.F.; Zeh, H.; Bartels, C.J.; Lee, K.K.W.; Bartlett, D.L. C-Stage in Colon Cancer: Implications of Carcinoembryonic Antigen Biomarker in Staging, Prognosis, and Management. *J. Natl. Cancer Inst.* **2011**, *103*, 689–697. [CrossRef]
24. Claus, C.; Ferrara, C.; Xu, W.; Sam, J.; Lang, S.; Uhlenbrock, F.; Albrecht, R.; Herter, S.; Schlenker, R.; Hüsser, T.; et al. Tumor-Targeted 4-1BB Agonists for Combination with T Cell Bispecific Antibodies as off-the-Shelf Therapy. *Sci. Transl. Med.* **2019**, *11*, eaav5989. [CrossRef] [PubMed]
25. Trüb, M.; Uhlenbrock, F.; Claus, C.; Herzig, P.; Thelen, M.; Karanikas, V.; Bacac, M.; Amann, M.; Albrecht, R.; Ferrara-Koller, C.; et al. Fibroblast Activation Protein-Targeted-4-1BB Ligand Agonist Amplifies Effector Functions of Intratumoral T Cells in Human Cancer. *J. Immunother. Cancer* **2020**, *8*, e000238. [CrossRef] [PubMed]
26. Bacac, M.; Colombetti, S.; Herter, S.; Sam, J.; Perro, M.; Chen, S.; Bianchi, R.; Richard, M.; Schoenle, A.; Nicolini, V.; et al. CD20-TCB with Obinutuzumab Pretreatment as Next-Generation Treatment of Hematologic Malignancies. *Clin. Cancer Res.* **2018**, *24*, 4785–4797. [CrossRef]
27. Bacac, M.; Fauti, T.; Sam, J.; Colombetti, S.; Weinzierl, T.; Ouaret, D.; Bodmer, W.; Lehmann, S.; Hofer, T.; Hosse, R.J.; et al. A Novel Carcinoembryonic Antigen T-Cell Bispecific Antibody (CEA TCB) for the Treatment of Solid Tumors. *Clin. Cancer Res.* **2016**, *22*, 3286–3297. [CrossRef] [PubMed]
28. Lehmann, S.; Perera, R.; Grimm, H.-P.; Sam, J.; Colombetti, S.; Fauti, T.; Fahrni, L.; Schaller, T.; Freimoser-Grundschober, A.; Zielonka, J.; et al. In Vivo Fluorescence Imaging of the Activity of CEA TCB, a Novel T-Cell Bispecific Antibody, Reveals Highly Specific Tumor Targeting and Fast Induction of T-Cell-Mediated Tumor Killing. *Clin. Cancer Res.* **2016**, *22*, 4417–4427. [CrossRef]
29. Griessinger, C.M.; Olafsen, T.; Mascioni, A.; Jiang, Z.K.; Zamilpa, C.; Jia, F.; Torgov, M.; Romero, J.M.; Marchioni, F.; Satpayev, D.; et al. The PET-Tracer 89Zr-Df-IAB22M2C Enables Monitoring of Intratumoral CD8 T-Cell Infiltrates in Tumor-Bearing Humanized Mice after T-Cell Bispecific Antibody Treatment. *Cancer Res.* **2020**, *80*, 2903–2913. [CrossRef]
30. Otano, I.; Azpilikueta, A.; Glez-Vaz, J.; Alvarez, M.; Medina-Echeverz, J.; Cortés-Domínguez, I.; Ortiz-de-Solorzano, C.; Ellmark, P.; Fritzell, S.; Hernandez-Hoyos, G.; et al. CD137 (4-1BB) Costimulation of CD8$^+$ T Cells Is More Potent When Provided in Cis than in Trans with Respect to CD3-TCR Stimulation. *Nat. Commun.* **2021**, *12*, 7296. [CrossRef]
31. Sun, Y.; Zhou, L.; Gu, X.; Zhao, J.; Bi, J.; Pan, L. Leveraging T Cell Co-Stimulation for Enhanced Therapeutic Efficacy of Trispecific Antibodies Targeting Prostate Cancer. *J. Immunother. Cancer* **2025**, *13*, e010140. [CrossRef] [PubMed]

32. Kamikawa, T.; Kimura, N.; Ishii, S.; Muraoka, M.; Kodama, T.; Taniguchi, K.; Yoshimoto, M.; Miura-Okuda, M.; Uchikawa, R.; Kato, C.; et al. SAIL66, a next Generation CLDN6-Targeting T-Cell Engager, Demonstrates Potent Antitumor Efficacy through Dual Binding to CD3/CD137. *J. Immunother. Cancer* **2024**, *12*, e009563. [CrossRef]
33. Mikami, H.; Feng, S.; Matsuda, Y.; Ishii, S.; Naoi, S.; Azuma, Y.; Nagano, H.; Asanuma, K.; Kayukawa, Y.; Tsunenari, T.; et al. Engineering CD3/CD137 Dual Specificity into a DLL3-Targeted T-Cell Engager Enhances T-Cell Infiltration and Efficacy against Small-Cell Lung Cancer. *Cancer Immunol. Res.* **2024**, *12*, 719–730. [CrossRef]
34. Carter, P. Bispecific Human IgG by Design. *J. Immunol. Methods* **2001**, *248*, 7–15. [CrossRef] [PubMed]
35. Schaefer, W.; Regula, J.T.; Bähner, M.; Schanzer, J.; Croasdale, R.; Dürr, H.; Gassner, C.; Georges, G.; Kettenberger, H.; Imhof-Jung, S.; et al. Immunoglobulin Domain Crossover as a Generic Approach for the Production of Bispecific IgG Antibodies. *Proc. Natl. Acad. Sci. USA* **2011**, *108*, 11187–11192. [CrossRef]
36. Regula, J.T.; Imhof-Jung, S.; Mølhøj, M.; Benz, J.; Ehler, A.; Bujotzek, A.; Schaefer, W.; Klein, C. Variable Heavy-Variable Light Domain and Fab-Arm CrossMabs with Charged Residue Exchanges to Enforce Correct Light Chain Assembly. *Protein Eng. Des. Sel.* **2018**, *31*, 289–299. [CrossRef] [PubMed]
37. Schlothauer, T.; Herter, S.; Koller, C.F.; Grau-Richards, S.; Steinhart, V.; Spick, C.; Kubbies, M.; Klein, C.; Umaña, P.; Mössner, E. Novel Human IgG1 and IgG4 Fc-Engineered Antibodies with Completely Abolished Immune Effector Functions. *Protein Eng. Des. Sel.* **2016**, *29*, 457–466. [CrossRef] [PubMed]
38. Jumper, J.; Evans, R.; Pritzel, A.; Green, T.; Figurnov, M.; Ronneberger, O.; Tunyasuvunakool, K.; Bates, R.; Žídek, A.; Potapenko, A.; et al. Highly Accurate Protein Structure Prediction with AlphaFold. *Nature* **2021**, *596*, 583–589. [CrossRef] [PubMed]
39. Fleming, J.; Magana, P.; Nair, S.; Tsenkov, M.; Bertoni, D.; Pidruchna, I.; Lima Afonso, M.Q.; Midlik, A.; Paramval, U.; Žídek, A.; et al. AlphaFold Protein Structure Database and 3D-Beacons: New Data and Capabilities. *J. Mol. Biol.* **2025**, *437*, 168967. [CrossRef]
40. Korotkova, N.; Yang, Y.; Le Trong, I.; Cota, E.; Demeler, B.; Marchant, J.; Thomas, W.E.; Stenkamp, R.E.; Moseley, S.L.; Matthews, S. Binding of Dr Adhesins of Escherichia Coli to Carcinoembryonic Antigen Triggers Receptor Dissociation. *Mol. Microbiol.* **2008**, *67*, 420–434. [CrossRef] [PubMed]
41. Kumar, A.; Duffieux, F.; Gagnaire, M.; Rapisarda, C.; Bertrand, T.; Rak, A. Structural Insights into Epitope-Paratope Interactions of a Monoclonal Antibody Targeting CEACAM5-Expressing Tumors. *Nat. Commun.* **2024**, *15*, 9377. [CrossRef]
42. Bacac, M.; Klein, C.; Umana, P. CEA TCB: A Novel Head-to-Tail 2:1 T Cell Bispecific Antibody for Treatment of CEA-Positive Solid Tumors. *Oncoimmunology* **2016**, *5*, e1203498. [CrossRef] [PubMed]
43. Teijeira, A.; Migueliz, I.; Garasa, S.; Karanikas, V.; Luri, C.; Cirella, A.; Olivera, I.; Cañamero, M.; Alvarez, M.; Ochoa, M.C.; et al. Three-Dimensional Colon Cancer Organoids Model the Response to CEA-CD3 T-Cell Engagers. *Theranostics* **2022**, *12*, 1373–1387. [CrossRef] [PubMed]
44. Neumaier, M.; Shively, L.; Chen, F.S.; Gaida, F.J.; Ilgen, C.; Paxton, R.J.; Shively, J.E.; Riggs, A.D. Cloning of the Genes for T84.66, an Antibody That Has a High Specificity and Affinity for Carcinoembryonic Antigen, and Expression of Chimeric Human/Mouse T84.66 Genes in Myeloma and Chinese Hamster Ovary Cells. *Cancer Res.* **1990**, *50*, 2128–2134.
45. Yazaki, P.J.; Sherman, M.A.; Shively, J.E.; Ikle, D.; Williams, L.E.; Wong, J.Y.C.; Colcher, D.; Wu, A.M.; Raubitschek, A.A. Humanization of the Anti-CEA T84.66 Antibody Based on Crystal Structure Data. *Protein Eng. Des. Sel.* **2004**, *17*, 481–489. [CrossRef]
46. Harwood, P.J.; Britton, D.W.; Southall, P.J.; Boxer, G.M.; Rawlins, G.; Rogers, G.T. Mapping Epitope Characteristics on Carcinoembryonic Antigen. *Br. J. Cancer* **1986**, *54*, 75–82. [CrossRef]
47. Chester, K.A.; Robson, L.; Keep, P.A.; Pedley, R.B.; Boden, J.A.; Boxer, G.M.; Hawkins, R.E.; Begent, R.H. Production and Tumour-Binding Characterization of a Chimeric Anti-CEA Fab Expressed in Escherichia Coli. *Int. J. Cancer* **1994**, *57*, 67–72. [CrossRef] [PubMed]
48. Boehm, M.K.; Corper, A.L.; Wan, T.; Sohi, M.K.; Sutton, B.J.; Thornton, J.D.; Keep, P.A.; Chester, K.A.; Begent, R.H.; Perkins, S.J. Crystal Structure of the Anti-(Carcinoembryonic Antigen) Single-Chain Fv Antibody MFE-23 and a Model for Antigen Binding Based on Intermolecular Contacts. *Biochem. J.* **2000**, *346*, 519–528. [CrossRef] [PubMed]
49. Chester, K.A.; Begent, R.H.; Robson, L.; Keep, P.; Pedley, R.B.; Boden, J.A.; Boxer, G.; Green, A.; Winter, G.; Cochet, O. Phage Libraries for Generation of Clinically Useful Antibodies. *Lancet* **1994**, *343*, 455–456. [CrossRef] [PubMed]
50. Graff, C.P.; Chester, K.; Begent, R.; Wittrup, K.D. Directed Evolution of an Anti-Carcinoembryonic Antigen scFv with a 4-Day Monovalent Dissociation Half-Time at 37 Degrees C. *Protein Eng. Des. Sel.* **2004**, *17*, 293–304. [CrossRef]
51. Yamamoto, Y.; Hirakawa, E.; Mori, S.; Hamada, Y.; Kawaguchi, N.; Matsuura, N. Cleavage of Carcinoembryonic Antigen Induces Metastatic Potential in Colorectal Carcinoma. *Biochem. Biophys. Res. Commun.* **2005**, *333*, 223–229. [CrossRef]
52. Conaghan, P.; Ashraf, S.; Tytherleigh, M.; Wilding, J.; Tchilian, E.; Bicknell, D.; Mortensen, N.J.; Bodmer, W. Targeted Killing of Colorectal Cancer Cell Lines by a Humanised IgG1 Monoclonal Antibody That Binds to Membrane-Bound Carcinoembryonic Antigen. *Br. J. Cancer* **2008**, *98*, 1217–1225. [CrossRef] [PubMed]

53. Wild, N.; Andres, H.; Rollinger, W.; Krause, F.; Dilba, P.; Tacke, M.; Karl, J. A Combination of Serum Markers for the Early Detection of Colorectal Cancer. *Clin. Cancer Res.* **2010**, *16*, 6111–6121. [CrossRef] [PubMed]
54. Oberst, M.D.; Fuhrmann, S.; Mulgrew, K.; Amann, M.; Cheng, L.; Lutterbuese, P.; Richman, L.; Coats, S.; Baeuerle, P.A.; Hammond, S.A. CEA/CD3 Bispecific Antibody MEDI-565/AMG 211 Activation of T Cells and Subsequent Killing of Human Tumors Is Independent of Mutations Commonly Found in Colorectal Adenocarcinomas. *mAbs* **2014**, *6*, 1571–1584. [CrossRef]
55. Segal, N.H.; Melero, I.; Moreno, V.; Steeghs, N.; Marabelle, A.; Rohrberg, K.; Rodriguez-Ruiz, M.E.; Eder, J.P.; Eng, C.; Manji, G.A.; et al. CEA-CD3 Bispecific Antibody Cibisatamab with or without Atezolizumab in Patients with CEA-Positive Solid Tumours: Results of Two Multi-Institutional Phase 1 Trials. *Nat. Commun.* **2024**, *15*, 4091. [CrossRef]
56. Kopetz, S.; Boni, V.; Kato, K.; Raghav, K.P.S.; Vieito, M.; Pallis, A.; Habermehl, C.; Siddiqui, A.; Courlet, P.; Sloot, W.; et al. Precemtabart Tocentecan, an Anti-CEACAM5 Antibody-Drug Conjugate, in Metastatic Colorectal Cancer: A Phase 1 Trial. *Nat. Med.* **2025**, *31*, 3504–3513. [CrossRef] [PubMed]
57. Stern, N.; Markel, G.; Arnon, T.I.; Gruda, R.; Wong, H.; Gray-Owen, S.D.; Mandelboim, O. Carcinoembryonic Antigen (CEA) Inhibits NK Killing via Interaction with CEA-Related Cell Adhesion Molecule 1. *J. Immunol.* **2005**, *174*, 6692–6701. [CrossRef]
58. Roda, G.; Jianyu, X.; Park, M.S.; DeMarte, L.; Hovhannisyan, Z.; Couri, R.; Stanners, C.P.; Yeretssian, G.; Mayer, L. Characterizing CEACAM5 Interaction with CD8α and CD1d in Intestinal Homeostasis. *Mucosal Immunol.* **2014**, *7*, 615–624. [CrossRef] [PubMed]
59. Shi, H.; Tsang, Y.; Yang, Y. Identification of CEACAM5 as a Stemness-Related Inhibitory Immune Checkpoint in Pancreatic Cancer. *BMC Cancer* **2022**, *22*, 1291. [CrossRef] [PubMed]
60. Bluemel, C.; Hausmann, S.; Fluhr, P.; Sriskandarajah, M.; Stallcup, W.B.; Baeuerle, P.A.; Kufer, P. Epitope Distance to the Target Cell Membrane and Antigen Size Determine the Potency of T Cell-Mediated Lysis by BiTE Antibodies Specific for a Large Melanoma Surface Antigen. *Cancer Immunol. Immunother.* **2010**, *59*, 1197–1209. [CrossRef]
61. Li, J.; Stagg, N.J.; Johnston, J.; Harris, M.J.; Menzies, S.A.; DiCara, D.; Clark, V.; Hristopoulos, M.; Cook, R.; Slaga, D.; et al. Membrane-Proximal Epitope Facilitates Efficient T Cell Synapse Formation by Anti-FcRH5/CD3 and Is a Requirement for Myeloma Cell Killing. *Cancer Cell* **2017**, *31*, 383–395. [CrossRef] [PubMed]
62. Chen, W.; Yang, F.; Wang, C.; Narula, J.; Pascua, E.; Ni, I.; Ding, S.; Deng, X.; Chu, M.L.-H.; Pham, A.; et al. One Size Does Not Fit All: Navigating the Multi-Dimensional Space to Optimize T-Cell Engaging Protein Therapeutics. *mAbs* **2021**, *13*, 1871171. [CrossRef] [PubMed]

antibodies

MDPI

Review

Breakthrough for Anticancer Immunotherapy: Current Advances in Manufacturing Protocols of Chimeric Antigen Receptor-Based Therapies

Yuxin Qian *, Weiwei Ma and Xiao-Ning Xu *

Department of Infectious Disease, Faculty of Medicine, Imperial College London, London AW7 2AZ, UK; weiwei.ma@hotmail.com

* Correspondence: yuxin.qian19@imperial.ac.uk (Y.Q.); x.xu@imperial.ac.uk (X.-N.X.)

Abstract: Chimeric antigen receptor (CAR)-based immunotherapy has emerged as a transformative strategy in anticancer treatment, driven by advances in CAR construct design, manufacturing platforms, and expansion to diverse immune cell types. The landmark success of CD19-targeted CAR-T cell therapy in B cell malignancies has paved the way for broader clinical applications. As of 2025, the U.S. FDA has approved multiple autologous CAR-T products, underscoring their therapeutic promise. However, challenges persist, including cytokine release syndrome (CRS), neurotoxicity, product inconsistency, and the high cost and complexity of cell manufacturing. Variations in cell source, gene delivery methods, expansion protocols, and CAR design significantly influence the safety, efficacy, and scalability of these therapies. In this review, we comprehensively examine the current advances in manufacturing protocols for CAR-modified T cells, natural killer (NK) cells, and unconventional T cell subsets, including $\gamma\delta$ T, invariant natural killer T (iNKT), and mucosal-associated invariant T (MAIT) cells. We also highlight emerging innovations such as in vivo CAR-T generation and off-the-shelf allogeneic approaches. By integrating updated strategies with a critical evaluation of current limitations, this review aims to support the development of standardized, robust, and accessible CAR-based immunotherapies.

Keywords: cancer immunotherapy; chimeric antigen receptor (CAR); cell manufacturing; antibody-derived scFv

1. Introduction

For decades, cancer treatment has primarily relied on surgery, radiation, and chemotherapy. However, the advent of immunotherapy has ushered in a new era of precision oncology by harnessing the host immune system to selectively target malignant cells. Among these strategies, adoptive cell therapy, particularly the genetic engineering of T cells to express chimeric antigen receptor (CAR), has emerged as one of the most transformative breakthroughs in hematologic oncology. CARs are synthetic fusion proteins that redirect immune cell specificity toward tumour-associated antigens. They typically consist of an extracellular single-chain variable fragment (scFv) derived from a monoclonal antibody, a hinge/transmembrane domain, and intracellular signaling motifs that activate immune effector functions upon antigen engagement [1]. This modular structure enables CAR-T cells to recognize tumour antigens in an HLA-independent manner and exert potent cytotoxicity.

Clinical studies have demonstrated exceptional efficacy of CAR-T cell therapy, particularly against B cell malignancies. Anti-CD19 CAR-T cells have achieved complete response (CR) rates ranging from approximately 50% to over 90% in patients with relapsed or refractory (r/r) B cell acute lymphoblastic leukemia (B-ALL) and non-Hodgkin lymphomas [2,3]. As of 2025, the U.S. Food and Drug Administration (FDA) has authorized six autologous CAR-T products: four targeting CD19—Kymriah® (Tisagenlecleucel) [4] for r/r ALL, Yescarta® (Axicabtagene Ciloleucel) [5], Breyanzi® (Lisocabtagene Maraleucel) [6] for r/r diffuse large B cell lymphoma (DLBCL), Tecartus® (Brexucabtagene Autoleucel) [7] for r/r mantle cell lymphoma (MCL), alongside two B cell maturation antigen (BCMA)-directed therapies for multiple myeloma (MM), Abecma® (Idecabtagene Vicleucel) [8] and Carvykti® (Ciltacabtagene Autoleucel) [9] (Table 1). All six products have also received marketing authorization from the European Medicines Agency (EMA) [10]. In China, the National Medical Products Administration (NMPA) has approved six CAR-T therapies, including the globally marketed Yescarta® and Carvykti®, as well as four domestically developed products, Carteyva® (relmacabtagene autoleucel) [11] for r/r follicular lymphoma (FL), Fucaso® (equecabtagene autoleucel) [12] and Zevor-cel® (zevorcabtagene autoleucel) [13] for r/r MM; and Yorwida® (inaticabtagene autoleucel) for r/r B-ALL [14]. As of June 2025, over 6000 interventional cell therapy trials were registered worldwide, including 1580 CAR-T studies listed on ClinicalTrials.gov [15,16]. It reflects the unprecedented growth of the CAR field and its expanding therapeutic footprint across hematologic cancers and emerging solid tumour indications.

Table 1. Overview of representative CAR-T cell products on the market for cancers.

Target Antigen	Diagnosis	Company/Institution	Year	Approval
CD19	B-ALL DLBCL	Novartis	2017 (FDA) 2018 (EMA)	FDA/EMA-approved Kymriah® (Tisagenlecleucel)
CD19	LBCL DLBCL FL	Kite Pharma	2017 (FDA) 2018 (EMA) 2021 (NMPA)	FDA/EMA/NMPA-approved Yescarta® (Axicabtagene Ciloleucel)
CD19	LBCL DLBCL	Juno Therapeutics	2021 (FDA) 2022 (EMA)	FDA/EMA-approved Breyanzi® (Lisocabtagene Maraleucel)
CD19	MCL	Kite Pharma	2020 (FDA, EMA)	FDA/EMA-approved Tecartus® (Brexucabtagene Autoleucel)
CD19	LBCL FL MCL	JW Therapeutics	2021	NMPA-approved Carteyva®(Relmacabtagene Autoleucel)
CD19	B-ALL DLBCL	Juventas Cell Therapy/CASI Pharmaceuticals	2021	NMPA-approved Yorwida® (Inaticabtagene Autoleucel)
BCMA	MM	Celgene Corporation	2021 (FDA, EMA)	FDA/EMA-approved Abecma® (Idecabtagene Vicleucel)
BCMA	MM	Janssen Biotech	2022 (FDA, EMA) 2023 (NMPA)	FDA/EMA/NMPA-approved Carvykti® (Ciltacabtagene Autoleucel)
BCMA	r/r MM	IASO Bio/Innovent Biologics	2023	NMPA-approved Fucaso®(Equecabtagene Autoleucel)
BCMA	r/r MM	CARsgen Therapeutics	2024	NMPA-approved Zevor-cel® (Zevorcabtagene Autoleucel)

This rapid clinical expansion, however, also highlights the need for innovation to address scalability, cost, and accessibility. While CAR-T therapies have transformed outcomes for many patients, their complex, patient-specific manufacturing and intensive clinical management limit broader application. Variability in T cell quality, activation, transduction efficiency, and expansion kinetics can substantially affect potency and persistence. At the same time, severe toxicities such as cytokine release syndrome (CRS) and immune effector cell-associated neurotoxicity syndrome (ICANS) remain significant clinical concerns [17]. To overcome these barriers, the field is rapidly advancing toward more efficient, standardized, and scalable solutions. Current research focuses on developing allogeneic "off-the-shelf" CAR-T products, exploring alternative immune cell platforms, including γδ T cells, invariant natural killer T (iNKT) cells, mucosal-associated invariant T (MAIT) cells, and natural killer (NK) cells, and integrating automated, closed-system manufacturing technologies. In parallel, in vivo CAR-T generation using viral or non-viral delivery systems is emerging as a transformative approach that could eventually eliminate the need for ex vivo cell manipulation altogether.

In this review, we comprehensively examine the state of the art in CAR-based cell manufacturing, encompassing conventional and unconventional cell sources, gene delivery technologies, expansion platforms, and clinical-grade production systems. We further highlight ongoing innovations aimed at enhancing safety, efficacy, and manufacturing efficiency, setting the stage for the next generation of broadly applicable immunotherapies.

2. Manufacturing Protocols for Conventional T Cell Therapy

The dramatic clinical outcomes of CD19-specific CAR-T cell therapy have boosted research interests in the development of armed immune cells to combat tumours. Minor differences in the manufacturing procedures and the CAR construct design might result in substantial variability in the quality and performance of the final cell products in patients [18–20]. The general CAR-T cell production outline includes T cell collection, cultivation, gene transfer and CAR-T cell expansion. The final cell products could be stored through cryopreservation before transfusion to patients, allowing for shipping from manufacturing sites to clinical centres and flexibility in scheduling the adoptive transfer timing (Figure 1).

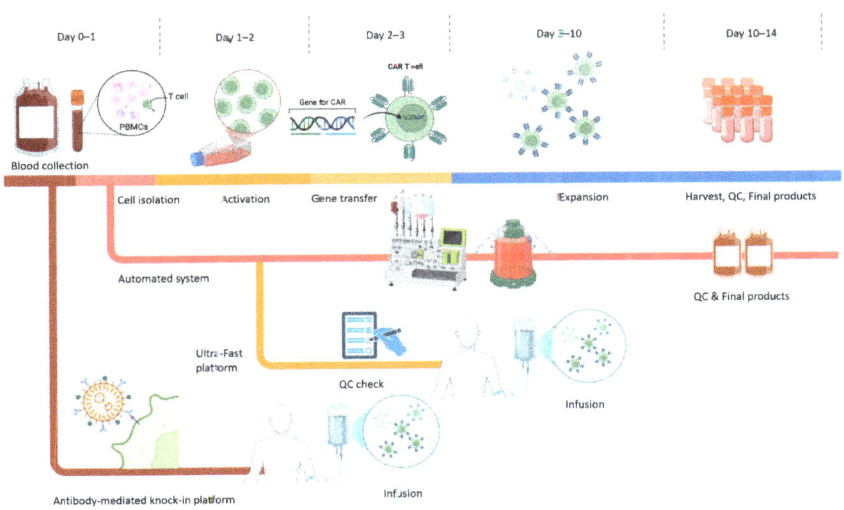

Figure 1. Schematic overview of CAR-T cell manufacturing workflow. The workflow depicts the general ex vivo CAR-T manufacturing process applicable to both autologous and allogeneic settings

from Day 0 to Day 14, highlighting conventional processes alongside next-generation rapid and automated platforms. Colours indicate different stages of the CAR-T manufacturing process: collection, activation, gene engineering, expansion, and infusion. Conventional workflows typically involve these steps, resulting in a final product in approximately 10–14 days. In parallel, automated, closed-system platforms integrate activation, transduction, and expansion in a single device to improve standardization, reduce operator variability, and support decentralized GMP manufacturing. Emerging rapid and ultra-fast manufacturing strategies shorten the activation and culture phases, enabling gene transfer as early as Days 1–2 and significantly compressing the overall timeline. Additionally, antibody-mediated knock-in systems (e.g., retargeted lentiviral platforms such as PACK-IT) enable selective T cell gene delivery without the need for prior enrichment or activation and support both ultra-rapid ex vivo production and in vivo CAR-T generation.

2.1. T Cell Collection and Enrichment

Although the U.S. FDA has approved multiple CAR-T cell products, there is still no universally standardized manufacturing protocol for CAR-T cell therapy. Variability in cell source, processing, and enrichment strategies remains a key determinant of product quality and clinical outcomes. In autologous manufacturing, production typically begins with leukapheresis to isolate white blood cells from the patient's peripheral blood. Traditionally, lymphocytes were separated using Ficoll density gradient centrifugation [21]. Currently, automated cell-washer machines such as Cytiva® Sepax S-100 (Cytiva, USA), Fresenius Kabi LOVO® (Fresenius Kabi, Germany) and Haemonetics® Cell Saver (Haemonetics Corporation, MA, USA) are used to streamline cell separation, reduce contamination risk, and remove red blood cells, granulocytes, and platelets [21,22]. Approximately 10^6 to 10^7 viable T cells are required for downstream CAR modification [23]. Many protocols now incorporate immunomagnetic bead-based selection to enrich specific T cell subsets (e.g., $CD4^+$, $CD8^+$, or naïve/memory fractions) for improved product consistency. Defined $CD4^+$:$CD8^+$ ratios, such as the fixed 1:1 ratio described by Turtle et al., have been associated with improved expansion kinetics, persistence, and clinical remission rates in ALL [24]. In certain disease settings, such as chronic lymphocytic leukaemia (CLL), where circulating tumour cells can contaminate leukapheresis products, additional $CD3^+$ T cell enrichment may be required to improve manufacturing efficiency [20]. Negative selection is also used to deplete unwanted populations. For example, $CD56^+$ bead depletion can remove excessive NK cells (>10%), which may otherwise compete with T cells during culture [25]. The quality of the apheresis product is strongly influenced by patient-related factors, including age, tumour burden, prior therapies, and the hostile tumour microenvironment (TME). For instance, ALL diagnosis and higher platelet count have been linked to lower lymphocyte collection efficiency [26]. Patients heavily pretreated with chemotherapy or radiotherapy often exhibit lymphocytopenia [27], while T cell exhaustion and senescence are common, leading to reduced expansion potential and diminished in vivo persistence [19,28]. These factors must be evaluated at the collection stage to assess the feasibility of CAR-T manufacturing, anticipate expansion challenges, and predict product potency. Standardized apheresis quality assessments and pre-manufacturing fitness assays are increasingly being implemented at clinical sites to optimize product yield and ensure therapeutic-grade cell quality.

2.2. T Cell Activation

Robust activation signals are required to initiate T cell proliferation and differentiation during CAR-T cell manufacturing. Several techniques have been developed to mimic physiological T cell activation in vitro, balancing expansion efficiency, phenotype preservation, and transduction competence. The most widely adopted clinical approach remains anti-CD3/CD28 antibody-coated magnetic beads (e.g., Dynabeads® CD3/CD28, Thermo Fisher

Scientific, USA), which accounted for the majority of 952 analysed manufacturing protocols in the systematic survey by Vormittag et al. [21]. This strategy offers practical advantages that the antibodies on the bead surface minimize the potential loss during media changes, and the beads can be removed post-expansion via strong magnets. Bead-based stimulation has been shown to induce 10–100-fold higher cytokine production and a less exhausted phenotype compared with OKT3 (anti-CD3) plus IL-2 stimulation [29,30]. Bead-free activator technologies are gaining popularity in recent manufacturing pipelines. Product like Miltenyi's TransAct™ CD3/CD28 reagent (Miltenyi Biotec, Germany) could provide soluble, biodegradable activation complexes that eliminate bead removal steps, reducing labour, contamination risk, and residual material in the final product [31,32]. Alternative activation systems are being increasingly explored to optimize scalability. Artificial antigen-presenting cells (aAPCs), particularly K562-derived platforms, can deliver sustained and more physiologic activation via engineered expression of co-stimulatory ligands. For example, Shrestha et al. engineered K562-based aAPCs co-expressing CD3, CD28, and CD137L, which supported prolonged expansion in a CAR-dependent manner [33]. Schmidts et al. further optimized this approach by knocking out the low-density lipoprotein receptor (LDLR) to prevent undesired lentiviral transduction, enabling direct co-culture with T cells without additional removal steps. In preclinical ALL models, 2:1 aAPC:T cell co-culture supported robust CAR-T expansion and potent antitumour activity [34]. Activation duration remains a critical process parameter. Stimulation is typically performed for 48 to 72 h, with evidence that shorter ex vivo culture times preserve a less differentiated T cell phenotype, enhanced in vitro effector function, and superior in vivo persistence compared with prolonged culture [35]. Activation extended beyond 5 days has been associated with decreased CD3 expression, reduced viability, lower expansion, and diminished transduction efficiency relative to 3-day activation [36]. Therefore, a balanced activation strategy with appropriate platform and optimized stimulation duration is crucial for generating a CAR-T cell product with high potency, viability, and consistent clinical performance.

2.3. CAR Structure and Generations

The CAR is embedded in the T cell to determine the specificity and activity of the cell products. A proper CAR construct design is critical for the therapeutic success of CAR-based immunotherapy. Since the introduction of CAR technology, the CAR construct has experienced several generations of renovation, each aimed at overcoming the shortcomings of the previous designs (Figure 2).

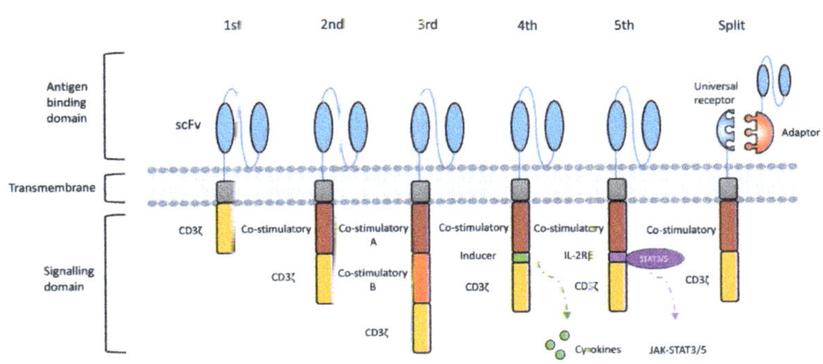

Figure 2. Evolutions of CAR designs. The first-generation CAR consists of an extracellular antigen-binding domain linked to an intracellular CD3ζ signaling module. The second- and third-generation CARs incorporate one or two co-stimulatory domains, respectively, to enhance proliferation, persistence,

and anti-tumour activity. Fourth-generation CARs (TRUCKs) are engineered to release immune-modulating molecules, such as proinflammatory cytokines, upon antigen recognition to remodel the tumour microenvironment. Fifth-generation CARs integrate a truncated cytokine receptor domain (e.g., IL-2Rβ) with a STAT3/5-binding motif, enabling antigen-dependent cytokine signaling to further support T cell expansion and function. Novel modular designs, such as split or adaptor CAR systems, separate antigen recognition from signaling by pairing a universal receptor with interchangeable, target-specific adaptors, allowing for adjustable specificity and tuneable activity.

First-generation CARs consist of an extracellular scFv for HLA-independent antigen recognition, a transmembrane domain for structural stability, and an intracellular CD3ζ signaling domain for T cell activation and cytotoxic function [20]. Although capable of antigen-specific recognition, these early CD3ζ-based CAR-T cells demonstrated poor in vivo persistence and suboptimal anti-tumour activity due to the absence of co-stimulatory signaling [37,38]. Second-generation CARs incorporate a single co-stimulatory domain, most often CD28 or 4-1BB, alongside CD3ζ, resulting in enhanced proliferation, prolonged survival, and delayed exhaustion [39]. Other co-stimulatory modules, such as OX40, ICOS and CD27, have also been explored to fine-tune T cell persistence and activity in vivo [20]. In patients with r/r B cell malignancies, second-generation CAR-T cells have achieved markedly improved clinical outcomes compared to the first-generation products [40]. Third-generation CARs combine two co-stimulatory domains (e.g., CD28 + 4-1BB or CD28 + OX40) with CD3ζ to deliver stronger and more sustained activation signals. Early-phase clinical studies have shown favourable safety profiles and enhanced anti-tumour efficacy, with more durable responses in certain B-cell lymphoma cohorts [41,42].

The evolution from first- to third-generation CARs was primarily aimed at augmenting T-cell activation, survival, and cytotoxicity. Fourth-generation CARs (TRUCKs) further expanded functionality by enabling T cells to release immune-modulating factors upon antigen encounter, allowing for active reshaping of TME [20]. For instance, CAR-T cells engineered to co-express IL-7 and CCL19 enhanced proliferation, cytotoxicity, and intratumoural recruitment of endogenous immune cells r/r MM (NCT03778346) [43]. Additional cytokines such as IL-12, IL-15, and IL-18 are being incorporated into TRUCK platforms to strengthen T cell function within immunosuppressive niches, although precise control of cytokine release remains essential to prevent systemic toxicity [44]. Fifth-generation CARs build on this concept by introducing truncated cytokine receptor elements (e.g., IL-2Rβ–STAT3/5 motifs) that couple antigen recognition to self-amplifying proliferative and survival signals, potentially reducing reliance on exogenous cytokines [45].

Despite these advances, single-antigen CAR designs remain highly vulnerable to tumour antigen loss or downregulation, one of the most common mechanisms of relapse following CAR-T therapy. Several next-generation CAR architectures have been developed to mitigate antigen escape by broadening tumour coverage. One approach is bispecific CAR format, in which two antigen-binding domains are incorporated into a single receptor. Clinical studies using BCMA/CD19 bispecific CAR-T cells in r/r MM have reported high complete response rates and durable remissions [46]. Beyond fixed dual-target receptors, adaptor-based CAR platforms offer a modular solution to antigen escape by enabling on-demand and reversible retargeting. In the SUPRA CAR system (split, universal, programmable), a universal "zipCAR" receptor on the T cell binds interchangeable "zipFv" adaptors through engineered leucine-zipper interactions, allowing the same CAR-T population to be redirected to new antigens as tumours evolve (Figure 2). This configuration enables rapid switching between antigens using a single cellular product, and competitive zippers can be used as tunable OFF-switches to modulate activity and mitigate toxicity [47]. Similarly, the SpyTag/SpyCatcher platform links a universal CAR scaffold to tumour-specific binders via covalent isopeptide bonds, enabling rapid retargeting, controlled dosing of adaptor molecules, and switching between antigens in response to tumour heterogeneity [48,49].

In addition to antigen escape, next-generation CAR designs increasingly address the toxicity risks inherent to conventional CAR-T therapies, particularly CRS and ICANS. These toxicities arise from excessive cytokine release, myeloid-cell activation, and uncontrolled T cell expansion within the tumour microenvironment. To mitigate these risks, multiple safety-switch mechanisms have been incorporated into CAR constructs. The most clinically advanced is the inducible caspase-9 (iC9) suicide switch, which triggers rapid apoptosis upon administration of a small-molecule dimerizer and has shown effective clinical performance in early trials [50]. Surface elimination markers, such as truncated EGFR (EGFRt), permit antibody-mediated depletion with cetuximab when severe toxicity occurs [51]. More recently, small-molecule, regulated ON/OFF switches, including synthetic Notch receptors, drug-controlled heterodimerization domains, and protease-regulated CARs, have been developed to allow for reversible, dose-dependent control of CAR-T activity [52,53].

These advanced CAR formats represent important developments toward overcoming antigenic heterogeneity and CAR functions. Selecting the appropriate CAR generation and architecture must be tailored to the intended effector cell type, clinical indication, and manufacturing platform to optimise therapeutic potency, durability, and safety.

2.4. Gene Delivery

The transfer of the CAR transgene into immune effector cells is a pivotal step in CAR-T cell manufacturing. Current clinical practice relies primarily on viral or non-viral systems to achieve stable CAR expression in patient T cells. Viral vectors remain the most widely used delivery platform, with retroviral and lentiviral systems dominating the field due to their high transduction efficiency, ability to stably integrate the CAR cassette, and safety profile demonstrated in multiple clinical trials [22,54]. For clinical application, all viral vectors must meet stringent GMP standards, necessitating dedicated production facilities and extensive biosafety testing to exclude replication-competent viruses [21]. These requirements contribute significantly to manufacturing costs. Efforts to streamline vector production have included the development of stable packaging cell lines to reduce batch-to-batch variability and cost. For example, Parente-Pereira et al. developed a two-step procedure yielding a fully positive retroviral packaging master cell bank with stable vector expression and high production consistency [23]. In addition, process optimisations such as the use of X-Vivo 15 medium, G-Rex culture vessels for improved nutrient exchange, and RetroNectin-coated surfaces for enhanced viral colocalization with T cells have improved CAR-T yield and transduction efficiency [55].

Alternatives are becoming available to support gene delivery work. In particular, the transposon/transposase system is one of the non-viral gene delivery approaches. Although it is a relatively new approach in the field of cell therapies, it can provide great economic benefits over viral transduction. The transposon/transposase system allows for integration of CAR transgene into the T cell genome with satisfactory efficiency, simple manufacturing procedure and relatively low cost compared to viral transduction. The desired CAR transgene is incorporated into a transposon sequence on the plasmid, encoding the required transposase for later transposon excision and insertion into the T-cell genome. Electroporation is applied to introduce the plasmids to target T cells before activation [56]. Clinical studies using the Sleeping Beauty platform to generate anti-CD19 CAR-T cells in r/r B-cell malignancies have reported durable expression with low risk of insertional mutagenesis [57,58]. Although more trials and research are warranted to confirm the efficacy and safety of this non-viral gene transfer methodology, the early success in the clinical trials has shown its potential as a competent and economical tool for both research and medical use.

2.5. T Cell Expansion

Scaling up the cell production to a required infusion level is a challenge in CAR-T manufacturing. Cytokine supplementation plays a central role in shaping the expansion kinetics, composition, and functional phenotype of the final product. The most widely used regimens include interleukin (IL)-2, IL-15, and IL-21, either alone or in combination, to support ex vivo proliferation [20]. However, excessive supplementation with cytokines, especially IL-2 exposure, may lead to a more differentiated and exhausted phenotype of T cells that impairs the killing functions towards cancer cells and reduces the persistence [59]. Comparative studies have shown that supplementation with IL-15 preferentially maintains T cells with a stem cell memory (Tscm)-like phenotype, partly by limiting mTORC1-driven glycolytic metabolism. CAR-T cells expanded with IL-15 alone have demonstrated enhanced proliferative capacity, reduced apoptotic susceptibility, and potent anti-tumour activity in preclinical models [59]. IL-21 has also emerged as a valuable cytokine in expansion protocols. In addition to boosting proliferation, IL-21 can inhibit terminal differentiation into exhausted late-memory phenotypes, thus preserving long-lived, functionally competent T cells [60].

As manufacturing volumes increase, he choice of expansion platform becomes critical. Manual flask cultures are labour-intensive, contamination-prone, and incompatible with large-scale GMP production. Automated, closed-system bioreactors, such as the CliniMACS Prodigy® (Miltenyi Biotec, Germany), Lonza Cocoon® (Lonza, Switzerland), and G-Rex® (Wilson Wolf Manufacturing, USA), provide sterile, tightly controlled environments that minimize operator intervention [22]. These systems have made CAR-T manufacturing more consistent and accessible across clinical centres. Typically, approximately 14 days are required to produce infusion-ready CAR-T products. Comparative studies indicate that CAR-T cells expanded using the CliniMACS Prodigy® system maintain phenotypic and functional equivalence to manually produced products, confirming the feasibility of automated, large-scale production [61,62]. Quality control and in-process monitoring remain essential for GMP compliance, and continued optimization of culture formulations is necessary. Importantly, the successful in-human trial using CliniMACS Prodigy® to generate bispecific anti-CD20/anti-CD19 CAR-T cells (NCT03019055) achieved robust antitumour activity with low toxicity, validating the system's potential for decentralized, on-site CAR-T manufacturing [63].

Recent technological advances aim to further shortened expansion timelines. Short-cycle manufacturing platforms, such as the PACK-IT (Programmable Antibody-mediated Cellular Knock-In of T cells) system, enable faster CAR-T generation [64]. It uses a mutated vesicular stomatitis virus glycoprotein (VSV-G) envelope that lacks native tropism and is fused with a T cell-specific scFv (e.g., anti-CD3), enabling targeted gene integration directly into T cells. This approach eliminates the need for cell purification or activation and completes ex vivo manufacturing within four days, yielding CAR-T cells with superior potency compared with conventional lentiviral products. Moreover, intravenous administration of anti-CD3 PACK-IT particles in humanized mice successfully generated functional CAR-T cells in vivo, highlighting its potential to unify rapid ex vivo and direct in vivo manufacturing [64].

Building upon these developments, ultra-fast systems such as the Ultra-Fast CAR (UF-CAR) and FasT CAR-T (F-CAR-T) platforms have further compressed production to under 48 h. These one-day protocols optimize activation signals, transduction efficiency, and early-memory phenotype retention to deliver functional CAR-T cells in a single manufacturing cycle [65–67]. In a clinical comparison, FasT CAR-T demonstrated encouraging responses and manageable safety in patients with r/r B-ALL while retaining higher proportions of naïve and Tscm-like cells than conventional products [66]. By substantially reducing vein-to-vein time and manufacturing costs, these platforms address logistical delays that can otherwise preclude timely therapy for patients with rapidly progressing malignancies.

Together, these ultra-rapid and programmable systems mark a major step toward next-generation CAR manufacturing, one that is faster, safer, and more compatible with global clinical deployment. Nevertheless, even with the advent of these rapid and automated manufacturing platforms, autologous CAR-T production remains constrained by patient-to-patient variability, high cost, and complex logistics. These ongoing challenges have spurred growing interest in universal, donor-derived allogeneic CAR approaches designed to deliver standardized, readily available therapies.

3. From Autologous to Allogeneic: Off-the-Shelf CAR-T Development

Conventional CAR-T therapies are primarily produced in the autologous setting, in which a patient's own T cells are collected, genetically modified, expanded, and reinfused. While clinically effective, this personalized approach is costly, time-consuming, and susceptible to production failure, particularly in patients with lymphocytopenia or poor T-cell fitness, resulting in variable product quality. To overcome these challenges, researchers are developing platforms that use healthy donor–derived T cells to create off-the-shelf allogeneic CAR-T products, enabling large-scale manufacture of therapeutic doses from a single donor source. This strategy provides a more economical and readily available alternative to patient-specific therapies. The manufacturing workflow closely parallels that of autologous production but incorporates donor screening to ensure optimal T-cell quality and standardized procedures (Figure 3). By generating multiple doses in advance, allogeneic manufacturing can shorten treatment timelines for patients with urgent clinical needs while reducing overall costs through economies of scale [68,69].

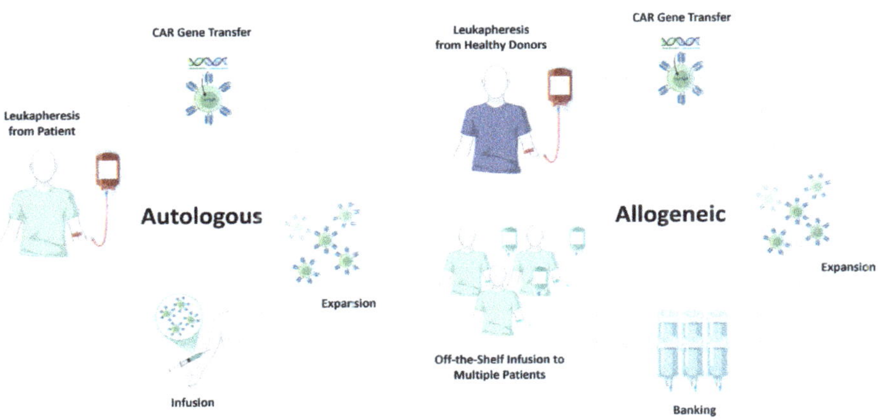

Figure 3. Comparison of autologous and allogeneic CAR cell therapy platforms. Schematic overview highlighting key differences between autologous and allogeneic CAR cell manufacturing. Autologous products are derived from the patient's own leukapheresis material and reinfused after CAR engineering, offering minimal GVHD risk but prolonged manufacturing times and variable product quality. Allogeneic products originate from healthy donors or engineered cell sources, enabling large-scale, off-the-shelf manufacturing and rapid availability, but requiring gene editing to mitigate alloreactivity and facing potential host immune rejection.

The principal challenge of allogeneic CAR-T therapy lies in preventing graft-versus-host disease (GVHD) and avoiding immune-mediated rejection. The most direct strategy to mitigate GVHD is disruption of the αβ T cell receptor (TCR), as TCR–alloantigen interactions drive alloreactivity [70]. Knockout of either the α or β constant chain prevents functional TCR assembly [71]. Early work by Torikai et al. in 2012 used zinc-finger nucleases to ablate αβ TCR expression, producing anti-CD19 CAR-T cells with preserved anti-tumour activity and no clinically significant GVHD [72]. More recently, CRISPR/Cas9, TALENs, and ARCUS nucleases have been applied to achieve TCR disruption [73,74]. Eyquem et al. demonstrated that targeted CAR insertion into the TCR alpha constant (*TRAC*) locus via CRISPR/Cas9 not only inactivated TCR expression but also improved CAR-T potency and reduced exhaustion in an ALL model [75]. To enhance persistence and resist host immune clearance, additional edits can be introduced. Deletion of CD52 renders CAR-T cells resistant to alemtuzumab-mediated depletion, allowing for selective removal of host $CD52^+$ lymphocytes before infusion [76,77]. In the first-in-human trial of αβ TCR– and CD52–deficient allogeneic CD19 CAR-T cells, the product showed a manageable safety profile and preliminary efficacy in r/r LBCL [78]. Alternative immune-conditioning strategies include engineering resistance to nucleoside analogues by disrupting deoxycytidine kinase (dCK), enabling CAR-T cell survival in the presence of purine nucleoside analogue chemotherapy while abrogating host lymphocytes [79].

Building on these foundational studies, CRISPR-based allogeneic CAR-T platforms have progressed rapidly from concept to clinical evaluation, establishing the feasibility of multiplex genome editing for off-the-shelf cell therapies. The CTX110 product from CRISPR Therapeutics employs targeted disruption of *TRAC* and *β2-microglobulin (B2M)* to generate CD19-specific CAR-T cells with reduced GVHD risk and diminished immunogenicity. In the phase I CARBON trial (NCT04035434), CTX110 induced durable remissions in patients with LBCL and a manageable safety profile [80]. The next-generation candidate CTX112 introduces additional edits in *TGFBR2, ZC3H12A* to enhance CAR expression, persistence and functional fitness, with preliminary results indicating clinical activity in r/r B cell malignancies (NCT05643742) [81]. Other gene-edited allogeneic CAR-T products, including CB-010 (Caribou Biosciences, USA, NCT04637763) and ALLO-501A (Allogene Therapeutics, USA, NCT04416984), further exemplify how multiplexed genome editing can generate off-the-shelf, TCR-deficient CAR-T cells with reduced host immunogenicity and improved in vivo persistence [82,83]. Across early-phase trials, gene-edited allogeneic CAR-T products have achieved overall response rates of roughly 50–70% in r/r B cell lymphomas, including complete responses in heavily pretreated patients and generally manageable rates of CRS, ICANS, and GVHD [84].

Although still at an early clinical stage, allogeneic CAR-T platforms hold considerable promise as scalable, rapidly deployable cancer immunotherapies. Continued optimisation of gene-editing strategies to mitigate GVHD, evade host rejection, and preserve effector function will be critical for realizing the potential of universal, off-the-shelf CAR-T products.

4. Manufacturing Protocols for Unconventional T Cell Therapy

To date, most clinically approved CAR products are based on conventional αβ T cells. While effective, this platform faces several inherent limitations that restrict its broader application. Manufacturing autologous αβ CAR-T cells remains complex and time-consuming, with variability in product quality due to patient-specific factors. Moreover, these cells carry the risk of serious, sometimes life-threatening toxicities, and their activity can be compromised by the diverse human leukocyte antigen (HLA) backgrounds present in genetically heterogeneous populations. From a therapeutic perspective, αβ T cell–based CAR products have demonstrated remarkable efficacy in B cell malignancies but only modest activity in solid tumours [3]. The unfavourable immunosuppressive TME and the heterogeneity of antigen expression in solid cancers present formidable barriers to effective targeting [2]. These challenges highlight the need for next-generation CAR platforms that can overcome antigen escape, function within suppressive tumour niches, and be manufactured efficiently at scale.

In parallel with optimizing conventional αβ CAR-T cell manufacturing, increasing attention has turned to unconventional T cell subsets, including γδ T cells, iNKT cells, and MAIT cells, as alternative immune effectors. These cell types offer unique biological advantages, such as MHC-independent antigen recognition, intrinsic anti-tumour properties, and potentially lower risks of GVHD, making them attractive candidates for next generation of CAR-based therapies (Figure 4).

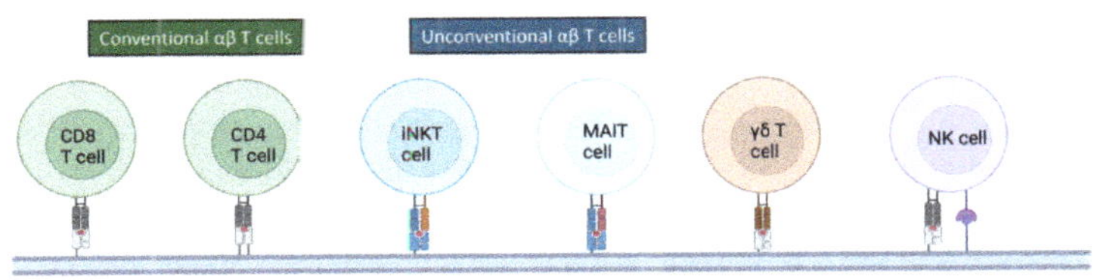

	MHC I	MHC II	CD1d	MR1	MHC I like	MHC I & receptors
Tumour infiltration	✓	✓	✓	✓	✓	✓
Antitumour efficacy	✓	✓	✓	✓	✓	✓
Low risk of GVHD	X	X	✓	✓	✓	✓
Transduction efficiency	High	High	Medium	Medium	Medium	Low
Ease of *ex vivo* expansion	High	High	Medium	Medium	Medium-Low	Medium
In vivo persistence	High	High	Medium	Unclear	Medium	Low-Medium
Scalability of starting material	High	High	Low	Low-Medium	Medium	Medium-High
Clinical maturity	FDA-approved	FDA-approved	Phase I	Preclinical	Phase I/II	Phase I/II

Figure 4. Comparative characteristics of conventional αβ T cells and emerging unconventional immune cell subsets used as platforms for CAR engineering. Key characteristics of conventional αβ T cells and unconventional effector populations, including iNKT, MAIT, γδ T, and NK cells, relevant to CAR-based therapy. These distinctions highlight the unique opportunities and limitations of each platform for autologous and allogeneic CAR product development.

4.1. Gamma-Delta (γδ) T Cells

γδ T cells are a unique subset of T lymphocytes characterized by their expression of γ and δ T cell receptor chains, in contrast to the conventional αβ T cells. These cells play a critical role in tissue homeostasis and tumour immunosurveillance and have recently gained attention as a promising platform for CAR engineering [85]. The intrinsic properties of γδ T cells offer distinct advantages for cellular immunotherapy, particularly in the context of allogeneic and off-the-shelf CAR-based therapies. One of the most significant features of γδ T cells is their ability to recognize antigens in an MHC-independent manner. This innate-like mode of antigen detection substantially reduces the risk of alloreactivity and GVHD, making donor matching unnecessary and supporting the development of universally compatible cell products [86]. In addition to γδ TCR signaling, these cells respond to a wide array of tumour-associated stress ligands through activating receptors such as Toll-like receptors (TLRs) and natural cytotoxicity receptors (NCRs) [87]. Their non-reliance on classical peptide-HLA complexes expands the scope of targetable tumour antigens and mitigates the impact of tumour antigen heterogeneity and immune escape.

The unique features of γδ T cells highlight their capacity to become a safer and more promising platform of CAR-T therapy in the allogeneic setting. Several groups have developed processing protocols using donor-derived γδ T cells as a CAR backbone to assess their therapeutic potential. Rozenbaum et al. established a robust 14-day manufacturing protocol that begins with peripheral blood mononuclear cells (PBMCs) isolation, followed by stimulation with IL-2 and OKT3 in the presence of zoledronic acid to selectively expand the Vγ9Vδ2 subset. CAR genes were introduced using retroviral transduction, and αβ T cells were removed by magnetic depletion to ensure population purity. Final products exhibited >99% γδ T cell purity and achieved a median 185-fold expansion. These CD19-targeting CAR-γδ T cells demonstrated cytotoxic activity against both CD19-positive and CD19-negative leukaemia cells in murine models, although their anti-tumour potency and persistence were inferior to conventional CAR-T cells, likely due to limited in vivo longevity. Strategies such as repeated infusion or cytokine support may help overcome this limitation [88]. To further optimize γδ T cell manufacturing, Wang et al. explored alternative activation agents including isopentenyl pyrophosphate (IPP) and bromohydrin pyrophosphate (BrHPP), both of which enhanced expansion and lentiviral transduction efficiency. High levels of CAR expression were achieved after 8–10 days of culture, followed by 5 additional days for effective gene transfer. These approaches have laid the foundation for scalable CAR-γδ T cell production [89]. In another example, Capsomidis et al. constructed GD2-specific CARs in donor-derived γδ T cells using a CD28–CD3ζ endodomain. This product demonstrated efficacy in preclinical models of neuroblastoma, highlighting the potential application of γδ CAR-T cells in solid tumours [90]. Given the poor efficacy of conventional CAR-T cells in solid cancers due to the immunosuppressive microenvironment and tumour antigen heterogeneity, the MHC-independent cytotoxicity of γδ T cells could represent a critical advantage.

The clinical translation of CAR-engineered γδ T cells has progressed rapidly in recent years, with several early-phase clinical trials now underway to assess their safety, feasibility, and antitumour potential in both haematologic and solid malignancies (Table 2). The most clinically advanced γδ CAR-T product to date is ADI-001, an allogeneic Vδ1 γδ T cell therapy expressing a CD20-specific CAR. In a phase I trial (NCT04735471), ADI-001 was administered to patients with relapsed or refractory B-cell non-Hodgkin lymphoma and achieved an overall response rate of 75%, with complete remissions in 69% of treated patients, including those who had failed prior CD19-targeted CAR-T therapy [91]. Treatment was well tolerated, with no dose-limiting toxicities, no graft-versus-host disease, and only low-grade cytokine release syndrome observed. Importantly, no TCR gene editing was required to mitigate alloreactivity, owing to the MHC-independent antigen recognition of γδ T cells [91]. A related long-term follow-up study (NCT04911478) is currently monitoring safety and persistence in patients treated with ADI-001 and other γδ CAR-T cell products. Parallel efforts are exploring the use of γδ CAR-T therapies in solid tumours. NCT04107142, a phase I dose-escalation trial sponsored by CytoMed Therapeutics, is evaluating CTM-N2D, an allogeneic Vδ2 γδ T cell product targeting NKG2D ligands, in patients with advanced colorectal, breast, sarcoma, and other refractory solid tumours. Preliminary findings from the first patient cohort indicated good tolerability, enabling dose escalation to the second cohort [92]. In a related approach, Ever Supreme Bio has launched a first-in-human phase I/II study (NCT06150885) evaluating donor-derived Vδ2 γδ T cells transiently co-engineered via mRNA electroporation to express an anti-HLA-G CAR and a bispecific T cell engager targeting PD-L1. This dual-targeting strategy is designed to overcome HLA-G–mediated immune evasion and recruit endogenous $CD3^+$ T cells for enhanced tumour clearance. The trial opened in late 2024 for patients with relapsed or refractory solid malignancies; enrolment is ongoing, though clinical outcomes have not yet been reported.

Table 2. Clinical Trials of Allogeneic CAR-Modified Innate and Unconventional Immune Cell Therapies.

Immune Cell Subtype	Target	No.	Company/Institution	Phase	Date
γδ T Cells	CD20	NCT04735471 NCT04911478	Adicet Therapeutics	Phase I/ Long term observational	2021–2025 2022–2039
	NKG2DL	NCT04107142	CytoMed Therapeutics Pte Ltd.	Phase I	2019–2021
	HLA-G	NCT06150885	Ever Supreme Biotechnology Co., Ltd.	Phase I/II	2024–2027
iNKT Cells	CD19	NCT03774654	Carlos Ramos, Baylor College of Medicine	Phase I	2020–2035
	CD19	NCT05487651	Athenex, Inc.	Phase I	2022–2024
	GD2	NCT03294954	Andras Heczey, Baylor College of Medicine	Phase I	2020–2035
NK Cells	CD19	NCT05020678	Nkarta, Inc	Phase I	2021–2038
	CD19	NCT03056339	M.D. Anderson Cancer Center	Phase I/II	2017–2023
	CD19	NCT04245722	Fate Therapeutics	Phase I	2020–2023
	HER2	NCT03383978	Johann Wolfgang Goethe University Hospital	Phase I	2017–2026
	PD-L1	NCT04847466	National Cancer Institute (NCI), US	Phase II	2021–2027
	Claudin-6, GPC3, Mesothelin, AXL	NCT05410717	Second Affiliated Hospital of Guangzhou Medical University	Phase I	2022–2036

Together, these early studies underscore the feasibility of γδ T cells as an allogeneic backbone for CAR-T therapy. Their innate-like, MHC-independent targeting reduces the risk of GVHD and may facilitate scalable, off-the-shelf production. As clinical evidence accumulates, γδ CAR-T platforms may offer a promising alternative or complement to conventional αβ CAR-T cells, particularly for patients with solid tumours or those ineligible for autologous therapy.

4.2. Invariant Natural Killer T (iNKT) Cells

iNKT cells represent a rare subset of innate-like T lymphocytes that express a semi-invariant TCR, composed of a Vα24-Jα18 α-chain paired with a Vβ11 β-chain. In contrast to conventional αβ T cells, iNKT cells recognize glycolipid antigens presented by the non-polymorphic MHC class I-like molecule CD1d [93]. This conserved antigen-presentation pathway, shared across mammals, significantly reduces the risk of alloreactivity and makes iNKT cells inherently safer for use in allogeneic adoptive cell therapies [94]. Functionally, iNKT cells exhibit NK-like cytotoxicity, rapid cytokine release, and tumour infiltration capacity. In preclinical models, they have demonstrated the ability to kill tumour cells directly, modulate myeloid populations, and eliminate tumour-associated macrophages [95,96]. These features, combined with their resistance to GVHD induction and their ability to traffic to inflamed tumour tissues, position iNKT cells as a promising platform for CAR engineering.

Several studies have developed CAR-modified iNKT cells targeting antigens such as CD19, GD2, CD38, and BCMA to explore their therapeutic applicability [97–100]. Due to their very low natural frequency in human peripheral blood (<0.1%), specialized enrichment protocols are required to enable clinical-scale manufacturing. After PBMC isolation, magnetic bead-based kits targeting the invariant Vα24 TCR are commonly used for iNKT cell selection [96,98]. Activation protocols vary across groups depending on target antigen and expansion goals. Poels et al. utilized IL-2, IL-7, IL-15, and α-galactosylceramide (α-GalCer) to achieve long-term expansion, incorporating 4-1BB signaling to support persistence and reach 300 to 1000-fold expansion within five weeks [100]. Heczey et al. demonstrated that IL-2 and IL-21 in combination with α-GalCer enhanced CD62L expression and iNKT proliferation. Additional OX-40 costimulation further improved antitumour activity [98]. Similarly, Rotolo et al. employed IL-15 and α-GalCer, combined with CD3/CD28 co-stimulation, for robust expansion and CAR transduction [97]. Gene delivery is typically achieved via lentiviral or retroviral vectors over 72 h. Given the rarity of iNKT cells in peripheral blood, restimulation and secondary transduction cycles are often required to reach therapeutic cell numbers [97,98].

In vitro studies have demonstrated that GD2-specific CAR-iNKT cells are capable of killing GD2-positive tumour cells as well as CD1d-expressing tumour-associated macrophages, enhancing tumour control across both direct and indirect mechanisms [99]. The attempts that equipped iNKT cells with CD38-CAR and/or BCMA-CAR also displayed a low risk to cause significant off-tumour toxicity toward normal hematopoietic cells ex vivo [100] and did not induce xenogeneic GVHD responses in murine models [101]. Their Th1-like cytokine profile and CAR-redirected killer functions toward multiple myeloma cell lines highlight their potential as an adoptive immunotherapy strategy [100,101]. In the first-in-human trial (NCT03294954), autologous GD2-targeting CAR-iNKT cells were administered to patients with relapsed or refractory neuroblastoma. The infused cells expanded in vivo, trafficked to bone marrow metastases, and induced signs of tumour regression. No cytokine release syndrome or GVHD was observed, supporting their safety and feasibility [98]. Following these early successes, efforts have turned to allogeneic iNKT cell therapies. A phase I trial (NCT03774654) is currently evaluating healthy donor–derived CD19-directed CAR-iNKT cells engineered with CD28 and IL-15 domains in patients with relapsed or refractory B-cell malignancies. This study aims to assess the persistence, safety, and immune evasion capacity of iNKT cells in an off-the-shelf format. Another trial (NCT05487651), launched in 2022, also investigates allogeneic CD19-CAR-iNKT cells in adult patients with CD19^{+} hematologic malignancies, focusing on dose escalation and treatment tolerability. Together, these trials represent the next phase of translation for iNKT-based CAR therapies and will be instrumental in validating their role as an allogeneic, GVHD-resistant platform for the treatment of lymphoid cancers (Table 2).

4.3. Mucosal-Associated Invariant T (MAIT) Cells

MAIT cells are a distinctive subset of innate-like T lymphocytes that express a semi-invariant TCR (Vα7.2-Jα33 with Vβ2 or Vβ13). Unlike conventional αβ T cells, MAIT cells recognize microbial riboflavin metabolites presented by the monomorphic MR1 molecule, rather than polymorphic HLA class I [102,103]. This MR1 restriction fundamentally limits their alloreactive potential and positions MAIT cells as highly suitable for allogeneic, off-the-shelf cellular therapies. A pivotal study by Tourret et al. demonstrated that human MAIT cells are intrinsically non-alloreactive. In both xenogeneic and fully MHC-mismatched allogeneic transplant models, MAIT cells failed to induce GVHD, in stark contrast to conventional αβ T cells [104]. Their inability to recognize alloantigens allows MAIT cells to persist in allogeneic hosts without provoking host-versus-graft or graft-versus-host immune responses, providing a strong biological rationale for developing MAIT cells as a universal donor platform.

MAIT cells are relatively abundant compared with other unconventional lymphocyte subsets. They can comprise up to 10% of circulating T cells and 30–50% of T cells in mucosal organs such as the liver and lung [105]. Their natural tissue residency reflects high expression of homing receptors including CCR6 and CXCR6, enabling efficient trafficking to peripheral tissues and tumours [106,107]. Upon activation, MAIT cells exert potent cytotoxicity through perforin and granzyme release, and secrete pro-inflammatory cytokines such as IFN-γ, TNF-α, IL-17, and IL-2 [108]. They also respond robustly to IL-12, IL-15, and IL-18 in a TCR-independent manner, further amplifying their effector capacity even in settings of limited antigen presentation [102].

The development of MAIT cells as a CAR backbone is still in an early stage, but significant progress has been made in establishing scalable isolation and culture workflows. MR1–5-OP-RU tetramers conjugated to fluorochromes enable highly specific enrichment of MAIT cells and can provide mild TCR stimulation during isolation [109,110]. Flow cytometric sorting based on Vα7.2 and CD161 co-expression remains another reliable approach for obtaining highly pure populations. Standard CD3/CD28 bead stimulation is compatible with MAIT expansion, but cytokine-enriched regimens can yield superior proliferation. Parrot et al. demonstrated over 300-fold expansion using IL-2 and irradiated autologous feeder PBMCs at a 1:10 ratio, providing sufficient output for downstream engineering [109]. Recent preclinical studies have confirmed the feasibility of generating CAR-modified MAIT cells. Lentiviral transduction following MR1-ligand stimulation and cytokine support (IL-2, IL-7, IL-15) enabled stable CAR expression targeting CD19, HER2, or mesothelin [111,112]. CAR-MAIT cells showed potent cytotoxicity comparable to conventional CAR-T cells while producing lower levels of inflammatory cytokines, suggesting a potentially improved safety profile [111]. Notably, MAIT cells could also target MR1-expressing tumour-associated macrophages, permitting dual mechanisms of tumour control and improved performance in immunosuppressive organoid models [112].

Although clinical trials of CAR-MAIT cells have not yet begun as of 2025, translational interest is accelerating. A small but growing patent landscape describes MAIT-cell purification methods, MR1-tetramer applications, and genetic engineering strategies for CAR or TCR modification. These filings remain early and largely academic, but they reflect increased recognition of MAIT cells as a promising allogeneic platform. Given their absence of alloreactivity, natural tissue tropism, dual innate–CAR killing capacity, and resilience within immunosuppressive microenvironments, MAIT cells represent a compelling foundation for next-generation universal cellular therapies. Rigorous preclinical evaluation, optimization of culture and gene-delivery methods, and standardized potency assays will be essential to advance CAR-MAIT cells toward future clinical translation.

5. Manufacturing Protocols for Natural Killer (NK) Cell Therapy

NK cells are cytotoxic innate lymphocytes that contribute to tumour immune surveillance by directly targeting transformed or infected cells without prior sensitization. Their function is regulated by a dynamic balance of activating and inhibitory receptors, rather than antigen-specific recognition. NK cell infiltration into tumours has been associated with improved clinical outcomes in several cancers, including lung carcinoma, melanoma, and hepatocellular carcinoma [113,114]. Given their favourable immunobiology and safety profile, NK cells have emerged as an attractive platform for CAR engineering.

Compared to T cells, CAR-NK cells offer several advantages. They are naturally alloreactive-tolerant and do not require strict HLA matching, enabling their use in off-the-shelf settings without inducing GVHD [115]. Furthermore, their limited production of pro-inflammatory cytokines, such as IL-6 and TNF-α, reduces the risk of CRS and neurotoxicity, two major toxicities observed in CAR-T therapy [18,116]. Importantly, NK cells retain intrinsic cytotoxic mechanisms, which allows them to eliminate tumour cells independently of CAR engagement, thereby mitigating the risk of tumour antigen escape [115].

Multiple NK cell sources have been exploited for CAR engineering, including peripheral blood, umbilical cord blood, induced pluripotent stem cells (iPSCs), and the NK-92 cell line [117]. Peripheral and cord blood NK cells are immediately accessible but exhibit donor-to-donor variability and limited proliferative capacity. By contrast, iPSCs provide a renewable, genetically defined, and highly standardized starting population. CAR insertion can be performed at the pluripotent stage, followed by directed differentiation into NK cells under GMP conditions, yielding uniform batches with consistent phenotype and function [118,119]. This top-down manufacturing architecture enables long-term master cell banking and supports true off-the-shelf scalability. Recent iPSC-derived platform, FT596, integrates a CD19 CAR, enhanced CD16 receptor, and IL-15 receptor fusion, showed excellent tolerability with no GVHD and low rates of CRS/ICANS in a phase I trial (NCT04245722) while achieving durable responses in r/r B cell lymphomas [120]. Similar iPSC-based frameworks are now being evaluated for CAR-T generation, offering a theoretically inexhaustible supply of rejuvenated T cells with reduced donor heterogeneity [121]. The NK-92 cell line represents another widely utilized source due to its ease of expansion, high cytotoxic potential, and efficient gene transfer. Clinical studies evaluating NK-92 or CAR-NK-92 products, including HER2-CAR-NK-92 for glioblastoma (NCT03383978), have demonstrated favourable safety profiles without GVHD or high-grade CRS [122,123]. However, because NK-92 cells originate from a malignant clone, they require irradiation prior to infusion to prevent engraftment, which limits persistence and restricts their use predominantly to solid tumour indications or repeated-dose regimens [124,125]. These characteristics distinguish NK-92 from primary and iPSC-derived CAR-NK products and highlight important considerations for platform selection.

Ex vivo NK cell expansion is typically supported by cytokine stimulation with IL-2, IL-15, or IL-21, often in combination with irradiated feeder cells or artificial antigen-presenting cells. Engineered K562 feeder cells expressing membrane-bound IL-15 and IL-21 can dramatically enhance expansion efficiency and cytotoxicity in both primary and iPSC-derived NK cells [126,127]. A major bottleneck in CAR engineering is the intrinsic resistance of NK cells to viral transduction. Conventional retroviral methods achieve only modest efficiencies (22–66%) [117,119]. In contrast, lentiviral vectors pseudotyped with baboon endogenous retrovirus (BaEV) envelope proteins markedly improve gene delivery, enabling transduction efficiencies exceeding 80% in some platforms [127].

To enhance function in immunosuppressive TME, several groups have applied targeted gene editing to improve NK-cell resilience. Deletion of *PD-1* or *CISH* using CRISPR/Cas9 enhances cytotoxicity, metabolic fitness, and in vivo persistence [128,129]. Similarly, knockout of *TGFβR2* confers resistance to TGF-β–mediated suppression without impairing cytolytic activity [130,131]. These modifications aim to equip CAR-NK cells with enhanced resilience against immunosuppressive signals and are being incorporated into next-generation constructs.

Scalable manufacturing is also advancing through automated, closed-system bioreactors. The CliniMACS Prodigy® platform has been adapted for NK-cell processing, enabling GMP-compliant generation of CAR-NK cells with minimal hands-on time. Albinger et al. produced functional CD33-CAR-NK cells using this system, achieving cytotoxicity comparable to manually manufactured batches in preclinical models [132]. Automation is likely to be particularly impactful for iPSC-derived platforms, where large, uniform cell banks are well aligned with fully closed manufacturing workflows. Clinical translation continues to accelerate (Table 2). One of the earliest and most prominent trials (NCT03056339) evaluated UCB-derived CD19-targeted CAR-NK cells engineered to co-express IL-15 and a safety switch, the results demonstrated encouraging remission rates with minimal toxicity [133]. Additional ongoing phase I/II studies, including NCT04847466 and NCT05410717, are evaluating emerging CAR-NK constructs targeting antigens such as Claudin-6, GPC3, mesothelin, and AXL, as well as innovative design strategies for the treatment of solid tumours [134].

Taken together, the favourable immunobiology of NK cells, combined with engineering flexibility, scalable iPSC-based manufacturing, and encouraging early clinical results, highlights CAR-NK therapy as a promising and increasingly mature alternative to CAR-T platforms in adoptive immunotherapy.

6. In Vivo CAR-T Cell Therapy

The clinical and logistical challenges inherent to conventional ex vivo CAR-T cell manufacturing have driven the development of in vivo CAR-T strategies. Instead of isolating, activating, and genetically modifying patient T cells outside the body, this approach delivers CAR-encoding constructs directly to endogenous T cells, enabling their reprogramming within the patient. In vivo CAR-T strategies offer the potential to eliminate patient-specific production lines, accelerate treatment timelines from weeks to days, and expand access to cellular immunotherapy beyond specialized manufacturing centres [135]. In vivo CAR-T therapy requires efficient, targeted delivery of CAR-encoding genes to peripheral T cells, and several delivery technologies are under active investigation (Table 3).

Table 3. Key features of delivery systems enabling in vivo CAR-T engineering.

Platform	Composition	Cargo	Mechanism of Action	Advantages	Limitations
Lipid Nanoparticles	Ionizable lipid + cholesterol + PEG-lipid; may include antibody–lipid conjugate for targeting.	mRNA DNA	Endocytosis → mRNA translation → transient CAR expression	• Safe • Non-viral, scalable (vaccine-like manufacturing) • Subset-specific targeting	• Transient expression • Innate immune sensing • Hepatic accumulation risk
Polymeric Nanoparticles	Biodegradable or cationic polymers (PLGA, PEI, chitosan)	mRNA mcDNA	Uptake by endocytosis → proton-sponge–mediated endosomal escape → cytoplasmic/nuclear delivery	• Highly customizable (surface charge, degradation rate). • Can co-deliver adjuvants or cytokines.	• Lower T cell efficiency • Risk of insertional mutagenesis • Variable biocompatibility Reproducibility issues
Lentiviral Delivery System	Enveloped RNA virus (HIV-1–derived; VSV-G pseudotyped)	RNA	Reverse transcription → genomic integration → stable CAR expression	• Long-term expression • Clinically validated • High efficiency	• Integration risk (mutagenesis) • Costly GMP manufacturing • Limited redosing

Table 3. *Cont.*

Platform	Composition	Cargo	Mechanism of Action	Advantages	Limitations
AAV-Based Delivery System	Non-enveloped capsid with single-stranded or self-complementary DNA genome (AAV2, AAV6, AAV9)	DNA	Nuclear entry → episomal persistence (non-integrating) → transient/semi-stable expression	• High efficiency • Low genotoxic risk	• Small cargo capacity (~4.7 kb) • Anti-AAV immunity • Expensive
Bioinstructive Implantable Scaffold	Biodegradable hydrogel/ polymer matrix (alginate, PEG) with viral vectors, antibodies, cytokines	Viral or mRNA CAR constructs	PBMCs infiltrate scaffold → in situ transduction & expansion → systemic CAR-T release	• Localized CAR induction • Controlled cytokine milieu • Suitable for solid tumours	• Invasive implantation • Scalability • Preclinical stage

6.1. Lipid Nanoparticle (LNP)

Among non-viral systems, lipid nanoparticles (LNPs) have emerged as the leading platform for in vivo CAR-T generation due to their clinical precedent in mRNA vaccines, high biocompatibility, and scalable manufacturing. LNPs consist of ionizable lipids, cholesterol, phospholipids, and PEG–lipids and can be conjugated with antibodies or scFvs that recognize T-cell surface markers such as CD3, CD4, CD5, CD7, or CD8. After intravenous administration, these particles are taken up by T cells and release their cargo into the cytoplasm following endosomal escape [136,137]. This transient CAR expression profile without genomic integration provides an excellent safety margin and supports repeated dosing for controlled therapeutic exposure. However, it also limits persistence, necessitating redosing to maintain efficacy. Moreover, innate immune activation through TLR7/8 or RIG-I signaling and hepatic uptake can restrict systemic delivery efficiency [137]. Optimization of lipid composition and targeting ligands is therefore critical to achieve selective, sustained T cell transfection in humans.

The feasibility of this approach has been evaluated in multiple early studies. In a preclinical model, CD3-targeted LNPs carrying CD19-CAR mRNA and IL-6 short hairpin RNA achieved efficient in vivo T-cell transfection, resulting in sustained CAR expression for up to 90 days, potent B cell depletion, and attenuated CRS-associated inflammation [138]. Billingsley et al. optimized ionizable LNP chemistry and achieved extra-hepatic targeting of T cells in vivo, leading to up to 90% B cell depletion [139]. A more advanced mRNA-LNP system to date is the L829 targeted LNP (tLNP) described by Hunter et al. This formulation delivered CAR mRNA directly to CD8^{+} T cells in humanized mice and non-human primates while minimizing hepatic accumulation. The therapy induced rapid and profound B cell depletion, followed by the repopulation of predominantly naïve B cells, reflecting an effective immune system reset. The treatment was well tolerated and caused only mild, transient cytokine elevations [136].

Recently, Bimbo et al. expanded this technology to a DNA-based targeted LNP system termed NCtx, which carries minicircle DNA and transposase mRNA instead of only CAR-mRNA. The particles were dual-targeted to CD3 and CD7, allowing for uptake by both resting and activated T cells and eliminating the need for prior stimulation. This design enabled stable genomic integration of the CAR construct, generating long-lived CAR-T cells directly in the host. In humanized mouse models, a single injection of NCtx-CD19 led to complete leukemia regression and durable survival. Moreover, dual-specific constructs encoding CD19/CD22 CARs maintained CAR expression for more than 40 days [140].

LNP-mediated delivery systems offer a compelling balance of precision, safety, and manufacturability. Transient mRNA formulations minimize genotoxic risk and enable repeat dosing, whereas DNA-based or transposase-assisted designs can extend expression durability. Ongoing refinements in lipid chemistry, targeting ligands, and nucleic acid payload engineering are rapidly advancing these platforms toward clinically viable in vivo CAR-T manufacturing.

6.2. Polymeric Nanoparticle

Polymeric and hybrid lipid–polymer nanoparticles provide an alternative non-viral route for in vivo CAR delivery, particularly for DNA-based payloads designed to support longer-term CAR expression. A seminal study by Smith et al. used synthetic poly (β-amino ester) (PBAE) nanoparticles loaded with plasmid DNA encoding CD19-specific CAR to program circulating T cells in mice. It proved that polymeric DNA nanocarriers can mediate clinically relevant anti-tumour responses without ex vivo manipulation [141]. Mechanistically, these cationic polymers electrostatically complex with negatively charged DNA, protect it from degradation, and facilitate endosomal escape and nuclear entry. Subsequent work and recent reviews have highlighted PBAE and related biodegradable polymers as promising vectors for in vivo immune-cell reprogramming, given their tuneable chemistry, relatively low cost, and ability to co-deliver multiple genetic elements [137]. However, polymeric systems face several limitations. DNA requires nuclear access, which can limit transfection efficiency compared with mRNA-based LNPs. When combined with transposase systems, carries a theoretical risk of insertional mutagenesis. In addition, formulation reproducibility and in vivo biodistribution can vary between batches [142]. Thus, while polymeric DNA nanocarriers offer a flexible, non-viral platform with promising preclinical efficacy, their translation will depend on improved control over targeting, integration, and long-term safety.

6.3. Viral Vectors

Viral vectors provided the earliest and most robust demonstrations that CAR-T cells can be generated directly inside the host. Recent work by Nicolai et al. described VivoVec, a T-cell-targeted lentiviral platform designed for in vivo CAR-T generation in non-human primates. The vector displays a multi-domain fusion ligand comprising CD80 and CD58 to engage CD3 and costimulatory receptors on T cells. A single infusion of VivoVec encoding a CD20 CAR generated substantial frequencies of CAR^+ T cells in blood and secondary lymphoid organs, leading to deep and sustained B cell depletion without the need for lymphodepleting chemotherapy [143]. In parallel, AAV-based systems have also been explored for the generation of in vivo CAR platform. AAV-based systems have similarly shown efficacy in generating functional CAR-T cells in vivo, leading to successful tumour regression [144,145].

These viral approaches offer very high transduction efficiency and long-term expression, supported by mature GMP manufacturing pipelines. However, both platforms face important challenges, including high production cost, pre-existing anti-vector immunity that can impair efficacy or prevent re-dosing, and, for lentivirus, a residual risk of insertional mutagenesis. These constraints are driving the rapid development of non-viral and hybrid systems that aim to retain the potency of viral vectors while improving safety and scalability.

6.4. Bioinstructive Implantable Scaffolds

Beyond systemic nanoparticles and viral vectors, bioinstructive implantable scaffolds offer a localized strategy to generate and release CAR-T cells directly within the patient. In a key study, Agarwalla et al. introduced MASTER (Multifunctional Alginate Scaffold for T-cell Engineering and Release), an alginate-based, antibody- and cytokine-functionalized scaffold, in which autologous PBMCs were briefly mixed ex vivo with CAR-encoding viral particles and loaded into MASTER, which was then implanted subcutaneously. Within hours, T cells were activated, transduced, and expanded inside the scaffold and subsequently released into circulation as CAR-T cells. In murine lymphoma models, MASTER-generated CAR-T cells showed superior engraftment, a less exhausted, more stem-like phenotype, and improved tumour control compared with conventionally manufactured CAR-T cells while reducing the overall production timeline to roughly one day [146].

Subsequent work has extended the scaffold concept to acellular systems that recruit and program host T cells directly in situ. Dandia et al. reported a three-dimensional, cell-free scaffold implanted adjacent to solid tumours, loaded with viral vectors and immunostimulatory materials, which locally attracted endogenous T cells, mediated CAR gene transfer, and promoted their expansion within the tumour bed, resulting in marked tumour growth inhibition in preclinical models [146]. Similarly, Pandit et al. developed "Drydux," a macroporous biomaterial "CAR-T factory" that can be implanted near solid tumours to generate and release tumour-specific CAR-T cells in situ [147].

Bioinstructive scaffolds thus offer several conceptual advantages. They favour early memory phenotypes by minimizing prolonged ex vivo culture and concentrate CAR-T activity at or near tumour sites, which may lower systemic toxicity and improve solid tumour responses. At the same time, they introduce new challenges, including the requirement for surgical or interventional delivery, added complexity in device-level manufacturing and regulation as combination products. Rather than replacing systemic in vivo CAR-T platforms, scaffold-based in situ programming is likely to serve as a complementary strategy, illustrating how the convergence of bioinstructive materials and gene delivery technologies can shape the future of CAR-T manufacturing and clinical translation.

7. Conclusions and Perspectives

Over the past decades, CAR technology has evolved into a foundational platform in cancer immunotherapy. By redirecting immune cell specificity through antibody-derived scFvs, CAR-engineered lymphocytes have produced transformative clinical outcomes in relapsed or refractory haematological malignancies. The approval of CD19-targeted products, such as tisagenlecleucel and axicabtagene ciloleucel, has validated this therapeutic concept and paved the way for rapid innovation [4,5]. Despite these achievements, conventional CAR-T therapies remain constrained by manufacturing bottlenecks, high cost, variable product quality, and limited efficacy in solid tumours.

Manufacturing is central to these limitations. The autologous model requires individualized leukapheresis, ex vivo activation, gene transfer, and expansion. These steps are logistically complex, costly, and difficult to scale. Product quality is further influenced by patient-intrinsic factors such as lymphopenia, prior chemotherapy, or immune exhaustion. To address these challenges, manufacturing protocols are evolving rapidly. Automated and closed systems such as the CliniMACS Prodigy®, Lonza Cocoon®, and G-Rex® now enable GMP-compliant, standardized production with reduced operator input and lower contamination risk [61,63]. Improvements in cytokine support, starting cell phenotype, and activation conditions have further refined product consistency. In parallel, ultrarapid manufacturing platforms, including UF-CAR, FasT CAR-T, and PACK-IT, have reduced production time from weeks to days while preserving early-memory phenotypes and functional potency [64,66,67]. Together, these developments signal a shift toward rapid, decentralized, and more accessible manufacturing models.

Beyond conventional $\alpha\beta$ T cells, a new generation of CAR platforms harnesses unconventional immune populations such as $\gamma\delta$ T, iNKT, MAIT, and NK cells. These subsets offer intrinsic advantages, including MHC-independent recognition, reduced risk of GVHD, natural tissue tropism, and robust baseline cytotoxicity [98,104,111]. Although most remain at preclinical or early clinical stages, advances in isolation, expansion, and transduction, such as MR1-tetramer sorting, feeder-cell systems, and optimized lentiviral pseudotypes, are accelerating their translational readiness [127].

Another transformative development is in vivo CAR-T engineering, where circulating T cells are genetically programmed inside the patient using lipid nanoparticles or polymeric carriers that deliver CAR-encoding mRNA or DNA [136,140]. This approach bypasses ex vivo manipulation and has the potential to reduce cost and broaden global access. Preclinical studies have shown efficient generation of functional CAR-T cells in vivo with robust activity in models of B cell depletion and cardiac fibrosis [136,148], and early-phase clinical evaluation is underway.

Despite these technological advances, solid tumours remain the most formidable barrier. Antigen heterogeneity, physical stromal barriers, and a highly immunosuppressive TME collectively limit CAR-T cell infiltration, persistence and effector function. To overcome these obstacles, several complementary engineering strategies are under active investigation. Metabolic reprogramming, such as overexpressing the mitochondrial regulator PGC1α or disrupting *REGNASE-1*, aims to enhance oxidative metabolism and nutrient utilisation to improve T cell fitness within the nutrient-depleted TME. Checkpoint modulation, including PD-1 knockout or PD-1/CD28 switch-receptor designs, can restore T cell activation in the presence of inhibitory ligands. Enzymatic strategies, such as ADA1-mediated degradation of adenosine, counteract metabolite-driven suppression in the tumour niche [149–151]. In parallel, nanocarrier-delivered siRNA targeting PD-L1 or TGF-β pathways, together with cytokine-armoured or matrix-remodelling CAR-T constructs, enhance tumour infiltration and functional resilience [152,153]. These complementary approaches demonstrate how manufacturing innovation and cellular engineering are converging to expand CAR-T applications beyond haematologic cancers.

While these innovations broaden the therapeutic landscape, several cross-cutting challenges must be resolved to ensure scalable, standardized, and clinically robust CAR-based therapies. Large-scale expansion of rare populations such as MAIT and iNKT cells is technically demanding. Robust, harmonized potency assays and real-time quality control will be essential to ensure inter-site consistency as automated systems become more prevalent. Gene-transfer platforms, such as viral, non-viral, and CRISPR-based, must continue to balance efficiency, cost, scalability, and genomic safety. Meanwhile, combinatorial engineering strategies, including logic-gated CARs, suicide switches, and cytokine armouring, are being refined to enhance precision and mitigate toxicity [44,154]. Integration of synthetic biology, single-cell analytics, and machine-learning–supported optimisation is expected to accelerate innovation across the CAR manufacturing ecosystem.

In conclusion, the CAR field is transitioning from custom-built, labour-intensive therapies toward programmable, scalable, and globally deployable platforms. The convergence of manufacturing innovation, novel immune cell sources, and in vivo CAR engineering is poised to broaden the reach of CAR-based immunotherapy across both haematologic and solid tumours. As these technologies mature, the establishment of harmonized regulatory frameworks, standardized potency metrics, and international quality guidelines will be critical to translating their full clinical and societal impact.

Author Contributions: Conceptualization, Y.Q. and X.-N.X.; literature search, Y.Q.; writing—original draft preparation, Y.Q.; writing—review and editing, Y.Q., W.M. and X.-N.X.; visualization, Y.Q.; supervision, X.-N.X.; project administration, X.-N.X.; funding acquisition, X.-N.X. All authors have read and agreed to the published version of the manuscript.

Funding: This research was funded by Medical Research Council: MR/Y011910/1.

Institutional Review Board Statement: Not applicable.

Informed Consent Statement: Not applicable.

Data Availability Statement: No new data were created or analyzed in this study.

Conflicts of Interest: The authors declare no conflict of interest.

References

1. Mchayleh, W.; Bedi, P.; Sehgal, R.; Solh, M. Chimeric Antigen Receptor T-Cells: The Future is Now. *J. Clin. Med.* **2019**, *8*, 207. [CrossRef] [PubMed]
2. Sadelain, M.; Brentjens, R.; Rivière, I. The Basic Principles of Chimeric Antigen Receptor Design. *Cancer Discov.* **2013**, *3*, 388–398. [CrossRef] [PubMed]
3. Labanieh, L.; Majzner, R.G.; Mackall, C.L. Programming CAR-T cells to kill cancer. *Nat. Biomed. Eng.* **2018**, *2*, 377–391. [CrossRef]
4. Maude, S.L.; Laetsch, T.W.; Buechner, J.; Rives, S.; Boyer, M.; Bittencourt, H.; Bader, P.; Verneris, M.R.; Stefanski, H.E.; Myers, G.D.; et al. Tisagenlecleucel in Children and Young Adults with B-Cell Lymphoblastic Leukemia. *N. Engl. J. Med.* **2018**, *378*, 439–448. [CrossRef]
5. Neelapu, S.S.; Locke, F.L.; Bartlett, N.L.; Lekakis, L.J.; Miklos, D.B.; Jacobson, C.A.; Braunschweig, I.; Oluwole, O.O.; Siddiqi, T.; Lin, Y.; et al. Axicabtagene Ciloleucel CAR T-Cell Therapy in Refractory Large B-Cell Lymphoma. *N. Engl. J. Med.* **2017**, *377*, 2531–2544. [CrossRef]
6. FDA. FDA Approves New Treatment for Adults with Relapsed or Refractory Large-B-Cell Lymphoma [Internet]. FDA NEWS RELEASE. 2021. Available online: https://www.fda.gov/vaccines-blood-biologics/cellular-gene-therapy-products/breyanzi (accessed on 20 January 2021).
7. FDA. FDA Approves First Cell-Based Gene Therapy for Adult Patients with Relapsed or Refractory MCL [Internet]. FDA NEWS RELEASE. 2020. Available online: https://www.fda.gov/drugs/resources-information-approved-drugs/fda-approves-brexucabtagene-autoleucel-relapsed-or-refractory-mantle-cell-lymphoma (accessed on 12 October 2025).
8. FDA. ABECMA (Idecabtagene Vicleucel) [Internet]. U.S. Food and Drug Administration. 2021. Available online: https://www.fda.gov/vaccines-blood-biologics/abecma-idecabtagene-vicleucel (accessed on 30 March 2021).
9. FDA. Carvykti. 2023. Available online: https://www.fda.gov/vaccines-blood-biologics/cellular-gene-therapy-products/carvykti (accessed on 12 October 2025).
10. Bellino, S.; La Salvia, A.; Cometa, M.F.; Botta, R. Cell-based medicinal products approved in the European Union: Current evidence and perspectives. *Front. Pharmacol.* **2023**, *14*, 1200808. [CrossRef]
11. JW Therapeutics. JW Therapeutics Announces NMPA Approval of Relmacabtagene Autoleucel Injection in Patients with Relapsed or Refractory Follicular Lymphoma. *PR Newswire*, 10 October 2022. Available online: https://www.prnewswire.com/news-releases/jw-therapeutics-announces-nmpa-approval-of-relmacabtagene-autoleucel-injection-in-patients-with-relapsed-or-refractory-follicular-lymphoma-301644493.html (accessed on 12 October 2025).
12. Keam, S.J. Equecabtagene Autoleucel: First Approval. *Mol. Diagn. Ther.* **2023**, *27*, 781–787. [CrossRef]
13. Chen, W.; Fu, C.; Fang, B.; Liang, A.; Xia, Z.; He, Y.; Lu, J.; Liu, H.; Hou, M.; Cai, Z.; et al. Phase II study of zevorcabtagene autoleucel, a fully human BCMA-targeting CAR T cell therapy, in patients with relapsed/refractory multiple myeloma. *Exp. Hematol. Oncol.* **2025**, *14*, 119. [CrossRef]
14. Wang, Y.; Lv, L.; Song, Y.; Wei, X.; Zhou, H.; Liu, Q.; Xu, K.; Yan, D.; Zhang, C.; Liu, S.; et al. Inaticabtagene autoleucel in adult relapsed or refractory B-cell acute lymphoblastic leukemia. *Blood Adv.* **2025**, *9*, 836–843. [CrossRef]
15. Benthani, F.; Upadhaya, S.; Zhou, A. Cancer cell therapies: Global clinical trial trends and emerging directions. *Nat. Rev. Drug Discov.* **2025**, *24*, 898–899. [CrossRef] [PubMed]
16. Cao, L.-Y.; Zhao, Y.; Chen, Y.; Ma, P.; Xie, J.-C.; Pan, X.-M.; Zhang, X.; Chen, Y.-C.; Wang, Q.; Xie, L.-L. CAR-T cell therapy clinical trials: Global progress, challenges, and future directions from ClinicalTrials.gov insights. *Front. Immunol.* **2025**, *16*, 1583116. [CrossRef] [PubMed]
17. Gust, J.; Taraseviciute, A.; Turtle, C.J. Neurotoxicity Associated with CD19-Targeted CAR-T Cell Therapies. *CNS Drugs* **2018**, *32*, 1091–1101. [CrossRef]

18. Frigault, M.J.; Maus, M.V. State of the art in CAR T cell therapy for CD19+ B cell malignancies. *J. Clin. Investig.* **2020**, *130*, 1586–1594. [CrossRef] [PubMed]
19. Arcangeli, S.; Falcone, L.; Camisa, B.; De Girardi, F.; Biondi, M.; Giglio, F.; Ciceri, F.; Bonini, C.; Bondanza, A.; Casucci, M. Next-Generation Manufacturing Protocols Enriching TSCM CAR T Cells Can Overcome Disease-Specific T Cell Defects in Cancer Patients. *Front. Immunol.* **2020**, *11*, 1217. [CrossRef]
20. Stock, S.; Schmitt, M.; Sellner, L. Optimizing Manufacturing Protocols of Chimeric Antigen Receptor T Cells for Improved Anticancer Immunotherapy. *Int. J. Mol. Sci.* **2019**, *20*, 6223. [CrossRef]
21. Vormittag, P.; Gunn, R.; Ghorashian, S.; Veraitch, F.S. A guide to manufacturing CAR T cell therapies. *Curr. Opin. Biotechnol.* **2018**, *53*, 164–181. [CrossRef]
22. Wang, X.; Rivière, I. Clinical manufacturing of CAR T cells: Foundation of a promising therapy. *Mol. Ther. Oncolytics* **2016**, *3*, 16015. [CrossRef]
23. Parente-Pereira, A.C.; Wilkie, S.; van der Stegen, S.J.; Davies, D.M.; Maher, J. Use of retroviral-mediated gene transfer to deliver and test function of chimeric antigen receptors in human T-cells. *J. Biol. Methods* **2014**, *1*, e7. [CrossRef]
24. Turtle, C.J.; Hanafi, L.-A.; Berger, C.; Gooley, T.A.; Cherian, S.; Hudecek, M.; Sommermeyer, D.; Melville, K.; Pender, B.; Budiarto, T.M.; et al. CD19 CAR-T cells of defined CD4+:CD8+ composition in adult B cell ALL patients. *J. Clin. Investig.* **2016**, *126*, 2123–2138. [CrossRef]
25. Singh, H.; Figliola, M.J.; Dawson, M.J.; Olivares, S.; Zhang, L.; Yang, G.; Maiti, S.; Manuri, P.; Senyukov, V.; Jena, B.; et al. Manufacture of clinical-grade CD19-specific T cells stably expressing chimeric antigen receptor using Sleeping Beauty system and artificial antigen presenting cells. *PLoS ONE* **2013**, *8*, e64138. [CrossRef] [PubMed]
26. Tuazon, S.A.; Li, A.; Gooley, T.; Eunson, T.W.; Maloney, D.G.; Turtle, C.J.; Linenberger, M.L.; Connelly-Smith, L.S. Factors affecting lymphocyte collection efficiency for the manufacture of chimeric antigen receptor T cells in adults with B-cell malignancies. *Transfusion* **2019**, *59*, 1773–1780. [CrossRef] [PubMed]
27. Rafiq, S.; Hackett, C.S.; Brentjens, R.J. Engineering strategies to overcome the current roadblocks in CAR T cell therapy. *Nat. Rev. Clin. Oncol.* **2020**, *17*, 147–167. [CrossRef] [PubMed]
28. Majzner, R.G.; Mackall, C.L. Clinical lessons learned from the first leg of the CAR T cell journey. *Nat. Med.* **2019**, *25*, 1341–1355. [CrossRef]
29. Barrett, D.M.; Singh, N.; Liu, X.; Jiang, S.; June, C.H.; Grupp, S.A.; Zhao, Y. Relation of clinical culture method to T-cell memory status and efficacy in xenograft models of adoptive immunotherapy. *Cytotherapy* **2014**, *16*, 619–630. [CrossRef]
30. Casati, A.; Varghaei-Nahvi, A.; Feldman, S.A.; Assenmacher, M.; Rosenberg, S.A.; Dudley, M.E.; Scheffold, A. Clinical-scale selection and viral transduction of human naïve and central memory CD8+ T cells for adoptive cell therapy of cancer patients. *Cancer Immunol. Immunother.* **2013**, *62*, 1563–1573. [CrossRef] [PubMed]
31. Kamdar, M.; Adams, C.; Bair, S.; Dropulic, B.; Gutman, J.; Haverkos, B.; Jordan, K.; Mallo, R.; Marians, R.; Mast, F.; et al. Abstract CT522: Feasibility and safety of a novel CD19 CAR T cell therapy in adults with R/R B-NHL. *Cancer Res.* **2022**, *32*, CT522. [CrossRef]
32. Wang, X.; Qu, J.; Stefanski, J.; Borquez-Ojeda, O.; Hack, A.; He, Q.; Wasielewska, T.; Du, F.; Sadelain, M.; Rivière, I. 459. Evaluation of Miltenyi ExpAct and TransAct CD3/28 Beads for CAR-T Cell Manufacturing. *Mol. Ther.* **2016**, *24*, S182. [CrossRef]
33. Shrestha, B.; Zhang, Y.; Yu, B.; Li, G.; Boucher, J.C.; Beatty, N.J.; Tsai, H.-C.; Wang, X.; Mishra, A.; Sweet, K.; et al. Generation of Antitumor T Cells for Adoptive Cell Therapy with Artificial Antigen Presenting Cells. *J. Immunother.* **2020**, *43*, 79–88. [CrossRef]
34. Schmidts, A.; Marsh, L.C.; A Srivastava, A.; A Bouffard, A.; Boroughs, A.C.; Scarfò, I.; Larson, R.C.; Bedoya, F.; Choi, B.D.; Frigault, M.J.; et al. Cell-based artificial APC resistant to lentiviral transduction for efficient generation of CAR-T cells from various cell sources. *J. Immunother. Cancer* **2020**, *8*, e000990. [CrossRef]
35. Ghassemi, S.; Nunez-Cruz, S.; O'Connor, R.S.; Fraietta, J.A.; Patel, P.R.; Scholler, J.; Barrett, D.M.; Lundh, S.M.; Davis, M.M.; Bedoya, F.; et al. Reducing Ex Vivo Culture Improves the Antileukemic Activity of Chimeric Antigen Receptor (CAR) T Cells. *Cancer Immunol. Res.* **2018**, *6*, 1100–1109. [CrossRef]
36. Reddy, O.L.; Stroncek, D.F.; Panch, S.R. Improving CAR T cell therapy by optimizing critical quality attributes. *Semin. Hematol.* **2020**, *57*, 33–38. [CrossRef]
37. Abate-Daga, D.; Davila, M.L. CAR models: Next-generation CAR modifications for enhanced T-cell function. *Mol. Ther. Oncolytics* **2016**, *3*, 16014. [CrossRef]
38. Thistlethwaite, F.C.; Gilham, D.E.; Guest, R.D.; Rothwell, D.G.; Pillai, M.; Burt, D.J.; Byatte, A.J.; Kirillova, N.; Valle, J.W.; Sharma, S.K.; et al. The clinical efficacy of first-generation carcinoembryonic antigen (CEACAM5)-specific CAR T cells is limited by poor persistence and transient pre-conditioning-dependent respiratory toxicity. *Cancer Immunol. Immunother.* **2017**, *66*, 1425–1436. [CrossRef]
39. Mohanty, R.; Chowdhury, C.R.; Arega, S.; Sen, P.; Ganguly, P.; Ganguly, N. CAR T cell therapy: A new era for cancer treatment (Review). *Oncol. Rep.* **2019**, *42*, 2183–2195. [CrossRef]

40. Al-Mansour, M.; Al-Foheidi, M.; Ibrahim, E. Efficacy and safety of second-generation CAR T-cell therapy in diffuse large B-cell lymphoma: A meta-analysis. *Mol. Clin. Oncol.* **2020**, *13*, 33. [CrossRef]
41. Shadman, M.; Yeung, C.; Redman, M.W.; Lee, S.Y.; Lee, D.H.; Ramachandran, A.; Ra, S.; A Marzbani, E.; Graf, S.A.; Warren, E.H.; et al. Third Generation CD20 Targeted CAR T-Cell Therapy (MB-106) for Treatment of Patients with Relapsed/Refractory B-Cell Non-Hodgkin Lymphoma. *Blood* **2020**, *136*, 38–39. [CrossRef]
42. Enblad, G.; Karlsson, H.; Gammelgård, G.; Wenthe, J.; Lövgren, T.; Amini, R.M.; Wikstrom, K.I.; Essand, M.; Savoldo, B.; Hallböök, H.; et al. A Phase I/IIa Trial Using CD19-Targeted Third-Generation CAR T Cells for Lymphoma and Leukemia. *Clin. Cancer Res.* **2018**, *24*, 6185–6194. [CrossRef] [PubMed]
43. Duan, D.; Wang, K.; Wei, C.; Feng, D.; Liu, Y.; He, Q.; Xu, X.; Wang, C.; Zhao, S.; Lv, L.; et al. The BCMA-Targeted Fourth-Generation CAR-T Cells Secreting IL-7 and CCL19 for Therapy of Refractory/Recurrent Multiple Myeloma. *Front. Immunol.* **2021**, *12*, 561. [CrossRef] [PubMed]
44. Chmielewski, M.; Abken, H. TRUCKS, the fourth-generation CAR T cells: Current developments and clinical translation. *Adv. Cell Gene Ther.* **2020**, *3*, e84. [CrossRef]
45. Kagoya, Y.; Tanaka, S.; Guo, T.; Anczurowski, M.; Wang, C.H.; Saso, K.; O Butler, M.; Minden, M.D.; Hirano, N. A novel chimeric antigen receptor containing a JAK-STAT signaling domain mediates superior antitumor effects. *Nat. Med.* **2018**, *24*, 352–359. [CrossRef]
46. Shi, M.; Wang, J.; Huang, H.; Liu, D.; Cheng, H.; Wang, X.; Chen, W.; Yan, Z.; Sang, W.; Qi, K.; et al. Bispecific CAR T cell therapy targeting BCMA and CD19 in relapsed/refractory multiple myeloma: A phase I/II trial. *Nat. Commun.* **2024**, *15*, 3371. [CrossRef]
47. Cho, J.H.; Collins, J.J.; Wong, W.W. Universal Chimeric Antigen Receptors for Multiplexed and Logical Control of T Cell Responses. *Cell* **2018**, *173*, 1426–1438.e11. [CrossRef]
48. Liu, X.; Wen, J.; Yi, H.; Hou, X.; Yin, Y.; Ye, G.; Wu, X.; Jiang, X. Split chimeric antigen receptor-modified T cells targeting glypican-3 suppress hepatocellular carcinoma growth with reduced cytokine release. *Ther. Adv. Med. Oncol.* **2020**, *12*, 1758835920910347. [CrossRef] [PubMed]
49. Minutolo, N.G.; Sharma, P.; Poussin, M.; Shaw, L.C.; Brown, D.P.; Hollander, E.E.; Smole, A.; Rodriguez-Garcia, A.; Hui, J.Z.; Zappala, F.; et al. Quantitative Control of Gene-Engineered T-Cell Activity through the Covalent Attachment of Targeting Ligands to a Universal Immune Receptor. *J. Am. Chem. Soc.* **2020**, *142*, 6554–6568. [CrossRef] [PubMed]
50. Diaconu, I.; Ballard, B.; Zhang, M.; Chen, Y.; West, J.; Dotti, G.; Savoldo, B. Inducible Caspase-9 Selectively Modulates the Toxicities of CD19-Specific Chimeric Antigen Receptor-Modified T Cells. *Mol. Ther.* **2017**, *25*, 580–592. [CrossRef] [PubMed]
51. Shabaneh, T.B.; Moffett, H.F.; Stull, S.M.; Derezes, T.; Tait, L.J.; Park, S.; Riddell, S.R.; Lajoie, M.J. Safety switch optimization enhances antibody-mediated elimination of CAR T cells. *Front. Mol. Med.* **2022**, *2*, 1026474. [CrossRef]
52. Lu, L.; Xie, M.; Yang, B.; Zhao, W.-B.; Cao, J. Enhancing the safety of CAR-T cell therapy: Synthetic genetic switch for spatiotemporal control. *Sci. Adv.* **2024**, *10*, eadj6251. [CrossRef]
53. Zugasti, I.; Espinosa-Aroca, L.; Fidyt, K.; Mulens-Arias, V.; Diaz-Beya, M.; Juan, M.; Urbano-Ispizua, Á.; Esteve, J.; Velasco-Hernandez, T.; Menéndez, P. CAR-T cell therapy for cancer: Current challenges and future directions. *Signal Transduct. Target. Ther.* **2025**, *10*, 210. [CrossRef]
54. Scholler, J.; Brady, T.L.; Binder-Scholl, G.; Hwang, W.-T.; Plesa, G.; Hege, K.M.; Vogel, A.N.; Kalos, M.; Riley, J.L.; Deeks, S.G.; et al. Decade-long safety and function of retroviral-modified chimeric antigen receptor T cells. *Sci. Transl. Med.* **2012**, *4*, 132ra53. [CrossRef]
55. Pampusch, M.S.; Haran, K.P.; Hart, G.T.; Rakasz, E.G.; Rendahl, A.K.; Berger, E.A.; Connick, E.; Skinner, P.J. Rapid Transduction and Expansion of Transduced T Cells with Maintenance of Central Memory Populations. *Mol. Ther. Methods Clin. Dev.* **2020**, *16*, 1–10. [CrossRef]
56. Narayanavari, S.A.; Chilkunda, S.S.; Ivics, Z.; Izsvák, Z. Sleeping Beauty transposition: From biology to applications. *Crit. Rev. Biochem. Mol. Biol.* **2017**, *52*, 18–44. [CrossRef]
57. Magnani, C.F.; Gaipa, G.; Lussana, F.; Belotti, D.; Gritti, G.; Napolitano, S.; Matera, G.; Cabiati, B.; Buracchi, C.; Borleri, G.; et al. Sleeping Beauty-engineered CAR T cells achieve antileukemic activity without severe toxicities. *J. Clin. Investig.* **2020**, *130*, 6021–6033. [CrossRef] [PubMed]
58. Srour, S.A.; Singh, H.; McCarty, J.; de Groot, E.; Huls, H.; Rondon, G.; Qazilbash, M.; Ciurea, S.; Bardelli, G.; Buck, J.; et al. Long-term outcomes of Sleeping Beauty-generated CD19-specific CAR T-cell therapy for relapsed-refractory B-cell lymphomas. *Blood* **2020**, *135*, 862–865. [CrossRef]
59. Alizadeh, D.; Wong, R.A.; Yang, X.; Wang, D.; Pecoraro, J.R.; Kuo, C.-F.; Aguilar, B.; Qi, Y.; Ann, D.K.; Starr, R.; et al. IL15 Enhances CAR-T Cell Antitumor Activity by Reducing mTORC1 Activity and Preserving Their Stem Cell Memory Phenotype. *Cancer Immunol. Res.* **2019**, *7*, 759–772. [CrossRef]
60. Štach, M.; Musil, J.; Cetkovsky, P.; Otahal, P. Interleukin 21 Enhances Survival and Expansion of CAR T Cells via Inhibition of Their Terminal Differentiation during Interaction with Tumor Target Cells. *Blood* **2018**, *132*, 4545. [CrossRef]

61. Zhu, F.; Shah, N.N.; Xu, H.; Schneider, D.; Orentas, R.; Dropulic, B.; Hari, P.; Keever-Taylor, C.A. CAR-T Cell Production Using the Clinimacs® Prodigy System. *Blood* **2016**, *128*, 5724. [CrossRef]
62. Shah, N.; Zhu, F.; Xu, H.; Schneider, D.; Orentas, R.; Dropulic, B.; Hari, P.; Keever-Taylor, C. Dual Targeted CD20/19 CAR-T Cell Production Using the ClinimacsÂ® Prodigy System. *Biol. Blood Marrow Transplant.* **2017**, *23*, S162. [CrossRef]
63. Shah, N.N.; Johnson, B.D.; Schneider, D.; Zhu, F.; Szabo, A.; Keever-Taylor, C.A.; Krueger, W.; Worden, A.A.; Kadan, M.J.; Yim, S.; et al. Bispecific anti-CD20, anti-CD19 CAR T cells for relapsed B cell malignancies: A phase 1 dose escalation and expansion trial. *Nat. Med.* **2020**, *26*, 1569–1575. [CrossRef]
64. Burdsall, K.A.; Xu, P.; Castro-Martinez, D.; Labanieh, L.; Ho, K.; Shi, Q.; Yu, B.; Sotillo, E.; Chang, H.Y.; Mackall, C.L. Abstract LB091: T cell targeted lentiviral gene delivery using the PACK-IT Platform generates CAR-T cells with superior potency compared to conventional lentivirus and enables in vivo generation of CD19-CAR T cells capable of controlling leukemia in preclinical models. *Cancer Res.* **2025**, *85*, LB091. [CrossRef]
65. Yang, J.; He, J.; Zhang, X.; Li, J.; Wang, Z.; Zhang, Y.; Qiu, L.; Wu, Q.; Sun, Z.; Ye, X.; et al. Next-day manufacture of a novel anti-CD19 CAR-T therapy for B-cell acute lymphoblastic leukemia: First-in-human clinical study. *Blood Cancer J.* **2022**, *12*, 104. [CrossRef] [PubMed]
66. Tan, X.; Wang, J.; Chen, S.; Liu, L.; Li, Y.; Tu, S.; Yi, H.; Zhou, J.; Wang, S.; Liu, L.; et al. Novel CD19 Fast-CAR-T cells vs. CD19 conventional CAR-T cells for the treatment of relapsed/refractory CD19-positive B-cell acute lymphoblastic leukemia. *Chin. Med. J.* **2025**, *138*, 2491–2497. [CrossRef] [PubMed]
67. Stadel, R.; Idippily, N.; Giraudo, M.F.; Caimi, P.; Van Besien, K.; Martin, J.; Wu, J.; Deng, C.; Wald, D. Development and Clinical Validation of an Ultra-Fast CAR-T Manufacturing Platform Enabling Production of CAR-T Cells in Less Than 1 Day. *Blood* **2023**, *142*, 4848. [CrossRef]
68. Depil, S.; Duchateau, P.; Grupp, S.A.; Mufti, G.; Poirot, L. 'Off-the-shelf' allogeneic CAR T cells: Development and challenges. *Nat. Rev. Drug Discov.* **2020**, *19*, 185–199. [CrossRef] [PubMed]
69. Caldwell, K.J.; Gottschalk, S.; Talleur, A.C. Allogeneic CAR Cell Therapy—More Than a Pipe Dream. *Front. Immunol.* **2021**, *11*, 3466. [CrossRef] [PubMed]
70. Brown, R.A.; Byersdorfer, C.A. Metabolic Pathways in Alloreactive T Cells. *Front. Immunol.* **2020**, *11*, 1517. [CrossRef]
71. Presti, V.L.; Buitenwerf, F.; van Til, N.P.; Nierkens, S. Gene Augmentation and Editing to Improve TCR Engineered T Cell Therapy against Solid Tumors. *Vaccines* **2020**, *8*, 733. [CrossRef]
72. Torikai, H.; Reik, A.; Liu, P.-Q.; Zhou, Y.; Zhang, L.; Maiti, S.; Huls, H.; Miller, J.C.; Kebriaei, P.; Rabinovitch, B.; et al. A foundation for universal T-cell based immunotherapy: T cells engineered to express a CD19-specific chimeric-antigen-receptor and eliminate expression of endogenous TCR. *Blood* **2012**, *119*, 5697–5705. [CrossRef]
73. Osborn, M.J.; Webber, B.R.; Knipping, F.; Lonetree, C.-L.; Tennis, N.; DeFeo, A.P.; McElroy, A.N.; Starker, C.G.; Lee, C.; Merkel, S.; et al. Evaluation of TCR Gene Editing Achieved by TALENs, CRISPR/Cas9, and megaTAL Nucleases. *Mol. Ther.* **2016**, *24*, 570–581. [CrossRef]
74. McCreedy, B.J.; Senyukov, V.V.; Nguyen, K.T. Off the shelf T cell therapies for hematologic malignancies. *Best Pr. Res. Clin. Haematol.* **2018**, *31*, 166–175. [CrossRef]
75. Eyquem, J.; Mansilla-Soto, J.; Giavridis, T.; van der Stegen, S.J.C.; Hamieh, M.; Cunanan, K.M.; Odak, A.; Gönen, M.; Sadelain, M. Targeting a CAR to the TRAC locus with CRISPR/Cas9 enhances tumour rejection. *Nature* **2017**, *543*, 113–117. [CrossRef]
76. Poirot, L.; Philip, B.; Schiffer-Mannioui, C.; Le Clerre, D.; Chion-Sotinel, I.; Derniame, S.; Potrel, P.; Bas, C.; Lemaire, L.; Galetto, R.; et al. Multiplex Genome-Edited T-cell Manufacturing Platform for "Off-the-Shelf" Adoptive T-cell Immunotherapies. *Cancer Res.* **2015**, *75*, 3853–3864. [CrossRef]
77. Qasim, W.; Zhan, H.; Samarasinghe, S.; Adams, S.; Amrolia, P.; Stafford, S.; Butler, K.; Rivat, C.; Wright, G.; Somana, K.; et al. Molecular remission of infant B-ALL after infusion of universal TALEN gene-edited CAR T cells. *Sci. Transl. Med.* **2017**, *9*, eaaj2013. [CrossRef]
78. Neelapu, S.S.; Munoz, J.; Locke, F.L.; Miklos, D.B.; Brown, R.; McDevitt, J.T.; Mardiros, A.; Demirhan, E.; Konto, C.; Tees, M.T. First-in-human data of ALLO-501 and ALLO-647 in relapsed/refractory large B-cell or follicular lymphoma (R/R LBCL/FL): ALPHA study. *J. Clin. Oncol.* **2020**, *38*, 8002. [CrossRef]
79. Valton, J.; Guyot, V.; Marechal, A.; Filhol, J.-M.; Juillerat, A.; Duclert, A.; Duchateau, P.; Poirot, L. A Multidrug-resistant Engineered CAR T Cell for Allogeneic Combination Immunotherapy. *Mol. Ther.* **2015**, *23*, 1507–1518. [CrossRef]
80. McGuirk, J.P.; Tam, C.S.; Kröger, N.; Riedell, P.A.; Murthy, H.S.; Ho, P.J.; Maakaron, J.E.; Waller, E.K.; Awan, F.T.; Shaughnessy, P.J.; et al. CTX110 Allogeneic CRISPR-Cas9-Engineered CAR T Cells in Patients (Pts) with Relapsed or Refractory (R/R) Large B-Cell Lymphoma (LBCL): Results from the Phase 1 Dose Escalation Carbon Study. *Blood* **2022**, *140*, 10303–10306. [CrossRef]
81. Ghobadi, A.; McGuirk, J.P.; Shaughnessy, P.; Tam, C.S.; Allen, M.; Pan, C.; Li, Z.; Stevens, W.N.; Weaver, A.; Cheah, C.Y. CTX112, a Next-Generation Allogeneic CRISPR-Cas9 Engineered CD19 CAR T Cell with Novel Potency Edits: Data from Phase 1 Dose Escalation Study in Patients with Relapsed or Refractory B-Cell Malignancies. *Blood* **2024**, *144*, 4829. [CrossRef]

82. Locke, F.L.; Lekakis, L.J.; Eradat, H.; Munoz, J.; Tees, M.T.; de Vos, S.; Nath, R.; Stevens, D.A.; Malik, S.; Popplewell, L.; et al. Phase 1 results with anti-CD19 allogeneic CAR T ALLO-501/501A in relapsed/refractory large B-cell lymphoma (r/r LBCL). *J. Clin. Oncol.* **2023**, *41*, 2517. [CrossRef]
83. Chatham, W.; Fiorenza, S.; Weinmann-Menke, J.; Sanchez-Pernaute, O.; Lafayette, R.; Askanase, A.; Gerber, M.; Alexander, L.; Mao, P.; Garner, E.; et al. A Phase 1, Multicenter, Open-Label Study of CB-010, a Next-Generation CRISPR-Edited Allogeneic Anti-CD19 CAR-T Cell Therapy, in Patients with Refractory Systemic Lupus Erythematosus (GALLOP). *J. Rheumatol.* **2025**, *52*, 147.3–148. [CrossRef]
84. Biederstädt, A.; Bassermann, F.; Hecker, J.S. Allogeneic CAR-engineered cellular therapy for relapsed and refractory large B cell lymphoma: A systematic review and meta-analysis. *Front. Immunol.* **2025**, *16*, 1585556. [CrossRef]
85. Zhao, Y.; Niu, C.; Cui, J.; Zhao, Y.; Niu, C.; Cui, J. Gamma-delta (γδ) T cells: Friend or foe in cancer development? *J. Transl. Med.* **2018**, *16*, 3. [CrossRef]
86. Legut, M.; Cole, D.K.; Sewell, A.K. The promise of γδ T cells and the γδ T cell receptor for cancer immunotherapy. *Cell. Mol. Immunol.* **2015**, *12*, 656–668. [CrossRef]
87. Scheper, W.; Sebestyen, Z.; Kuball, J. Cancer immunotherapy using γδT cells: Dealing with diversity. *Front. Immunol.* **2014**, *5*, 601. [CrossRef]
88. Rozenbaum, M.; Meir, A.; Aharony, Y.; Itzhaki, O.; Schachter, J.; Bank, I.; Jacoby, E.; Besser, M.J. Gamma-Delta CAR-T Cells Show CAR-Directed and Independent Activity Against Leukemia. *Front. Immunol.* **2020**, *11*, 1347. [CrossRef]
89. Wang, R.; Wen, Q.; He, W.; Yang, J.; Zhou, C.; Xiong, W.; Ma, L. Optimized protocols for γδ T cell expansion and lentiviral transduction. *Mol. Med. Rep.* **2019**, *19*, 1471–1480. [CrossRef]
90. Capsomidis, A.; Benthall, G.; Van Acker, H.H.; Fisher, J.; Kramer, A.M.; Abeln, Z.; Majani, Y.; Gileadi, T.; Wallace, R.; Gustafsson, K.; et al. Chimeric Antigen Receptor-Engineered Human Gamma Delta T Cells: Enhanced Cytotoxicity with Retention of Cross Presentation. *Mol. Ther.* **2018**, *26*, 354–365. [CrossRef]
91. Nishimoto, K.P.; Barca, T.; Azameera, A.; Makkouk, A.; Romero, J.M.; Bai, L.; Brodey, M.M.; Kennedy-Wilde, J.; Shao, H.; Papaioannou, S.; et al. Allogeneic CD20-targeted γδ T cells exhibit innate and adaptive antitumor activities in preclinical B-cell lymphoma models. *Clin. Transl. Immunol.* **2022**, *11*, e1373. [CrossRef] [PubMed]
92. Tan, E. *Update on Clinical Trial Milestone—CytoMed Therapeutics Achieves Completion of Dose Level 1 of Its ANGELICA TRIAL, and Proceeding to Dose Level 2 in Patients with Advanced Solid Tumors or Haematological Malignancies*; Cytomed Therapeutics Limited: Singapore, 2025.
93. Nair, S.; Dhodapkar, M.V. Natural Killer T Cells in Cancer Immunotherapy. *Front. Immunol.* **2017**, *8*, 1178. [CrossRef]
94. Brossay, L.; Chioda, M.; Burdin, N.; Koezuka, Y.; Casorati, G.; Dellabona, P.; Kronenberg, M. CD1d-mediated recognition of an alpha-galactosylceramide by natural killer T cells is highly conserved through mammalian evolution. *J. Exp. Med.* **1998**, *188*, 1521–1528. [CrossRef]
95. Cortesi, F.; Delfanti, G.; Grilli, A.; Calcinotto, A.; Gorini, F.; Pucci, F.; Lucianò, R.; Grioni, M.; Recchia, A.; Benigni, F.; et al. Bimodal CD40/Fas-Dependent Crosstalk between iNKT Cells and Tumor-Associated Macrophages Impairs Prostate Cancer Progression. *Cell Rep.* **2018**, *22*, 3006–3020. [CrossRef] [PubMed]
96. de Araújo, N.D.; Gama, F.M.; Barros, M.d.S.; Ribeiro, T.L.P.; Alves, F.S.; Xabregas, L.A.; Tarragô, A.M.; Malheiro, A.; Costa, A.G. Translating Unconventional T Cells and Their Roles in Leukemia Antitumor Immunity. *J. Immunol. Res.* **2021**, *2021*, 6633824. [CrossRef] [PubMed]
97. Rotolo, A.; Caputo, V.S.; Holubova, M.; Baxan, N.; Dubois, O.; Chaudhry, M.S.; Xiao, X.; Goudevenou, K.; Pitcher, D.S.; Petevi, K.; et al. Enhanced Anti-lymphoma Activity of CAR19-iNKT Cells Underpinned by Dual CD19 and CD1d Targeting. *Cancer Cell* **2018**, *34*, 596–610.e11. [CrossRef]
98. Heczey, A.; Courtney, A.N.; Montalbano, A.; Robinson, S.; Liu, K.; Li, M.; Ghatwai, N.; Dakhova, O.; Liu, B.; Raveh-Sadka, T.; et al. Anti-GD2 CAR-NKT cells in patients with relapsed or refractory neuroblastoma: An interim analysis. *Nat. Med.* **2020**, *26*, 1686–1690. [CrossRef]
99. Heczey, A.; Liu, D.; Tian, G.; Courtney, A.N.; Wei, J.; Marinova, E.; Gao, X.; Guo, L.; Yvon, E.; Hicks, J.; et al. Invariant NKT cells with chimeric antigen receptor provide a novel platform for safe and effective cancer immunotherapy. *Blood* **2014**, *124*, 2824–2833. [CrossRef]
100. Poels, R.; Drent, E.; Lameris, R.; Katsarou, A.; Themeli, M.; van der Vliet, H.J.; de Gruijl, T.D.; van de Donk, N.W.C.J.; Mutis, T. Preclinical Evaluation of Invariant Natural Killer T Cells Modified with CD38 or BCMA Chimeric Antigen Receptors for Multiple Myeloma. *Int. J. Mol. Sci.* **2021**, *22*, 1096. [CrossRef]
101. Sommer, C.; Boldajipour, B.; Kuo, T.C.; Bentley, T.; Sutton, J.; Chen, A.; Geng, T.; Dong, H.; Galetto, R.; Valton, J.; et al. Preclinical Evaluation of Allogeneic CAR T Cells Targeting BCMA for the Treatment of Multiple Myeloma. *Mol. Ther.* **2019**, *27*, 1126–1138. [CrossRef]
102. Hinks, T.S.; Zhang, X.W. MAIT cell activation and functions. *Front. Immunol.* **2020**, *11*, 1014. [CrossRef] [PubMed]

103. Riegert, P.; Wanner, V.; Bahram, S. Genomics, isoforms, expression, and phylogeny of the MHC class I-related MR1 gene. *J. Immunol.* **1998**, *161*, 4066–4077. [CrossRef] [PubMed]
104. Tourret, M.; Talvard-Balland, N.; Lambert, M.; Ben Youssef, G.; Chevalier, M.F.; Bohineust, A.; Yvorra, T.; Morin, F.; Azarnoush, S.; Lantz, O.; et al. Human MAIT cells are devoid of alloreactive potential: Prompting their use as universal cells for adoptive immune therapy. *J. Immunother. Cancer* **2021**, *9*, e003123. [CrossRef]
105. Kurioka, A.; Walker, L.J.; Klenerman, P.; Willberg, C.B. MAIT cells: New guardians of the liver. *Clin. Transl. Immunol.* **2016**, *5*, e98. [CrossRef]
106. Wang, Z.; Wang, M.; Chen, J.; Zhang, L.; Zhang, L.; Yu, L. MR1-restricted T cells: The new dawn of cancer immunotherapy. *Biosci. Rep.* **2020**, *40*, BSR20202962. [CrossRef] [PubMed]
107. Dusseaux, M.; Martin, E.; Serriari, N.; Péguillet, I.; Premel, V.; Louis, D.; Milder, M.; Le Bourhis, L.; Soudais, C.; Treiner, E.; et al. Human MAIT cells are xenobiotic-resistant, tissue-targeted, CD161hi IL-17-secreting T cells. *Blood* **2011**, *117*, 1250–1259. [CrossRef]
108. Vacchini, A.; Chancellor, A.; Spagnuolo, J.; Mori, L.; De Libero, G. MR1-Restricted T Cells Are Unprecedented Cancer Fighters. *Front. Immunol.* **2020**, *11*, 751. [CrossRef]
109. Parrot, T.; Healy, K.; Boulouis, C.; Sobkowiak, M.J.; Leeansyah, E.; Aleman, S.; Bertoletti, A.; Chen, M.S.; Sandberg, J.K. Expansion of donor-unrestricted MAIT cells with enhanced cytolytic function suitable for TCR redirection. *J. Clin. Investig.* **2021**, *6*, e140074. [CrossRef] [PubMed]
110. Souter, M.N.; Loh, L.; Li, S.; Meehan, B.S.; Gherardin, N.A.; Godfrey, D.I.; Rossjohn, J.; Fairlie, D.P.; Kedzierska, K.; Pellicci, D.G.; et al. Characterization of Human Mucosal-associated Invariant T (MAIT) Cells. *Curr. Protoc. Immunol.* **2019**, *127*, e90. [CrossRef] [PubMed]
111. Dogan, M.; Karhan, E.; Kozhaya, L.; Placek, L.; Chen, X.; Yigit, M.; Unutmaz, D. Engineering Human MAIT Cells with Chimeric Antigen Receptors for Cancer Immunotherapy. *J. Immunol.* **2022**, *209*, 1523–1531. [CrossRef] [PubMed]
112. Li, Y.-R.; Brown, J.; Yu, Y.; Lee, D.; Zhou, K.; Dunn, Z.S.; Hon, R.; Wilson, M.; Kramer, A.; Zhu, Y.; et al. Targeting Immunosuppressive Tumor-Associated Macrophages Using Innate T Cells for Enhanced Antitumor Reactivity. *Cancers* **2022**, *14*, 2749. [CrossRef]
113. Bald, T.; Krummel, M.F.; Smyth, M.J.; Barry, K.C. The NK cell-cancer cycle: Advances and new challenges in NK cell-based immunotherapies. *Nat. Immunol.* **2020**, *21*, 835–847. [CrossRef]
114. Wu, M.; Mei, F.; Liu, W.; Jiang, J. Comprehensive characterization of tumor infiltrating natural killer cells and clinical significance in hepatocellular carcinoma based on gene expression profiles. *Biomed. Pharmacother.* **2020**, *121*, 109637. [CrossRef]
115. Lupo, K.B.; Matosevic, S. Natural Killer Cells as Allogeneic Effectors in Adoptive Cancer Immunotherapy. *Cancers* **2019**, *11*, 769. [CrossRef]
116. Chow, V.A.; Gopal, A.K.; Maloney, D.G.; Turtle, C.J.; Smith, S.D.; Ujjani, C.S.; Shadman, M.; Cassaday, R.D.; Till, B.G.; Tseng, Y.D.; et al. Outcomes of patients with large B-cell lymphomas and progressive disease following CD19-specific CAR T-cell therapy. *Am. J. Hematol.* **2019**, *94*, E209–E213. [CrossRef]
117. Herrera, L.; Santos, S.; Vesga, M.A.; Anguita, J.; Martin-Ruiz, I.; Carrascosa, T.; Juan, M.; Eguizabal, C. Adult peripheral blood and umbilical cord blood NK cells are good sources for effective CAR therapy against CD19 positive leukemic cells. *Sci. Rep.* **2019**, *9*, 18729. [CrossRef] [PubMed]
118. Ueda, T.; Kaneko, S. In Vitro Differentiation of T Cell: From CAR-Modified T-iPSC. In *In Vitro Differentiation of T-Cells: Methods and Protocols*; Kaneko, S., Ed.; Springer: New York, NY, USA, 2019; pp. 85–91. [CrossRef]
119. Li, Y.; Hermanson, D.L.; Moriarity, B.S.; Kaufman, D.S. Human iPSC-Derived Natural Killer Cells Engineered with Chimeric Antigen Receptors Enhance Anti-Tumor Activity. *Cell Stem Cell* **2018**, *23*, 181–192.e5. [CrossRef]
120. Ghobadi, A.; Bachanova, V.; Patel, K.; Park, J.H.; Flinn, I.; A Riedell, P.; Bachier, C.; Diefenbach, C.S.; Wong, C.; Bickers, C.; et al. Induced pluripotent stem-cell-derived CD19-directed chimeric antigen receptor natural killer cells in B-cell lymphoma: A phase 1, first-in-human trial. *Lancet* **2025**, *405*, 127–136. [CrossRef]
121. Cerneckis, J.; Cai, H.; Shi, Y. Induced pluripotent stem cells (iPSCs): Molecular mechanisms of induction and applications. *Signal Transduct. Target. Ther.* **2024**, *9*, 112. [CrossRef]
122. Strassheimer, F.; Elleringmann, P.; Ludmirski, G.; Roller, B.; Macas, J.; Alekseeva, T.; Cakmak, P.; Aliraj, B.; Krenzlin, H.; Demes, M.C.; et al. CAR-NK cell therapy combined with checkpoint inhibition induces an NKT cell response in glioblastoma. *Br. J. Cancer* **2025**, *132*, 849–860. [CrossRef]
123. Burger, M.C.; Forster, M.-T.; Romanski, A.; Straßheimer, F.; Macas, J.; Zeiner, P.S.; Steidl, E.; Herkt, S.; Weber, K.J.; Schupp, J.; et al. Intracranial injection of natural killer cells engineered with a HER2-targeted chimeric antigen receptor in patients with recurrent glioblastoma. *Neuro-Oncol.* **2023**, *25*, 2058–2071. [CrossRef] [PubMed]
124. Suck, G.; Odendahl, M.; Nowakowska, P.; Seidl, C.; Wels, W.S.; Klingemann, H.G.; Tonn, T. NK-92: An "off-the-shelf therapeutic" for adoptive natural killer cell-based cancer immunotherapy. *Cancer Immunol. Immunother.* **2016**, *65*, 485–492. [CrossRef]

125. Montagner, I.M.; Penna, A.; Fracasso, G.; Carpanese, D.; Pietà, A.D.; Barbieri, V.; Zuccolotto, G.; Rosato, A. Anti-PSMA CAR-engineered NK-92 Cells: An Off-the-shelf Cell Therapy for Prostate Cancer. *Cells* **2020**, *9*, 1382. [CrossRef]
126. Liu, E.; Marin, D.; Banerjee, P.; Macapinlac, H.A.; Thompson, P.; Basar, R.; Kerbauy, L.N.; Overman, B.; Thall, P.; Kaplan, M.; et al. Use of CAR-Transduced Natural Killer Cells in CD19-Positive Lymphoid Tumors. *N. Engl. J. Med.* **2020**, *382*, 545–553. [CrossRef] [PubMed]
127. Colamartino, A.B.L.; Lemieux, W.; Bifsha, P.; Nicoletti, S.; Chakravarti, N.; Sanz, J.; Roméro, H.; Selleri, S.; Béland, K.; Guiot, M.; et al. Efficient and Robust NK-Cell Transduction with Baboon Envelope Pseudotyped Lentivector. *Front. Immunol.* **2019**, *10*, 2873. [CrossRef]
128. Afolabi, L.O.; Adeshakin, A.O.; Sani, M.M.; Bi, J.; Wan, X. Genetic reprogramming for NK cell cancer immunotherapy with CRISPR/Cas9. *Immunology* **2019**, *158*, 63–69. [CrossRef]
129. Kim, H. Overcoming Immune Barriers in Allogeneic CAR-NK Therapy: From Multiplex Gene Editing to AI-Driven Precision Design. *Biomolecules* **2025**, *15*, 935. [CrossRef]
130. Daher, M.; Basar, R.; Shaim, H.; Gokdemir, E.; Uprety, N.; Kontoyiannis, A.; Mendt, M.C.; Imahashi, N.; Kerbauy, L.N.; Wei Inng Lim, F.L.; et al. The TGF-β/SMAD Signaling Pathway as a Mediator of NK Cell Dysfunction and Immune Evasion in Myelodysplastic Syndrome. *Blood* **2017**, *130*, 53.
131. Dehghan, F.; Metanat, Y.; Askarizadeh, M.; Ahmadi, E.; Moradi, V. Novel gene manipulation approaches to unlock the existing bottlenecks of CAR-NK cell therapy. *Front. Cell Dev. Biol.* **2025**, *12*, 1511931. [CrossRef]
132. Albinger, N.; Müller, S.; Kostyra, J.; Kuska, J.; Mertlitz, S.; Penack, O.; Zhang, C.; Möker, N.; Ullrich, E. Manufacturing of primary CAR-NK cells in an automated system for the treatment of acute myeloid leukemia. *Bone Marrow Transplant.* **2024**, *59*, 489–495. [CrossRef] [PubMed]
133. Marin, D.; Li, Y.; Basar, R.; Rafei, H.; Daher, M.; Dou, J.; Mohanty, V.; Dede, M.; Nieto, Y.; Uprety, N.; et al. Safety, efficacy and determinants of response of allogeneic CD19-specific CAR-NK cells in CD19+ B cell tumors: A phase 1/2 trial. *Nat. Med.* **2024**, *30*, 772–784. [CrossRef] [PubMed]
134. Li, J.; Hu, H.; Lian, H.; Yang, S.; Liu, M.; He, J.; Cao, B.; Chen, D.; Hu, Y.; Zhi, C.; et al. NK-92MI Cells Engineered with Anti-claudin-6 Chimeric Antigen Receptors in Immunotherapy for Ovarian Cancer. *Int. J. Biol. Sci.* **2024**, *20*, 1578–1601. [CrossRef] [PubMed]
135. Bui, T.A.; Mei, H.; Sang, R.; Ortega, D.G.; Deng, W. Advancements and challenges in developing in vivo CAR T cell therapies for cancer treatment. *eBioMedicine* **2024**, *106*, 105266. [CrossRef]
136. Hunter, T.L.; Bao, Y.; Zhang, Y.; Matsuda, D.; Riener, R.; Wang, A.; Li, J.J.; Soldevila, F.; Chu, D.S.H.; Nguyen, D.P.; et al. In vivo CAR T cell generation to treat cancer and autoimmune disease. *Science* **2025**, *388*, 1311–1317. [CrossRef]
137. Li, Y.-R.; Zhu, Y.; Halladay, T.; Yang, L. In vivo CAR engineering for immunotherapy. *Nat. Rev. Immunol.* **2025**, *25*, 725–744. [CrossRef]
138. Zhou, J.-E.; Sun, L.; Jia, Y.; Wang, Z.; Luo, T.; Tan, J.; Fang, X.; Zhu, H.; Wang, J.; Yu, L.; et al. Lipid nanoparticles produce chimeric antigen receptor T cells with interleukin-6 knockdown in vivo. *J. Control. Release* **2022**, *350*, 298–307. [CrossRef] [PubMed]
139. Billingsley, M.M.; Gong, N.; Mukalel, A.J.; Thatte, A.S.; El-Mayta, R.; Patel, S.K.; Metzloff, A.E.; Swingle, K.L.; Han, X.; Xue, L.; et al. In Vivo mRNA CAR T Cell Engineering via Targeted Ionizable Lipid Nanoparticles with Extrahepatic Tropism. *Small* **2023**, *20*, e2304378. [CrossRef]
140. Bimbo, J.F.; van Diest, E.; E Murphy, D.; Ashoti, A.; Evers, M.J.W.; A Narayanavari, S.; Vaz, D.P.; Rijssemus, H.; Zotou, C.; Saber, N.; et al. T cell-specific non-viral DNA delivery and in vivo CAR-T generation using targeted lipid nanoparticles. *J. Immunother. Cancer* **2025**, *13*, e011759. [CrossRef]
141. Smith, T.T.; Stephan, S.B.; Moffett, H.F.; McKnight, L.E.; Ji, W.; Reiman, D.; Bonagofski, E.; Wohlfahrt, M.E.; Pillai, S.P.S.; Stephan, M.T. In situ programming of leukaemia-specific T cells using synthetic DNA nanocarriers. *Nat. Nanotechnol.* **2017**, *12*, 813–820. [CrossRef]
142. Short, L.; Holt, R.A.; Cullis, P.R.; Evgin, L. Direct in vivo CAR T cell engineering. *Trends Pharmacol. Sci.* **2024**, *45*, 406–418. [CrossRef]
143. Nicolai, C.J.; Parker, M.H.; Qin, J.; Tang, W.; Ulrich-Lewis, J.T.; Gottschalk, R.J.; Cooper, S.E.; Lopez, S.A.H.; Parrilla, D.; Mangio, R.S.; et al. In vivo CAR T-cell generation in nonhuman primates using lentiviral vectors displaying a multidomain fusion ligand. *Blood* **2024**, *144*, 977–987. [CrossRef]
144. Münch, R.C.; Muth, A.; Muik, A.; Friedel, T.; Schmatz, J.; Dreier, B.; Trkola, A.; Plückthun, A.; Büning, H.; Buchholz, C.J. Off-target-free gene delivery by affinity-purified receptor-targeted viral vectors. *Nat. Commun.* **2015**, *6*, 6246. [CrossRef] [PubMed]
145. Michels, A.; Ho, N.; Buchholz, C.J. Precision medicine: In vivo CAR therapy as a showcase for receptor-targeted vector platforms. *Mol. Ther.* **2022**, *30*, 2401–2415. [CrossRef] [PubMed]
146. Agarwalla, P.; Ogunnaike, E.A.; Ahn, S.; Froehlich, K.A.; Jansson, A.; Ligler, F.S.; Dotti, G.; Brudno, Y. Bioinstructive implantable scaffolds for rapid in vivo manufacture and release of CAR-T cells. *Nat. Biotechnol.* **2022**, *40*, 1250–1258. [CrossRef]

147. Pandit, S.; Agarwalla, P.; Song, F.; Jansson, A.; Dotti, G.; Brudno, Y. Implantable CAR T cell factories enhance solid tumor treatment. *Biomaterials* **2024**, *308*, 122580. [CrossRef]
148. Rurik, J.G.; Tombácz, I.; Yadegari, A.; Fernández, P.O.M.; Shewale, S.V.; Li, L.; Kimura, T.; Soliman, O.Y.; Papp, T.E.; Tam, Y.K.; et al. CAR T cells produced in vivo to treat cardiac injury. *Science* **2022**, *375*, 91–96. [CrossRef] [PubMed]
149. Ai, K.; Liu, B.; Chen, X.; Huang, C.; Yang, L.; Zhang, W.; Weng, J.; Du, X.; Wu, K.; Lai, P. Optimizing CAR-T cell therapy for solid tumors: Current challenges and potential strategies. *J. Hematol. Oncol.* **2024**, *17*, 105. [CrossRef]
150. Peng, J.-J.; Wang, L.; Li, Z.; Ku, C.-L.; Ho, P-C. Metabolic challenges and interventions in CAR T cell therapy. *Sci. Immunol.* **2023**, *8*, eabq3016. [CrossRef]
151. Hu, Y.; Sarkar, A.; Song, K.; Michael, S.; Hook, M.; Wang, R.; Heczey, A.; Song, X. Selective refueling of CAR T cells using ADA1 and CD26 boosts antitumor immunity. *Cell Rep. Med.* **2024**, *5*, 101530. [CrossRef]
152. Xia, X.; Yang, Z.; Lu, Q.; Liu, Z.; Wang, L.; Du, J.; Li, Y.; Yang, D.-H.; Wu, S. Reshaping the tumor immune microenvironment to improve CAR-T cell-based cancer immunotherapy. *Mol. Cancer* **2024**, *23*, 175. [CrossRef] [PubMed]
153. Choi, Y.; Seok, S.H.; Yoon, H.Y.; Ryu, J.H.; Kwon, I.C. Advancing cancer immunotherapy through siRNA-based gene silencing for immune checkpoint blockade. *Adv. Drug Deliv. Rev.* **2024**, *209*, 115306. [CrossRef] [PubMed]
154. Rafiq, S.; O Yeku, O.; Jackson, H.J.; Purdon, T.J.; van Leeuwen, D.G.; Drakes, D.J.; Song, M.; Miele, M.M.; Li, Z.; Wang, P.; et al. Targeted delivery of a PD-1-blocking scFv by CAR-T cells enhances anti-tumor efficacy in vivo. *Nat. Biotechnol.* **2018**, *36*, 847–856. [CrossRef]

MDPI AG
Grosspeteranlage 5
4052 Basel
Switzerland
Tel.: +41 61 683 77 34

Antibodies Editorial Office
E-mail: antibodies@mdpi.com
www.mdpi.com/journal/antibodies

www.ingramcontent.com/pod-product-compliance
Lightning Source LLC
Chambersburg PA
CBHW061601210926
51850CB00009B/2174